**Multiblock Grid
Generation
Results of the EC/BRITE-
EURAM Project
EUROMESH, 1990–1992**

Edited by
Nigel P. Weatherill,
Michael J. Marchant,
and D. A. King

Notes on Numerical Fluid Mechanics (NNFM)   Volume 44

Series Editors: Ernst Heinrich Hirschel, München
                 Kozo Fujii, Tokyo
                 Bram van Leer, Ann Arbor
                 Keith William Morton, Oxford
                 Maurizio Pandolfi, Torino
                 Arthur Rizzi, Stockholm
                 Bernard Roux, Marseille

Volume 26 Numerical Solution of Compressible Euler Flows (A. Dervieux / B. van Leer / J. Periaux / A. Rizzi, Eds.)

Volume 27 Numerical Simulation of Oscillatory Convection in Low-Pr Fluids (B. Roux, Ed.)

Volume 28 Vortical Solution of the Conical Euler Equations (K. G. Powell)

Volume 29 Proceedings of the Eighth GAMM-Conference on Numerical Methods in Fluid Mechanics (P. Wesseling, Ed.)

Volume 30 Numerical Treatment of the Navier-Stokes Equations (W. Hackbusch / R. Rannacher, Eds.)

Volume 31 Parallel Algorithms for Partial Differential Equations (W. Hackbusch, Ed.)

Volume 32 Adaptive Finite Element Solution Algorithm for the Euler Equations (R. A. Shapiro)

Volume 33 Numerical Techniques for Boundary Element Methods (W. Hackbusch, Ed.)

Volume 34 Numerical Solutions of the Euler Equations for Steady Flow Problems (A. Eberle / A. Rizzi / H. E. Hirschel)

Volume 35 Proceedings of the Ninth GAMM-Conference on Numerical Methods in Fluid Mechanics (J. B. Vos / A. Rizzi / I. L. Ryhming, Eds.)

Volume 36 Numerical Simulation of 3-D Incompressible Unsteady Viscous Laminar Flows (M. Deville / T.-H. Lê / Y. Morchoisne, Eds.)

Volume 37 Supercomputers and Their Performance in Computational Fluid Mechanics (K. Fujii, Ed.)

Volume 38 Flow Simulation on High-Performance Computers I (E. H. Hirschel, Ed.)

Volume 39 3-D Computation of Incompressible Internal Flows (G. Sottas / I. L. Ryhming, Eds.)

Volume 40 Physics of Separated Flow – Numerical, Experimental, and Theoretical Aspects (K. Gersten, Ed.)

Volume 41 Incomplete Decompositions (ILU) – Algorithms, Theory and Applications (W. Hackbusch / G. Wittum, Eds.)

Volume 42 EUROVAL – An European Initiative on Validation of CFD Codes (W. Haase / F. Brandsma / E. Elsholz / M. Leschziner / D. Schwamborn, Eds.)

Volume 43 Nonlinear Hyperbolic Problems: Theoretical, Applied, and Computational Aspects Proceedings of the Fourth International Conference on Hyperbolic Problems, Taormina, Italy, April 3 to 8, 1992 (A. Donato / F. Oliveri, Eds.)

Volume 44 Multiblock Grid Generation – Results of the EC/BRITE-EURAM Project EUROMESH, 1990–1992 (N. P. Weatherill / M. J. Marchant / D. A. King, Eds.)

Volume 45 Numerical Methods for Advection – Diffusion Problems (C. B. Vreugdenhil / B. Koren, Eds.)

Volumes 1 to 25 are out of print.
The addresses of the Editors and further titles of the series are listed at the end of the book.

Multiblock Grid Generation

Results of the EC/BRITE-EURAM
Project EUROMESH, 1990–1992

Edited by
Nigel P. Weatherill,
Michael J. Marchant,
and D. A. King

Die Deutsche Bibliothek – CIP-Einheitsaufnahme

Multiblock grid generation: results of the EC/BRITE-EURAM
project EUROMESH, 1990–1992 / ed. by Nigel P. Weatherill ... –
Braunschweig; Wiesbaden: Vieweg, 1993
   (Notes on numerical fluid mechanics; 44)
   ISBN 3-528-07644-5

NE: Weatherill, Nigel P. [Hrsg.]; Europäische Gemeinschaften; GT

All rights reserved
© Friedr. Vieweg & Sohn Verlagsgesellschaft mbH, Braunschweig/Wiesbaden, 1993

Vieweg ist a subsidiary company of the Bertelsmann Publishing Group International.

No part of this publication may be reproduced, stored in a retrieval
system or transmitted, mechanical, photocopying or otherwise,
without prior permission of the copyright holder.

Produced by W. Langelüddecke, Braunschweig
Printed on acid-free paper
Printed in Germany

ISSN 0179-9614
ISBN 3-528-07644-5

Foreword

Computational Fluid Dynamics research, especially for aeronautics, continues to be a rewarding and industrially relevant field of applied science in which to work. An enthusiastic international community of expert CFD workers continue to push forward the frontiers of knowledge in increasing number. Applications of CFD technology in many other sectors of industry are being successfully tackled. The aerospace industry has made significant investments and enjoys considerable benefits from the application of CFD to its products for the last two decades. This era began with the pioneering work of Murman and others that took us into the transonic (potential flow) regime for the first time in the early 1970's. We have also seen momentous developments of the digital computer in this period into vector and parallel supercomputing. Very significant advances in all aspects of the methodology have been made to the point where we are on the threshold of calculating solutions for the Reynolds-averaged Navier-Stokes equations for complete aircraft configurations.

However, significant problems and challenges remain in the areas of physical modelling, numerics and computing technology. The long term industrial requirements are captured in the U. S. Governments 'Grand Challenge' for 'Aerospace Vehicle Design' for the 1990's: 'Massively parallel computing systems and advanced parallel software technology and algorithms will enable the development and validation of multidisciplinary, coupled methods. These methods will allow the numerical simulation and design optimisation of complete aerospace vehicle systems throughout the flight envelope'.

This volume contains a set of papers describing work carried out during the EuroMesh project on 'Multi-Block Mesh Generation for Computational Fluid Dynamics'. The work was performed under a cost shared research contract (AERO 0018) within the programme BRITE/EURAM Area 5 Aeronautics of the Commission of the European Communities (CEC). EuroMesh was a pre-competitive research project lead by British Aerospace Regional Aircraft Ltd under the umbrella of the Aeronautics initiative managed and administered by the CEC DGXIIF. The project ran for two years with fourteen partners (6 from the aeronautics industry, 3 universities and 5 research institutes) from seven countries from the European Community and EFTA.

I would like to thank all those involved with EuroMesh for their enthusiasm and cooperation. In particular I would like to thank the Task Managers and Working Group Coordinators for their efforts. I would also like to offer my gratitude to Nigel Weatherill and Michael Marchant (University College of Swansea) for their assistance in the preparation of this publication and to Drietrich Knoerzer (CEC-DGXIIh) for his guidance on the running of the project.

D. A. King, BAe Woodford, February 1993.

Contents

	Page
I. Introduction	1

The EUROMESH Project 3
D. A. King

An introduction to grid generation using the multiblock approach . . . 6
N. P. Weatherill

II. Topology Generation 19

Topology generation within CAD systems 21
V. Treguer-Katossky, D. Bertin and E. Chaput

A topological modeller 27
R. Scateni

Advancing front technique used to generate block quadrilaterals 32
T. Schönfeld

III. Surface Grid Generation and Geometry Modelling 35

Surface mesh generation using projections 37
J. M. de la Viuda

Generation of surface grids using elliptic PDEs 45
P. Weinerfelt

Generation of structured meshes over complex surfaces 48
B. Morin and V. Tréguer-Katossky

Surface modelling using Coons multipatch and non-uniform rational surface . 55
E. Chaput

Reparametrization of block boundary surface grids 63
S. Farestam

Aircraft surface generation 71
H. Sobieczky

Contents (continued)

	Page
IV. Volume Grid Generation	77
Use of ONERA grid optimization method at CASA J. M. de la Viuda, J. J. Guerra and A. Abbas	79
Multi-block mesh generation for complete aircraft configurations K. Becker and S. Rill	86
Development of 3D multi-block mesh generation tools J. Oppelstrup, O. Runborg, P. Mineau, P. Weinerfelt, R. Lehtimäki and B. Arlinger	117
Multi-block mesh optimization T. Fol and V. Tréguer-Katossky	130
Smoothing of grid discontinuities across block boundaries P. Mineau	139
V. Grid Optimization and Adaption Methods	149
Grid adaption in computational aerodynamics R. Hagmeijer and K. M. J. de Cock	151
Embedding within structured multi-block computational fluid dynamics simulation S. N. Sheard and M. C. Fraisse	169
Adaptive mesh generation within a 2D CFD environment using optimisation techniques A. F. E. Horne	179
Two dimensional multi-block grid optimisation by variational techniques M. R. Morris	189
Local mesh enrichment for a block structured 3D Euler solver T. Schönfeld	199
The adaptation of two-dimensional multiblock structured grids using a PDE-based method D. Catherall	207
Contribution to the development of a multiblock grid optimization and adaption code O-P. Jacquotte, G. Coussement, F. Desbois and C. Gaillet	224
General grid adaptivity for flow simulation M. J. Marchant, N. P. Weatherill and J. Szmelter	263
Error estimates and mesh adaption for a cell vertex finite volume scheme J. A. Mackenzie, D. F. Mayers and A. J. Mayfield	290
Multigrid methods for the acceleration and the adaptation of the transonic flow problems A. E. Kanarachos, N. G. Pantelelis and I. P. Vournas	311

I. INTRODUCTION

The EUROMESH Project

D.A. King

Research Department
British Aerospace RAL
Woodford Aerodrome, Woodford, Cheshire, SK7 1QR - UK

Computational Fluid Dynamics (CFD) methods are now well established as an integral part of the aerodynamic design process throughout the civil aerospace industry. They have been successfully employed in the wing design for modern civil transport aircraft and executive jets over the last two decades. Some significant increments in wing performance have been associated with the introduction of new CFD methods. The design of the Airbus A310 saw the introduction of double curvature wings into the Airbus family in part through the introduction of transonic small perturbation (TSP) methods. The A330/340 wing was primarily designed with viscous-coupled full potential techniques. The next generation civil transports will be designed using methods both well established and those capable of modelling full aircraft configurations with Euler and Navier-Stokes flow solvers. These new methods will enable adverse aerodynamic interference effects to be designed out from an early stage in the product development through an integrated total aircraft approach. CFD techniques continue to underpin our ability to design aircraft with ever decreasing drag, emitting less pollution and consuming less fuel. In addition, the introduction of various new aerodynamic technologies and design concepts (eg laminar flow for lower drag, low cost for manufacture etc) will rely heavily on CFD to minimise high cost testing or prototyping. For high speed cruise design the traditional approach of designing in the wind tunnel has largely been replaced by design on the supercomputer with checking in the tunnel. This has yealded significant cost and performance benefits. However 3D CFD is not yet able to predict Clmax and so low speed design and optimisation is still carried out experimentaly. Advanced CFD methods have yet to make the same impact on dynamic design issues such as aeroelastic flutter and buffet.

The major European airframe manufacturers have made substantial investment in the development and calibration of state-of-the-art computational aerodynamics codes. Most companies have specialised teams dedicated to the development, integration and application of CFD to the design and analysis of their products. However, industry has traditionally relied on the support of universities and research establishments to carry out innovative basic research into new improved methods through both national governmental and direct industrial support. The success of CFD in Europe is due in no small part to the success of that partnership. The commercial sector has been slow to offer suitable proprietary CFD systems to the aerospace industry in part because of its unique transonic modelling requirements but also because of the relatively high accuracy and consistancy levels demanded for aircraft design. The development and marketing of commercial codes tends to have been concentrated on those industries whoose most basic requirement involves complex physical modelling (eg of combustion, radiation or two-phase flows).

Alongside transition, turbulence modelling and high performance computing one of the key pacing items for industrial CFD is the development of fast user friendly techniques for the generation of a suitable computational mesh for aircraft components or entire configurations. Aerodynamic designers generally require very high standards of flow prediction from CFD to enable them to progress the evolution of a design without regular recourse to the wind tunnel. This means that meshes should be of a high quality to facilitate accurate modelling of the complex air flows accross a wide speed range around the various configurations of interest. In addition the grid generation systems should be very flexible and easy to use to enable key geometric features to be modelled within realistic design time-frames. Effective interfaces are required between company CFD and CAD systems to ensure accurate geometrical representation and rapid model set-up times. Where geometric compromises are necessitated appropriate tools should be in place. A key design aim for a grid generation system should be to remove the need for the user to interact with the field (off surface) mesh. This has not generaly been a feature of current industrial (or any other) systems for modelling whole aircraft shapes. The need for use during an intensive design cycle rather than post design analysis yields distinct requirements for a system. Very rapid model set-up times for incremental adjustments to configurations are essential where daily or hourly turnaround is required.

Over the last 15 years a number of grid generation approaches have been proposed to meet these requirements. The use of irregular (unstructured) grids of tetrahedra with finite element solvers or non-aligned meshes with finite volume or difference solvers have been popular with some groups. In addition over the last 7-8 years considerable interest has evolved in regular (structured) grid multi-block methods. Many different implementations of multi-block grid generation systems have been developed across Europe. Subtley different block topology concepts have been adopted with a wide variation to the degree of automation to the generation. A number of approaches for grid node specification from simple algebraic teechniques to direct solution of partial differential equations or optimisation formulations have been implemented. Further work is required to bring these industrial systems up to a fully acceptable standard and there is considerable scope for benefits to be accrued from collaboration on development and assesment.

With an increasing focus of the aerospace business towards European companies and consortia competing with large North American companies and perhaps with the Japanese in the next century the time is right for coordinated European wide collaboration on CFD development. A framework for pre-competative industrial collaborative research in the European Community has been established through the CEC Aeronautics Programme managed and administered by the DGXIIF. A call for proposals was made in 1989 and a number of CFD projects were accepted and began early in the following year. The EuroMesh project (BRITE/EURAM AERO 0018) entitled 'Multi-Block Mesh Generation for Computational Fluid Dynamics' ran from February 1990 to March 1992 and involved fourteen partners (6 from the aeronautics industry,3 universities and 5 research institutes) from seven countries of the the European Community and EFTA. Informal links at a working level were established with the EuroVal Project (Validation of CFD Codes, BRITE/EURAM AERO 0014) which ran in parallel to EuroMesh.

The technical work of the EuroMesh programme fell into a number of areas:
Topology Generation - Various approaches to the specification of block connectivity were were investigated. For example a frontal technique was adapted to automatically generate a block structure.
Geometric Surface Modelling - The representation of geometric components was studied with a view to generating block structured surface meshes.
Mesh Generation Methods Several different approaches to generating mesh nodes were implemented in 2D and 3D. Methods have been produced based on simple algebraic formulations, on the direct solution of partial differential equations and the minimisation of a functional.
Mesh Embedding - Additional nodes may be added locally or on a block basis to enhance accuracy and reduce costs. Embedding techniques were investigated with the corresponding development of the flow algorithms.
Mesh Adaptation - As an aid to efficiency and accuracy nodes may be re-distributed within a block structured grid in an attempt to produce an even distribution of solution error. Various grid generation techniques were enhanced to accomdate adaptivity. Sensors based on some measure of flow activity and direct measures of numerical solution error were developed to drive the adaptation process. These techniques should be most beneficial for capturing leading edge suction peaks, shock waves and all forms of shear-layer without recourse to a globally fine grid.

Within the project several Working Groups were established to discuss and carry out joint exercises on the assesment of particular aspects of the methodology. One group considered mesh quality measures and another compared the results of mesh adaptation. In addition a pilot exercise was to begun to look at the issues of exchanging multi-block data sets between partners.

An Introduction to Grid Generation using the Multiblock Approach

N. P. Weatherill

Institute for Numerical Methods in Engineering,
Department of Civil Engineering,
University College of Swansea,
Singleton Park,
Swansea, SA2 8PP, UK

Summary

This introductory paper presents a tutorial on grid generation using the multiblock approach, highlights some of the key areas in its practical implementation and reviews the areas of work addressed by the BRITE-EURAM EUROMESH project.

1. Introduction

Techniques for the construction of structured quadrilateral and hexahedral grids (or mesh) have been studied now for many years[1-7]. Although easy to generate for very simple shapes which do not deviate far from Cartesian, box-like shapes, the task of constructing grids of reasonable quality for complicated shapes has proved to be a considerable technical challenge. Although some studies of grid generation techniques for general geometries had been investigated at an early date [8], modern structured grid generation began with the paper published in the Journal of Computational Physics by Joe Thompson, Frank Thames and Wayne Mastin [9]. This paper described how a system of coupled partial differential equations, based upon Poisson's equation, could be used to provide a means of constructing high quality structured grids with good control of grid point spacing. After this landmark paper, the subject sprang to life with papers appearing which described the more fine details of the basic approach. The topic received considerable attention, in part stimulated by the parallel efforts on solution techniques for the potential, Euler and Navier-Stokes equations, all of which, in general, require a computational grid for their numerical solution. The mathematics of structured grid generation became well understood and with the additional work on grid generation using algebraic techniques[10-15] the subject began to provide a basis for numerical applications to real engineering problems.

The early fundamental development work on grid generation techniques utilised a single block approach. This involves the region to be meshed being mapped to a single rectangular domain. The restrictions imposed on structured grid generation in a single blocked domain have been identified from an early stage in the development of computational fluid dynamics. It soon became apparent that for general geometries it would be necessary to subdivide the space into a set of blocks within which a structured grid could be generated. The grids within each block would then be connected to form a grid throughout the region of interest. So in this way the multiblock or composite grid was established. Today, multiblock grids are viewed as essential for the applications of structured grids to realistic problems. More recently research and development effort has been expended in developing grid techniques which are applicable to the blocked approach and overcoming some of the additional problems inherent to blocked grids.

The European Commission identified the importance to industry of automatic grid generation techniques based upon structured multiblock grid generation and, within the BRITE-EURAM programme, funded the work which is described in these proceedings. To a large extent the work programme targetted several areas which from work already undertaken by several groups in Europe was seen to be critical to the continued development of multiblock grid generation in computational fluid dynamics for industry.

This short introductory paper to these proceedings aims to provide a tutorial in multiblock grid generation, a short description of how the multiblock approach is used in practice and the topics of research and development covered by the BRITE-EURAM EUROMESH project.

2. Basic Multiblock Techniques

2.1 Topology Definition
Central to the multiblock concept is the block subdivision of a domain. Here, a block is defined as a geometrical domain which is topologically equivalent to a rectangle, in 2 dimensions, or a cube in 3 dimensions (Figure 1).

A trivial example of a block subdivision of the domain around an aerofoil is given in Figure 2. In this case, the mesh topology is a so-called 'C' mesh. The block subdivision for the mesh topology, shown in Figure 2a, is not unique. Two equally valid subdivisions are shown in Figure 2b and 2c. In the case of Figure 2b, the aerofoil boundary is restricted to one block side, whilst in the second case, the aerofoil boundary extends over one complete side and partially over 2 others. Either are possible subdivisions, but for discussion purposes it will be assumed that only one type of boundary can exist on any block side.

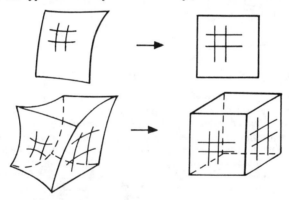

Figure 1. Regions in physical space are topologically equivalent to squares or cubes in the computational space.

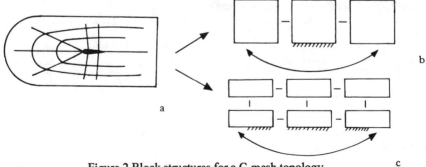

Figure 2 Block structures for a C-mesh topology.

Another 2 examples of block subdivision of the domain around an aerofoil are shown in Figures 3 and 4. The block structure for a polar mesh can be reduced to a single block in which 2 opposite sides map onto each other. The Cartesian or H-topology mesh around an aerofoil is given by 6 blocks suitably connected.

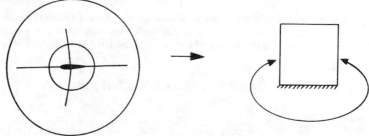

Figure 3 Block structure for an O (polar) mesh topology.

Figures 2-4 show the block structures for the construction of the 3 basic mesh topologies around an aerofoil. These are rather trivial examples of the multiblock concept but they help to illustrate the basic ideas.

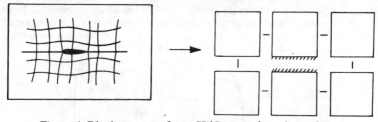

Figure 4. Block structure for an H (Cartesian) mesh topology.

Consider now the more complicated geometry of two aerofoils, with rounded leading edges, arranged in a tandem configuration. It is possible, in this case to generate a mesh which effectively has one global coordinate system. A Cartesian mesh could be constructed local to each aerofoil in a way which is schematically shown in Figure 5, or, alternatively, the forward aerofoil could be favoured with a C-mesh but the aft element would then have a Cartesian mesh (Figure 6.). On the grounds of mesh quality and flow sensitivity in the vicinity of the leading edges a Cartesian mesh local to the leading edge may be found to be unsuitable. Ideally, it can be argued, a C-mesh or a polar mesh is required to provide the necessary grid quality for accurate flow simulation. If this is the case, it is no longer possible to construct such mesh topologies and conserve a global curvilinear coordinate system. However, it is topologically possible to construct such meshes and the concept of a multiblock subdivision of the domain provides the appropriate mechanism.

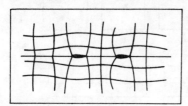

Figure 5. Cartesian mesh around both aerofoils.

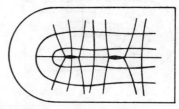

Figure 6. C-mesh around the leading aerofoil, H-mesh around the second aerofoil.

Consider first the construction of a C-mesh local to each aerofoil. Figure 7 shows, in schematic form, one type of mesh which would satisfy this basic requirement. This is not the only mesh topology which would produce C-meshes around aerofoils, but it is suitable for the discussion. An analysis of this construction indicates that the equivalent block structure is made of 18 blocks. (In fact, it is possible to reduce this to 16 blocks, but, in order to ensure symmetry about the aerofoils a cut has been made from the leading edge of the first aerofoil to the upstream boundary.) The arrangement of these blocks and their connectivity is shown in Figure 8.

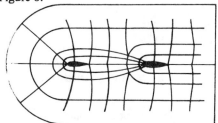
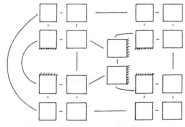

Figure 7. C-mesh topology constructed around both aerofoils

Figure 8. Block connectivity for the mesh topology shown in Figure 7.

Now consider a mesh which has polar meshes locally around the aerofoils. Such a mesh is show schematically in Figure 9, with the equivalent block topology given in Figure 10. Again, there is no unique way of decomposing the domain such that polar meshes are constructed.

These two examples illustrate that given a configuration, it is possible, using the concepts of multiblock domain decomposition, to construct mesh topologies which are ideally suited for each component of the configuration. The term 'component adaptive meshes' is often used to describe this approach.

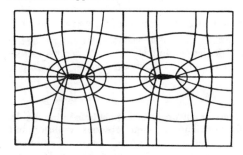
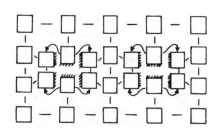

Figure 9. Polar mesh topology around each aerofoil.

Figure 10. Block connectivity for polar topology.

It is clear that the 2 mesh topologies shown in Figures 7 - 10 differ in the number of blocks and their arrangement or connectivities. Thus in a multiblocked environment it is necessary to define a way of describing the relative position of one block with respect to others ie. just as in an unstructured mesh it is necessary to define an element connectivity matrix, it is necessary here to define a block connectivity matrix. In fact, at a block level the topology of a multiblocked mesh is equivalent to an unstructured quadrilateral mesh in the computational space. This last observation is important. The block structure is completely independent of any coordinates in the physical space. The physical position of any computational block is not relevant in constructing the mesh topology. The format and information contained in the block connectivity data depends upon the type of boundary conditions to be imposed at the interface between blocks.

A typical format for the block connections is, for each face, for each block;-
    i) the number of points in each direction on the face, say (imax,jmax)
    ii) the type of boundary condition imposed,
and
    iii) if appropriate,   a) the adjacent block number,
                              b) the adjacent side of the adjacent block,
                              c) the orientation of the coordinate system of the adjacent block relative to that of the current block.

2.2 Solution Procedure within a block and at block boundaries

2.2.1 Mesh Generation within a block

From the discussion, it is clear that each block has its own curvilinear coordinate system. This system is independent of those in the adjacent blocks. Hence, to generate a mesh within a block it is necessary to construct an algorithm which operates on a rectangular computational domain. With a suitable treatment of boundary points, the generation of the global mesh then involves a loop over all blocks.

If the mesh within each block is to be generated using a system of elliptic partial differential equations, then it is clear that for a block defined by (ξ_i, η_j), i=2,....imax, j=1,2,....jmax in which the boundary points are assumed known, it is possible to solve for the interior points (x,y) i=2,...imax-1, j=2,..jmax-1, without the requirement for information from adjacent blocks. This assumes a second order accurate central difference formulation for the derivatives in the elliptic partial differential equations.

Thus, given a point (ξ_i, η_j) the computational molecule for the solution of the elliptic equations at this point involves all immediate neighbours to the point. Hence, assuming that the boundary points are known, (equivalent to a Dirichlet condition) the interior points can be determined. A similar situation arises if the grid points are to be computed using, for example, transfinite interpolation.

2.2.2 Point continuity at block boundaries

At the boundary between 2 blocks, the grid points in physical space must be continuous. In general, there cannot be holes or overlaps in the grid. This is a rather obvious condition, but one which should not be overlooked. Whatever treatment for the generation of points is designed for an edge it is necessary that the operations are commutative between common edges ie. the treatment of the points on the edges in block 1 edge 2 must be equivalent to the treatment of points on the block 2 edge 1 (Figure 11).

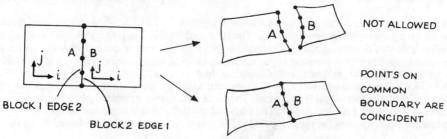

Figure 11. Point correspondence on block edges.

2.2.3 Slope continuity at block boundaries

With fixed data points on block boundaries, it is possible to form a well defined problem for the generation of grid points within a block. In some cases, fixed data points will correspond to points which define a geometrical object or boundary. In these cases, unless special treatments are applied, the grid lines extending from such boundaries will not possess special angular relationships with the boundary. A typical case is shown in Figure 12.

Figure 12. Grid points on the boundary are fixed. Angles of the grid lines to the boundary are arbitrary.

If a block boundary is adjacent to another boundary, and the boundary points are fixed, then the resulting mesh will, in general, be discontinuous in gradients (and higher derivatives) across that boundary (Figure 13).

Figure 13. Grid lines are discontinuous in slope across the boundary.

Clearly, in this case no information is provided to ensure any form of continuity and the mesh within each block is generated independently.

It is possible to achieve continuity of slope at block boundaries. Transfinite interpolation can be formulated to include slopes at boundaries. For elliptic systems used for grid generation again it is possible to formulate the control functions to control the slope of grid lines at boundaries[18].

2.2.4 Complete continuity at block boundaries

Complete continuity at block interfaces implies that the grid lines at block boundaries are as smooth, in the sense of differentiability, as lines interior to the blocks. Complete continuity at block interfaces can be achieved, but in this case the position of grid points on boundaries must, in general, evolve as part of the solution procedure for interior mesh points. The implication is that it is not possible to solve for grid points on a block-by-block basis, but rather the mesh in all blocks must be generated simultaneously.

Following this philosophy, the computational data necessary to generate the position of a point on a boundary must be of the same form as the information necessary to define the position of a point interior to a block. The position of an interior point, determined by the solution of elliptic equations, requires information from direct neighbours (Figure 14)

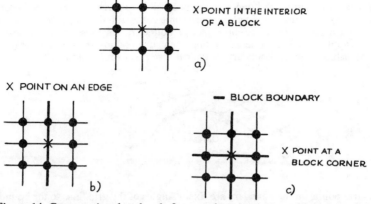

Figure 14. Computational molecule for complete continuity at block boundaries.

For a point on a block boundary the same computational molecule is applicable but, in this case, some of the points in this molecule are contained in adjacent blocks (Figure 15). For corner points the information is required from 3 adjacent blocks. Following this approach obviously leads to additional complexity in the data handling. No longer can a mesh in a block be derived from information contained only within that block. To ensure complete continuity the information related to a block must be augmented with information from adjacent blocks. This information is, in general, effectively the halo of points surrounding the block (Figure 15)

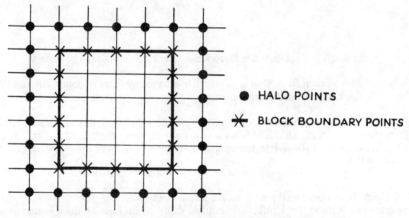

Figure 15. Block with halo of points obtained from adjacent blocks.

Any block of size i=1,....imax, j=1,.....jmax is increased to size i=1,....imax+2, j=1,.....jmax+2. This increase in size provides the necessary space to accommodate the appropriate points from the adjacent blocks. During the iterative solution routine the halo of points are constantly updated when new values of halo points are derived.

The disadvantage of the halo system is that it increases the memory requirements. For large blocks the overheads are small but for small blocks the overheads are high relative to the block size. The main advantage is that having filled the halo with current values the solution routine acts on a rectangular domain with no special implementation required for boundary points.

2.2.5 Singular points

Particular arrangements of blocks require more than or less than 4 blocks to join at a corner. Such arrangements lead to singular of special points in a grid a which more than or less than 4 grid lines meet. An example of two commonly occurring singularities is shown in Figure 16. In the grid generation procedure these points require special treatment. A common form of treatment is to average neighbour points.

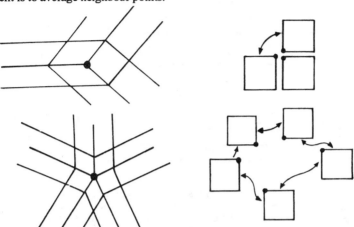

Figure 16. Special points which arise from block connections

2.2.6 Extension to 3 dimensions

The basic ideas outlined here can be extended in a natural way to 3 dimensions. Figure 17 shows the three basic block topologies for 'C', 'O', and 'H' meshes. These can be regarded as the basic block structures and combinations of such units are used to construct block toplogies for more complex shapes.

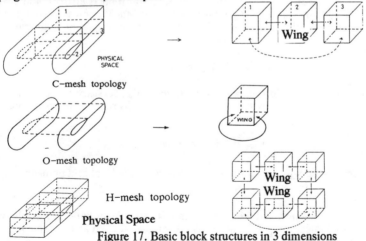

Figure 17. Basic block structures in 3 dimensions

3. Practical Implementation of the Multiblock Approach

The basic ideas discussed above can be applied to the generation of grids for realistic shapes and aerodynamic geometries. A computer system which incorporates such ideas will, in general, involve the topics outlined in Table 1.

Table 1. Key topics for multiblock grid generation and flow computer system

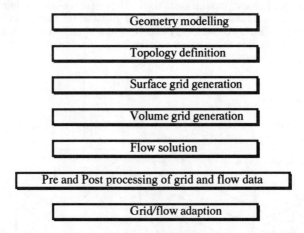

The approach has proved to very popular and successful and many impressive computations in aerospace engineering have been performed [19-27]. It has some disadvantages which are now widely recognised. Firstly, a criticism is the ease of use of such an approach and secondly, its applicability to all aerospace geometries and flow computations. The first difficulty primarily relates to the specification of the block connectivities. This is a difficult task, since it requires expert user effort and, furthermore, if an inappropriate grid topology is defined it is likely to lead to problems later in the grid or flow process. Automatic ways of subdividing a domain have been explored with limited success[28]. Now emphasis appears to be given to interactive specification using graphics workstations[29]. However, although this decreases the time for the task it still requires expert user effort and is prone to mistakes. The second problem is the application of the approach to all aerospace geometries. For some configurations, the specification of a suitable block decomposition is difficult and even if achieved can lead to a highly constrained grid of poor quality. Much is now understood of the mathematics of structured grid generation and it is likely that grid quality techniques will continue to improve.

4. Grid Adaptivity

To resolve features of a flowfield accurately it is, in general, necessary to introduce grid adaptivity techniques[30-34]. Adaptivity is based upon the equidistribution of errors principle, namely,

$$w_i \, ds_i = \text{constant}$$

where w_i is the error or activity indicator at node i and ds_i is the local grid point spacing at node i.

Central to adaptivity techniques, and the satisfaction of this equidistribution principle, is to define an appropriate indicator w_i. Adaptivity criteria are based upon an assessment of the error in the solution of the flow equations or are constructed to detect features of the flowfield. These estimators are intimately connected to the flow equations to be solved. For example, some of the main features of a solution of the Euler equations can be shock waves, stagnation points and vortices, and any indicator should accurately identify these flow characteristics.

However, for the Navier-Stokes equations, it is important not only to refine the mesh in order to capture these features but, in addition, to adequately resolve viscous dominated phenomena such as the boundary layers. Hence it seems likely, that certainly in the near future, adaptivity criteria will be a combination of measures each dependent upon some aspects of the flow and, in turn, upon the flow equations.

Once an adaptivity criterion has been established the equidistribution principle is achieved through a variety of methods, including point enrichment, point derefinement, node movement and remeshing, or combinations of these. It should be noted that an alternative approach to adaption is to suitably modify the flow algorithm, perhaps in the form of the interpolation used. This is often called p-refinement[35].

4.1 Grid refinement
Grid refinement, or h-refinement, involves the addition of points into regions where adaptivity is required. Such a procedure clearly provides additional resolution at the expense of increasing the number of points in the computation.

Grid refinement on a structured or multiblock grids is not straightforward The addition of points will, in general, break the regular array of points. The resulting distributed grid points no longer naturally fit into the elements of an array. Furthermore, some points will not 'conform' to the grid in that they have a different number of connections to other points. Hence grid refinement on structured grids requires a modification to the basic data structure and also the existence of so-called non-conforming nodes requires modifications to the flow solver. Clearly point enrichment on structured grids is not as natural a process as the method applied on unstructured grids and hence is not so widely employed. Work has been undertaken to implement point enrichment on structured grids and the results demonstrate the benefits to be gained from the additional effort in modifications to the data structure and flow solver.

4.2 Grid movement
Grid movement satisfies the equidistribution principle through the migration of points from regions of low activity into regions of high activity. The number of nodes in this case remains fixed. Traditionally, algorithms to move points involve some optimisation principle. Typically, expressions for smoothness, orthogonality and weighting according to the flowfield or errors are constructed[16] and then an optimisation is performed such that movement can be driven by a weight function, but not at the expense of loss of smoothness and orthogonality.

4.3 Combinations of node movement, point enrichment and derefinement
An optimum approach to adaptation is to combine node movement and point enrichment with derefinement. These procedures should be implemented in a dynamic way, i.e. applied at regular intervals within the flow simulation. Such an approach also provides the possibility of using movement and enrichment to independently capture different features of the flow.

5. BRITE-EURAM EUROMESH Topics

The key areas studied in the EUROMESH project are reflected in the topics outlined in Table 1. Issues related to topology generation, or how the domain is subdivided into blocks which then defines the grid topology, is covered in 3 papers. The generation of surface meshes, together with the associated problem of geometry modelling, is discussed in 6 papers, whilst volume grid generation is covered by 5 papers. Grid adaption techniques, which provide a mechanism for more accurate flow simulations is addressed by 10 papers, which indicates the anticipated importance of these techniques in future multiblock grid systems. Since adaptivity involves flow solution solvers, many of these papers also discuss issues related to flow algorithms.

References

1. Smith R. E. (Ed.) ' Numerical Grid Generation', NASA Conference Publication, CP-2166, 1980.
2. Eiseman P. R. 'Grid generation for fluid mechanics computation', Annual Review of Fluid Mechanics, Vol. 17, pp487-522, 1985.
3. Hauser J. and Taylor C (Eds) 'Numerical grid generation', Pineridge Press, Swansea, 1986.
4. Thompson J. F. and Steger J.F (Eds.) Three Dimensional Grid Generation for Complex Configurations - Recent Progress, AGARD-AG-309,1988.
5. Sengupta, S.(Eds.) 'Numerical Grid Generation in Computational Fluid Dynamics', Pineridge Press, Swansea, 1988.
6. Arcilla, A.S., Hauser, J. Eiseman, P.R. and Thompson J. F. (Eds.) 'Numerical Grid Generation in Computational Fluid Dynamics and related Fields', North-Holland, Amsterdam, 1991.
7. Applications of Mesh Generation to Complex 3-D Configurations', AGARD Conference Proceedings No. 464, May 1989.
8. Winslow A. M.,'Numerical solution of the quasi-linear Poisson's equation in a non-uniform triangle mesh', J. Comp. Phys., Vol. 1, p149-172, 1967.
9. Thompson J. F., Thames, F. and Mastin W.,' Automatic numerical grid generation of body-fitted curvilinear coordinate system of field containing any number of arbitrary 2-dimensional bodies', J. Comput. Phys, 15:299-319, 1974.
10. Gordon W. N. and Hall C. A. ' Construction of curvilinear coordinate systems and applications to mesh generation', Int. J. Num. Mthds in Eng, Vol. 7, pp461-477, 1973.
11. Eiseman P. R. 'A multi-surface method of coordinate generation', J. Comp. Phys, Vol. 33. No. 1, pp118-150, 1979.
12. L-E Eriksson, 'Generation of boundary-conforming grids around wing-body configurations using transfinite interpolation', AIAA Journal, Vol. 20, No. 10, p1313-1320, 1982.
13. Eiseman P. R.' Coordinate generation with precise control over mesh properties', J. Comp. Phys., Vol. 47, No. 3 pp331-351, 1982.
14. L-E Eriksson, 'Practical three-dimensional mesh generation using transfinite interpolation', SIAM J. Sci. Stat. Comput., Vol. 6, No. 3, pp712-741, July 1985.
15. Eiseman P. R. ' A control point form of algebraic grid generation', Int. J. Num. Mthds in Fluids, Vol. 8, pp1165-1181, 1988.
16. Brackbill J. U. and Saltzman J. S ' Adaptive zoning for singular problems in two dimensions' J. Comp. Phys., 46,342, 1982.
17. Thompson J. F., Warsi, C. W. Mastin, *Numerical grid generation: Foundations and applications,* North-Holland, 1985.
18. Thomas P. D. and Middlecoff J. F.' Direct control of the grid point distribution in meshes generated by elliptic equations', AIAA J. 18:652-656, 1980.
19. Roberts A. 'Automatic topology generation and generalised B-spline mapping', in Numerical Grid Generation , (Ed) J. F. Thompson, (North Holland, Amsterdam), 1982.
20. Weatherill N. P. and Forsey C. R., 'Grid generation and flow calculations for aircraft geometries', J. of Aircraft, Vol. 22, N. 10, p855-860, October 1985.
21. Thompson J. F. , 'A general 3-dimensional elliptic grid generation system on a composite block structure', Computer Methods in Applied Mechanics and Engineering, 64, p377-411, 1987.
22. Coleman, R. M. ' NUGGET: A program for three-dimensional grid generation', DTRC Report DTNSRDC - 87036, Sept. 1987.
23. Boerstoel J. W. 'Numerical grid generation in 3-dimensional Euler flow simulation', In Numerical methods in fluid dynamics', (Eds) K. W. Morton and M. J. Baines, Oxford University Press, 1988.
24. Thompson J. F. 'A composite grid generation code for general 3D regions - the EAGLE Code', AIAA Journal, Vol. 26, N. 3, pp271, March 1988.

25. Shaw, J. A., Georgala J. M. and Weatherill N. P., 'The construction of component-adaptive grids for aerodynamic geometries', Proc. of Int. Conf. on Numerical Grid Generation in CFD, Ed. Sengupta, Hauser, Eiseman, Thompson. Pub. Pineridge Press, Swansea, UK, 1988.
26. Seibert W., 'A graphic-interactive program system to generate composite grids for general configurations', in Proc. of Int. Conf. on Numerical Grid Generation in CFD, Ed. Sengupta, Hauser, Eiseman, Thompson. Pub. Pineridge Press, Swansea, UK, 1988.
27. Whitfield D. L. ' Unsteady Euler solutions on dynamic blocked grids for complex configurations', in Applications of Mesh Generation to Complex 3-D Configurations', AGARD Conference Proceedings No. 464, May 1989.
28. Shaw J. A. and Weatherill 'Automatic topology generation for aircraft geometries', To appear, Applied Mathematics and Computation, 1992.
29. Allwright, S. E. 'Techniques in multiblock domain decomposition and surface grid generation', in Numerical Grid Generation in CFD. (Proceedings of the Second International Conference), Pineridge Press, Swansea, UK, 1988)
30. Thompson J. F.' Review on the state of the art of adaptive grids', AIAA Paper 84-1606, 1984.
31. Nakahashi K. and Deiwert G. S.' A practical adaptive-grid method for complex fluid-flow problems' NASA TM 85989, June 1984.
32. Hyun Jin Kim and Thompson J. F.' Three dimensional adaptive grid generation on a composite block grid', AIAA Paper 88-0311, 1988.
33. Shephard M. S. and Weatherill N. P. (Eds.) 'Adaptive meshing', Special Issue of Int. J. Num. Mthds in Eng., Vol.32 No. 4, 1991.
34. Dannenhoffer J. F. ' A comparison of adaptive-grid redistribution and embedding for steady transonic flows', Int. J. Num. Mthds in Eng., Vol.32 No. 4, pp651-653, 1991.
35. Demkowicz L., Oden J. T., Rachowicz W. and Hardy O. ' An h-p Taylor Galerkin finite element method for compressible Euler equations', to appear Computer Methods Appl. Mech. Eng..

II. TOPOLOGY GENERATION

TOPOLOGY GENERATION WITHIN CAD SYSTEMS

V. Tréguer-Katossky*, D. Bertin*, E. Chaput**

* AEROSPATIALE, DIVISION AVIONS
316, Route de Bayonne
31060 Toulouse Cedex 03, FRANCE.

** AEROSPATIALE, DIVISION SYSTEMES
STRATEGIQUES ET SPATIAUX
Route de Verneuil BP 96
78133 Les Mureaux Cedex, FRANCE.

SUMMARY

The process of extracting reusable topology data from a 3D multi-block mesh created within a CAD system is analysed for two CAD systems: CATIA and ICEMDDN. A general method is derived, applicable to arbitrary CAD systems. For this purpose, the CAD system is modelled by a small kernal of high level capabilities. The principle of a general "Record/Replay system" for mesh generation is sketched.

INTRODUCTION

Non linear CFD methods are about to show their engineering utility within industrial design studies. However, "although solutions about complex configurations are being obtained, a discipline loosely called grid generation remains the key pacing item in making CFD useful to most engineering applications in aerodynamics" [1].

Great progress have been made recently, making effective interactive tools for multiblocks mesh generation available. But meshing a complex configuration, such as a complete airplane, costs still too much. The link between CAD data and the grid generation tools is now identified as the new bottleneck for most of the methods.

Brite Euram/Euromesh Sub-Task 6.1: Topology generation within CAD systems aimed at shortening the process [2].

1. MESH GENERATION WITHIN CAD SYSTEMS

1.1. Within CATIA

The Finite Element Modeler of CATIA (CATIA-FEM) can be used to generate 3D multiblock structured meshes.

Within modules CATIA 3D and CATIA Advanced Surfaces of CATIA, geometry data is first made avalaible for subsequent control of the block faces and block edges

Within module MESH1 of CATIA-FEM, block boundaries ans boundary meshes are specified with commands BLOCK+CREATE+POINT (or CURVE, or SURFACE). Since version V3R2, volumes meshes can be interpolated from block boundaries with command BLOCK+CREATE+VOLUME.

Mesh points can be made equivalent (condensed) at the interfaces between blocks with command NODES+COND. The transition through block interfaces can be smoothed, by moving nodes along a prescribed curve with command NODES+MOVE+CURVE.

Mesh points can be projected onto existing curves and surfaces with commands PROJECT+POINT (or CURVE, or SURFACE). The experience shows that, in highly constrained regions, projection is easier with meshes generated within existing surfaces, than with meshes interpolated from contours (however, the generation of the appropriate control surfaces may be also difficult). Picking up existing surfaces for projections is a tedious operation.

The Geometry interface of CATIA-FEM (CATGEO) is run interactively or in batch mode for extracting the coordinates of the mesh points. Until version V3R2, only block boundaries mesh points were extracted and the interior mesh points were computed by 3D transfinite interpolation.

An history ASCII file is stored, that reports the main commands executed during the interactive session. It contains topology and mesh data, that will be modified in a text editor, if needed. Geometry data are stored separately in a CATIA data base.

1.2. Within ICEMDDN

CAD system ICEMDDN has an environment for mesh generation, that comprises the fully integrated module MULCAD and the stand alone program PADAMM.

Within module Advanced Design of ICEMDDN, geometry data is made available for subsequent control of the block faces and block edges.

Within module MULCAD, the multi-block topology is specified with menu items "Sub-Faces" and "Domains". Commands "Create" allows respectivelly to pick up the existing entities that compound the 4 edges of a new sub-face or to select existing sub-faces that compound the 6 faces of a new domain. The curve entities can be partitionned into

Masters and Slaves with menu item "Connectivity Control" (the number of mesh points along Masters prescribes the number of mesh points along Slaves).

Mesh Data are specified with menu items "Edge Meshing" and "Patch/Sub-Face Association". Command "Sample Modify" allows to specify numbers of mesh points (along masters only) and to choose among 3 types of stretching functions. Commands "Dis/Mod Patches" allows to pick up any number of existing surfaces onto which the Sub-Face mesh points must be projected.

The interface of MULCAD allows to store topology data, mesh data, geometry data (ICEMMDDN ASCII file) and PADAMM input data. Mesh points are computed within PADAMM from input and geometry data.

1.3. Within an arbitrary CAD system

Within a CAD system, the mesh generation environment is meant for supporting the following activities:

- GENERATE GEOMETRY DATA FOR MESHES
- EDIT OR CAPTURE TOPOLOGY DATA
- EDIT MESH DATA
- GENERATE MESHES
- REUSE TOPOLOGY AND MESH DATA

Since the geometry definition of the aircraft is not part of the process of mesh generation, it is assumed that geometry data which define the aircraft are already stored within the CAD data base.

Geometry data generated for meshes are stored into the data base of the CAD system. They can be generated interactively within the CAD system. Since this step costs a lot, it is recommended to perform it only once, for recording the process of generation. Hence, geometry data can also be generated by replaying recorded CAD commands in batch mode. It is not recommended to modify files of recorded CAD commands within a text editor.

Very few CAD systems offer capabilities for storing explicit topology data within the CAD data base, hence topology data must be regarded as standing alone information. Topology data can be best generated within the CAD system, where existing geometry data provide a visual support to this abstract exercise. However, topological links must be made between topological entities, and not between geometrical entities. It is helpful to denote topological entities with logical names.

As for topology data, the CAD system provides a helpful visual support for the generation of mesh data. Associating mesh data with topology data is the key for reusing data: changing geometry data associated with topology data, without changing topology nor mesh data, allows to generate similar meshes with different geometries.

It is assumed that the mesh generator can be executed out of the CAD system. It is recommended to have also mesh generation facilities within the CAD system, though it is not mandatory. If so, mesh data can be interpreted immediately, and the resulting mesh points displayed, in order to give the user an early feed back (e.g. the display of mesh curves gives an insight in the resulting mesh surfaces).

2. RECORD/REPLAY SYSTEM

The following Record/Replay System deals with the automatic generation of geometry data for meshes. It can be implemented on top of any CAD system through a reduced interface.

It is recommended that the CAD system can support handles for picking up and manipulating groups of geometrical entities, because such handles would provide a convenient visual interface to topological data (see key and control data, below).

2.1. Object Model and Objects

An object model is topology information. It compounds a set of named topological points, edges and faces, and a set of topological links, denoted patching vertices and edges, that patch topological faces together. An object model describes a category of objects, such as Twin Engine Aircraft, Four Engine Aircraft... An object model can be compound of more simple object models, such as fuselage, wing ...

An object is an occurence of an object model. For example, A320 and A330 are occurences of Twin Engine Aircraft, and A340 is an occurence of Four Engine Aircraft.

An object is the association of geometry data (key entities) with topology data. Objects are represented through their boundaries. The intersections between objects of a compound object must be explicit.

2.2. Blocking Structure and Multiblocks Wireframe

A blocking structure is topological information. It compounds a set of named topological vertices, block edges, block faces, blocks and a set of patching vertices, edges and faces, that either patch blocks together, or patch blocks to the boundaries of the domain. Several blocking structures can be related to the same object model: "one C-block" or "two H-blocks" are 2 admissible blocking structures around a wing.

A multiblocks wireframe is an occurence of a blocking structure. It is the association of geometry data (control entities) with topology data.

2.3. Record a new Object Model

This function creates a new object model AND the first occurence of it. It allows to define interactively, to name and to record the topological entities and links that compound the object model, with the visual support of existing geometry data.

The user picks up one or more primary entities (existing geometrical entity) or secondary entities (result of a CAD command acting upon primary or secondary entities), in order to compose groups, denoted key geometry data. The definition of secondary geometry data is recorded.

Key entities are associated one to one, with topology data of the object model. The user picks up their handle for specifying topological links.

2.4. Replay an Object Model

This function requires only a reduced number of interventions from the user, in order to create a new occurence of an object model.

The user associates existing primary geometry data with topology data. The object is then completed automatically, by replaying the definition of secondary geometry data

2.5. Record a new Blocking Structure

This function creates a new blocking structure AND the first occurence of it. It allows to specify once and for all, how the multiblock wireframe can be controlled by geometry data.

The user picks up primary or secondary entities, in order to compose groups, denoted control geometry data. It is recommended to base secondary entities upon key entities, and to use as few primary entities as possible (because primary entities cannot be replayed).

Control entities are associated one to one, with topology data data of the blocking structure. The user picks up their handle, for specifying topological links.

The handles of control geometry data can be used for editing mesh data.

2.6. Replay a Blocking Structure

This function can rebuild automatically a new occurence of a blocking structure, after major changes in the object geometry, provided that control geometry data are completely based upon key geometry data.

CONCLUSION

A record/replay system [3] has been implemented, at Aerospatiale Division Avion, on top of ICEM-DDN. Functionalities Record Blocking Structure and Replay Blocking Structure are supported, partly by this system and partly by ICEM-MULCAD.

Minor changes in the geometry of the object, are taken into account by replacing geometry key data input in the mesh generator. A new mesh (of 200 000 nodes) can be produced within 3h to 4h on a workstation.

Major changes in the geometry of the object need first to run function Replay an Object Model and then function Replay a Blocking Structure. For example, being given a recorded 22 blocks mesh around an A320, rebuilding the key geometry data, for an A330, cost half a day, and rebuilding the multiblocks wireframe cost one and half day. The final mesh is obtained after 4 more hours on a workstation.

For a new mesh topology, run function Record new multiblocks structure costs 10 days, for a 40 blocks mesh around an A340.

For a complete new configuration, add 1 day for function Record new Object model.

REFERENCES

[1] J.L. Steger, Technical evaluation report on the Fluid Dynamics Panel Specialists' meeting on Application of Mesh Generation to Complex 3D Configurations, AGARD Advisory Report N° 268, 1991.

[2] D. Bertin, E. Chaput, "Brite Euram/Euromesh Sub-Task 6.1: Topology generation within CAD systems", Aerospatiale, Technical Report N°443.541/92, 1992.

[3] D. Bertin, J. Lordon, V. Moreux, "A new automatic grid generation environment for CFD applications", 10th AIAA Applied Aerodynamics Conference, 22-24 Juin 1992, Palo Alto, California.

A TOPOLOGICAL MODELLER

Riccardo Scateni

C.E.R.F.A.C.S.

European Centre for Research and Advanced Training in Scientific Computing
42 avenue Gustave Coriolis, F-31057 Toulouse Cedex, France

1 INTRODUCTION

In recent years, due to the increasing power of computers, it has become feasible to simulate fluid flows in complex geometries. This has introduced a new level of complexity in the field of Computational Fluid-Dynamics (CFD): users need to be able to describe much more complex computational spaces to flow solvers; for an efficient use of computational resources the computational domains have to be subdivided in possibly many sub-domains. Within each of these sub-domains the flow is then discretized depending on the flow solver used. The splitting of the computational domain is primarily done to maintain uniform precision in the computation, by creating smaller domains, and thus denser grids in the critical regions of the computational space. Moreover, one often has to consider a multi-domain computational space to be able to parallelize the code and/or use different flow solvers on different sub-domains.

In this task we addressed the problem of providing an interface between users and flow solvers. Ideally users should only be involved in the description of the physical geometry, and be relieved from the problem of domain splitting. Unfortunately, it is currently beyond the state of the art to fully automatize the subdivision of the computational domain (unless the method used for the flow simulation accepts fully unstructured meshes, but this is not generally the case). Thus, users should be provided with the ability to define both the computational geometry and how the multi-domain splitting should be done.

It is a solved problem to provide the first capability: it is a trivial extension to a CAD program. It is, however, not obvious how to provide a system that generates, in an interactive and user-friendly way, a data structure completely describing the split domain[1]. The minimal set of features that this system should have is the following:

- Interactive definition/editing of sub-domains.

- Automatic sub-domains readjustment after a user refinement or coarsening of a sub-domain (see Fig. 1).

- Merging domains with automatic readjustment of connected domains (see Fig. 2).

- Undo capabilities (keeping an operation log).

Some of these features are already present in currently available programs. However, these programs tend to be very much hard-wired on specific flow solvers, and to be designed and constructed in a bottom-up fashion.

We felt that the best way to approach this problem is to follow the object-oriented paradigm. This makes simpler to write code that allows easy manipulation of data structures and is

---
[1]Data that can then be fed into any flow solver via a specific filter

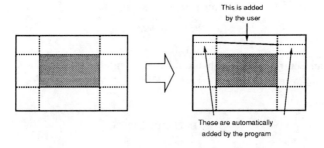

Figure 1: Automatic readjustment after a split

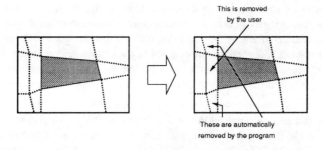

Figure 2: Automatic readjustment after a merge

robust, reusable and extendible. The latter quality is essential, since our top-down scheme implies a growth beyond the minimal implementation described above.

2 THE LAYOUT OF THE SYSTEM

This is the general layout of the object-oriented environment we designed. Consider it as a basic step of our proposed integrated environment for CFD applications.

2.1 Entities Needed in the Design of the System

First of all, following the object-oriented approach, we defined what were the types of objects (the *classes*) we needed to define for our purposes. To do this we considered what were the geometrical entities one has to deal with while operating in a CFD environment. We classified them as follows:

- There are 4 different categories of objects based on the number of geometrical dimensions they have: any geometrical entity can have 0, 1, 2 or 3 dimensions.

- We can have objects that are defined only by their *frontiers* or objects needing a larger amount of information.

- Each object can be a *simple object* or it can be defined in terms of simple objects.
- There can be more than one way to specify each object.

Based upon this classification we set the following rules:

- A **0D** geometrical object is a point belonging to an n-dimensional space ($n = 1, 2, 3$), that is a point described by n coordinates.

- A **1D** geometrical object is an entity defined at least by two **0D** objects. In the simplest case it is a segment; it can also be a straight line, a polyline (it is a *composite* object made by several segments connected at their endpoints), a chunk of cubic polynomial (since it is defined by more than two **0D** objects we can say it is an *over-defined* object) or a B-spline (it is, at the same time, *composite* and *over-defined*). For our purposes any of these objects is equivalent to a segment. In fact, in the interactive generation, any time we create a segment we can imagine that, at the same place, there could be any other object of the same dimension. Since we are generating a domain decomposition of the computational space, the most important informations we want to pass to the flow solver are informations about the connectivity of the domain. Fit the sub-domains to the real geometry is a minor task and an appropriate filter can perform it.

- A **2D** geometrical object is anything topologically equivalent to a square. So it can be a square, a rectangle or any other flat quadrilaterals but it can be as well a patch defined by two (or more, up to four) B-spline's as long as it keeps the property enunciated before, so it has four sides and four corners (in the case of a patch it will be an *over-defined* **2D** object).

- A **3D** geometrical object is anything topologically equivalent to a cube. Thus it can be a cube or any other kind of hexahedron.

Why only objects equivalent to square have been taken in account and not anything equivalent to any polygon with n sides? The assumption made is that we decided to model a tessellation of the plane inducted only by quadrilaterals, which are in turn only allowed to share a whole side. Under these conditions it is allowed to generate subdivisions like the one in Fig. 3a but not like the one in Fig. 3b.

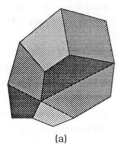
(a)
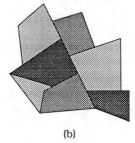
(b)

Figure 3: a) Legal subdivision of the domain. b) Illegal subdivision of the domain.

The only other reasonable choice would have been to generate only triangular sub-domains as it usually done when generated fully unstructured meshes with automatic or semi-automatic methods. It is quite obvious, however, that a mesh created in a triangular domain is not as well manageable as a mesh created in a quadrilateral domain. And, moreover, since the domain splitting is done by hand, it is not a problem to split in quadrilaterals rather than in triangles. The reasons to choose hexahedra in $3D$ are, of course, exactly the same than in the $2D$ case.

Each hexahedron is a *block* in the $3D$ domain and there is another very special kind of $3D$ object being the *domain* itself, considered as the juxtaposition of all the generated blocks. Obviously if the domain is not $3D$ but $2D$ the quadrilaterals are the blocks and the domain is a special type of $2D$ object.

Finally, as we said before, there can be more than one way to specify a certain object. Typically a $0D$ object (i.e. a point) can be defined in term of Cartesian coordinates, either in $2D$ or $3D$, in term of polar coordinates in $2D$ or in term of cylindrical or spherical coordinates in $3D$. What all of them have in common is the number of parameters: 2 for $2D$ Cartesian and polar, 3 for $3D$ Cartesian, cylindrical and spherical.

2.2 Classes

Once the *entities* involved in the project are defined the next step is to specify, for each entity, a layout of class reflecting its properties. This was accomplished specifying for each class the data it has to manage, the member functions it encapsulates and, eventually, the hierarchical relations with the other classes. A relation a class can have with another one can be either a *parent-child* relation or a reciprocal visibility of certain data.

The very first assumption we did at this level was to decide that only $0D$ objects carry geometrical informations, that is we record the description of the position in the domain only in the data structures encapsulated inside the object describing a point. All the other objects are carrying informations only on the *connections* they have with other objects. In other words changing the informations contained in the objects representing points we change the *geometry* of our scene, while changing informations contained in objects describing other entities (e.g. segments or rectangles) we change the *topology* of the scene.

2.3 Flexibility

How many ways there are to define a geometrical object, for instance a quadrilateral? It is quite clear that there is not only one. We can, in fact, specify it defining which are the four points at the corners, giving a segment and sweeping it in the plane, fixing two points that will be two opposite corners and maybe in other ways. It was important, from our point of view, to give to the user (or the programmer) as many ways as possible to generate new objects but at the same time not to confuse his ideas supplying too many functions, all named differently, and all doing (almost) the same things. In this case comes very useful the possibility to *overload* a function name. You overload a function when you actually define a *set* of functions having all the same name, but all operating with a different set of parameters. If this set of functions is associated to the name of the class *constructor*[2] both the previous conditions are respected since a single identifier is used to generate an object of a certain class in multiple ways.

---

[2]The class constructor is the function taking care to create a new instance of a certain class, a new object, allocating the needed memory, performing certain fixed operations and returning a pointer to the created object.

2.4 Maintenance of the Data Structures

The idea is to take advantage from the characteristics of the object-oriented design to keep totally separated the task of modifying the splitting of the computational domain and the task of modifying the data structures where the domain is mapped over. This means that a programmer of a CFD application will have the possibility to use a *tool-set* performing actions on the data structure maintaining the consistency of the connectivity tables.

Let us make an example. Consider that we have created two blocks sitting beside each other but not connected. When and if we decide to stick together these two blocks, the task of the application is to keep consistent the data structure eliminating a side (because the two blocks are going to share a side while before they didn't) and two points from the database. At the same time, it has to reorganize the system of connections of the two blocks reflecting this new condition (see Fig. 4).

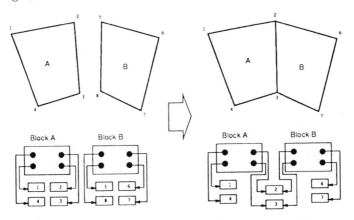

Figure 4: Two blocks stuck together

The availability of the object-oriented management system for the connectivity informations allows the programmer to say that he wants to *stick* the two blocks together while all the modifications are done implicitly by the piece of software sticking the two blocks together.

3 CURRENT IMPLEMENTATION

In the current implementation there is the availability of a library of classes and methods covering the **2D** space. The **3D** part is left as a future evolution.

Beside this there is a program available that can be considered as a proof of concepts of the system. It uses the library to implement an interactive domain editor. It allows the user to split a domain in multiple interconnected sub-domains. It is however at a prototype stage.

ADVANCING FRONT TECHNIQUE USED TO GENERATE BLOCK QUADRILATERALS

Thilo Schönfeld

C.E.R.F.A.C.S.

European Centre for Research and Advanced Training in Scientific Computing
42 avenue Gustave Coriolis, F-31057 Toulouse Cedex, France

SUMMARY

This short note presents an alternative approach of C.E.R.F.A.C.S. to sub-task 6.3.2 of the BRITE/EURAM EuroMesh project concerning the multi-block topology generation. The idea is to replace the interactive part of the common multi-block strategy by an automatic generation of the blocks. An unstructured mesh generation technique is used to automatically generate quadrangular blocks. In a separate step for each of the so-called macro-blocks a structured grid with rectangular cells is generated. Here we restrict ourselves to present some of the resulting grids. For a detailed description of the block generation algorithm we refer to the complete paper [1].

MACRO-BLOCK GENERATION

With the increasing demand to solve complex flow problems two principle strategies have been developed: the more classical *multi-block* technique and alternatively *unstructured* grid methods. The latter approach is distinguished by its nearly unlimited flexibility, since the generation of an unstructured grid is done locally and no special block topology needs to be defined for a given geometry. The major drawbacks to this method are the high computational costs and potentially poor accuracy due to distored elements. The regularity of quadrilateral grids is one of the main advantages of the multi-block technique. The aim of this work is to combine the advantages of both approaches.

The block generation procedure is based on the *advancing front technique* (AFT), a technique applied e.g. in [2] to generate small triangles that directly form the final grid cells. However, we use the AFT to generate large elements that serve as blocks and thus have to be covered by mesh lines to obtain the computational grid. The basic ideas behind the advancing front algorithm are retained, with the exception of two principal modifications. First, the generated *triangular* elements are replaced by *quadrilaterals*. As a consequence, two new nodes are generated simultaneously per "quadrangulation" step instead of only one point per triangulation step. Besides the common requirements for each new node, one must also resolve possible conflicts between these two points. The second major modification, motivated by the creation of *blocks* rather than small *cells*, is the generation of as large as possible elements. One framework is to combine adjacent faces to one large patch whenever the curvature is smooth.

Once the unstructured macro-blocks are generated, a mesh is then generated within each block. The block generator provides the input (in form of a set of registers) for both the mesh generator and the multi-block flow solver. The grids are generated by means of *transfinite interpolation* and an elliptic smoothing technique is applied across the block interfaces. The figures below show the blocks and the final meshes for the double ellipse, a multi-element airfoil, and a valve-cylinder assembly.

REFERENCES

[1] Schönfeld, T., Weinerfelt, P. : "The Automatic Generation of Quadrilateral Multi-Block Grids by the Advancing Front Technique", Proceedings of the *Third International Conference on Numerical Grid Generation* in Barcelona, pp. 743-754, Elsevier Science Publishers, North Holland, June 1991.

[2] Peraire, J. et al : "Adaptive Remeshing for Compressible Flow Computations", J. Comp. Phys. 72, pp. 449-466, 1987.

FIGURES

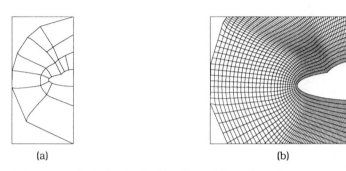

(a) (b)

Figure 1: a) Blocks for double-ellipse. b) Final mesh.

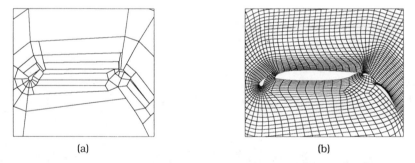

(a) (b)

Figure 2: a) Blocks for multi-element airfoil. b) Close-up of the mesh.

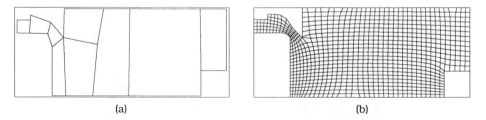

(a) (b)

Figure 3: a) Blocks for valve assembly. b) Final mesh.

III. SURFACE GRID GENERATION AND GEOMETRY MODELLING

SURFACE MESH GENERATION USING PROJECTIONS

José M. de la Viuda

Aerodynamics Dept., Dirección de Proyectos

CASA, Avda John Lennon s/n, GETAFE (SPAIN)

SUMMARY

This article describes a surface grid generation method developped in the EUROMESH project, a project sponsored by the Comission of the European Communities. The surface grid is generated in two steps. In the first one, an algebraic grid is generated from the edge definition of the grid. In the second one, the inner points of the algebraic grid are projected on the geometric patches. These geometric patches are modelled as Bezier surfaces. The process is performed by dividing the surface of the object into rectangles or triangles (sub-faces). The curves defining these rectangles or triangles are the edge mesh. Some applications are shown.

1. INTRODUCTION

In aerospace Industry, geometry definition is done within CAD systems, in most cases, using Bezier Surfaces. A solution to generate grids for CFD is to use the CAD definition as the starting point of the grid generation process. The objective of this work is to generate surface grids from the CAD definition without dependence of the CAD system in which the surfaces have been generated. The approach chosen is to generate an algebraic grid near the surfaces and to project its points on the CAD patches. The projection algorithm can be used in optimization and adaption methods.

2. DESCRIPTION OF THE PROCESS

The starting data are the edge mesh, the points defining the surfaces and the links between the edge mesh and the surface definition. There is no restriction on the origin of these data, although we retrieve them from our CAD system, ICEM from Control Data.

The surface of the object is divided into topologic squares, called "sub-faces". These "sub-faces" are defined by four edges, so that the edge mesh is made of all the points defining the "sub-faces" edges. All these points will be points of the final grid, and will control the density and the patterns of the surface grid.

The process of grid generation is performed by the use of "sub-faces". The process is divided into two steps: First, an algebraic grid is generated from the edge mesh. Second, the inner points of the algebraic grid are projected onto the surfaces of definition. To project the inner points of each "sub-face" grid, some aditional information is required : the patches on which the projections must be performed. This information is included in the structure of the surface data file. Since the edge definition is independent for each "sub-face" it is possible to have adjacent "sub-faces" with different number of points.

3. ALGEBRAIC SURFACE MESH

The initial grid is computed from the "sub-face" edge definition, using the following formulae :

$$s_{ij} = \text{(j-line length between i=1 to i)} / \text{(total j-line length)},$$
$$t_{ij} = \text{(i-line length between j=1 to j)} / \text{(total i-line length)}.$$

s_{ij} and t_{ij} are optimal distributions if :

$$s_{ij} = s_{i1}(1 - t_{ij}) + s_{ijm}t_{ij},$$
$$t_{ij} = t_{1j}(1 - s_{ij}) + t_{imj}s_{ij},$$

this system yields :

$$s_{ij} = \frac{s_{i1} + t_{1j}(s_{ijm} - s_{i1})}{w_{ij}},$$
$$r_{ij} = \frac{t_{1j} + s_{i1}(t_{imj} - t_{1j})}{w_{ij}},$$

where : $w_{ij} = 1 - (s_{ijm} - s_{i1})(t_{imj} - t_{1j})$.

The interpolation is performed in two steps. First, in the i direction (s_{ij}), between the edges 1 and 3. Second, in the j direction (t_{ij}), between the edges 2 and 4:

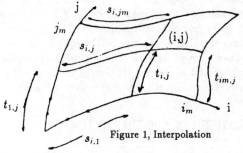

Figure 1, Interpolation

The edge points are kept as points of the final grid, so, only the inner points are projected. For sub-faces defined as part of revolution or ruled surfaces, the points obtained using this interpolation lay on the surface.

4. SURFACE MODEL

At the present moment the method can only project on Bezier surfaces. They have been selected beacuse of their importance for CAD systems. The mathematical model used is :

$$B(u,v) = \sum_{i=0}^{m} \sum_{j=0}^{n} P_{ij} \binom{m}{i} u^i (1-u)^{(m-i)} \binom{n}{j} v^j (1-v)^{(n-j)},$$

where : B(uv) = (x, y, z) ; $0 < u < 1; 0 < v < 1$,

and : $P_{ij} = (x_{i,j}, y_{i,j}, z_{i,j})$.

5. PROJECTION ALGORITHM

5.1 Mathematical Model

Given an initial mesh and a CAD patch (a Bezier surface), mesh points are projected on the patch following a fixed direction (direction for projection). This direction can be fixed or predetermined from the mesh points (the cross product of the vectors defined by the points (i,j), (i+1,j) and (i,j+1).

The equations representing this process are the following :

- surface model: (1)

$x = f_x(u,v)$,
$y = f_y(u,v)$,
$z = f_z(u,v)$.

- straight line equations (2)

$\frac{x-x_0}{a} = \frac{y-y_0}{b}$, $\frac{y-y_0}{b} = \frac{z-z_0}{c}$,

where x_0, y_0, z_0 are the coordinates of the mesh point and a, b, c the coordinates of the direction for projection.

(1) and (2) can be rewritten as :

$$\frac{f_x(u,v)-x_0}{a} = \frac{f_y(u,v)-y_0}{b} \quad , \quad (3)$$

$$\frac{f_y(u,v)-y_0}{b} = \frac{f_z(u,v)-z_0}{c} \quad . \quad (4)$$

(3) and (4) form a system of two non-linear equations with two unknowns (u,v).

The system is rewritten as :

$$F(u,v) = \frac{f_x(u,v)-x_0}{a} - \frac{f_y(u,v)-y_0}{b} \quad , \quad (5)$$

$$G(u,v) = \frac{f_y(u,v)-y_0}{b} - \frac{f_z(u,v)-z_0}{c} \quad . \quad (6)$$

The system is solved using a Newton method. This method requires good initial values; since, the search for initial values is a key part of the method.

5.2 Approximated Values

Several methods have been tested for obtaining good initial values, after many attempts, the approach described herein has been chosen.

The Bezier surfaces are defined by the polygon points, these polygon points can define a new "multiplane" surface by taking two consecutive points in u direction and two consecutive points in v direction. The resulting "multiplane" surface will be defined by (NU-1)x(NV-1) planes, each plane related to one rectangular region (u_i, u_{i+1}) (v_i, v_{i+1}) in the parameter plane (Figure 2).

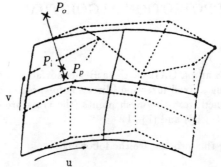

Figure 2, Search for initial values

For each plane, the intersection with the projection line is computed (pi), obtaining the coordinates in the x,y,z space. A box test is performed to determine if the intersection point lays in the rectangle defined by the region (ui,ui+1), (vi,vi+1). If the box test is positive, the u,v coordinates are obtained from the x,y,z coordinates using interpolation formulae :

$$ui= (\sum_{i=1}^{3} u1+(u2-u1)/(x2(i)-x1(i))*(xp(i)-x1(i)) + u1+(u2-u1)/(x4(i)-x3(i))*(xp(i)-x3(i)))/6 ,$$

$$vi= (\sum_{i=1}^{3} v1+(v2-v1)/(x3(i)-x1(i))*(xp(i)-x1(i)) + v1+(v2-v1)/(x4(i)-x2(i))*(xp(i)-x2(i)))/6 .$$

where :

x1 is the point defined by ui,vi ,
x2 is the point defined by ui+1,vi ,
x3 is the point defined by ui,vi+1 ,
x4 is the point defined by ui+1,vi+1 .

5.3 Newton Method

This method solves the following system of equations: (7)

$$\Delta u \frac{\partial F}{\partial u} + \Delta v \frac{\partial F}{\partial v} = u_0 ,$$

$$\Delta u \frac{\partial G}{\partial u} + \Delta v \frac{\partial G}{\partial v} = v_0 ,$$

$\frac{\partial G}{\partial u}$ and $\frac{\partial G}{\partial v}$ are computed at u_0, v_0 .

To obtain Δu and Δv, the new u and v values are obtained as $u_1 = u_0 + \Delta u$ and $v_1 = v_0 + \Delta v$. If $\frac{\Delta u}{u_0}$ and $\frac{\Delta v}{v_0}$ are less than a fixed tolerance the process stops and the solution is suppossed to have been found, if not, the values of u_1 and v_1 are introduced in 7 and the process starts again.

As a result of the test case computations described below, three criteria have been included to handle the cases of oscillating solutions :

1) The values of du and dv in the current step (n) are smaller than those of the step n-1 by a fixed tolerance , and the number of iteration is greater than a fixed value.

2) The values of du and dv in the current step (n) are smaller than those of the step n-2 by a fixed tolerance , and the number of iteration is greater than a fixed value.

3) The values u and v in the current step (n) are smaller than those of the step n-1 by a fixed tolerance , and the number of iteration is greater than a fixed value.

In any of these cases, the solution found is correct.

5.4 Multipatch Case

The program reads the polygon points coordinates for all the patches related to the treated "sub-face" and stores them in the memory; this process is repeated for each sub-face, so, if a patch is related to more than one "sub-face", its polygon points coordinates will be stored more than one time in the data file.

To project the inner points of a "sub-face", a loop on all the patches related to it is done. Within this loop an "intersection test" is performed for each patch: The intersection between the line defined by the point to project and the direction for projection, and a quadrilater defined by the four corner patch points is computed. If the intersection point lays in a box defined by the patch corner points the point projects on this patch and the test is positive. In this case the process described in (4.2) starts.

In some cases a patch can have the four corner points in a plane and the edge curves outside that plane. To handle these cases the box test has been relaxed so, there will be cases in which the result is positive but there is no projection.

The use of this "intersection test" reduces the number of calls to the TESIN routine, that searchs ui,vi values for a given patch. If the TESIN routine yields values for ui and vi, the system of equations is solved and, if there is solution the point is stored; if not a message will appear and the riginal point (of the algebraic grid) will be imposed in the projected grid.

If more than one projection points have been found, the program computes the distance between them and the original point, and the nearest point is chosen as the projected point.

6. RESULTS

Two main configurations have been tested : a fighter aircraft and a commuter aircraft.

6.1 Fighter Aircraft

In this case the surface grid of a canard aircraft has been generated, the number of patches is low and the computing time is also low. The caracteristics of this configuration are :

Number of sub-faces	:	74
Number of patches	:	623
Number of edge points	:	1454
Number of proj. points	:	1182
Total number of points	:	2736

All the sub-faces have been projected

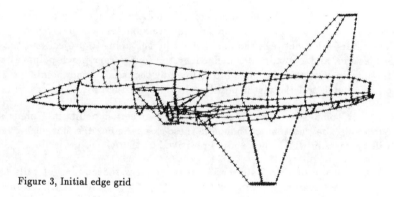

Figure 3, Initial edge grid

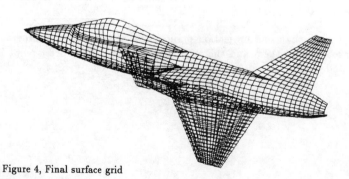

Figure 4, Final surface grid

6.2 Commuter Aircraft

This case is specially complex because of the following reasons :

-Geometry complex in some regions (wing-fuselage fairing, tailcone)

-Surface definition made of a great number of patches

-Combines regions with and without projections.

The caracteristics of this configuration are :

Number of sub-faces	:	79
Number of patches	:	4790
Number of edge points	:	754
Number of non-proj. points	:	554
Number of proj. points	:	561
Total number of points	:	2069

Only 37 sub-faces have been projected.

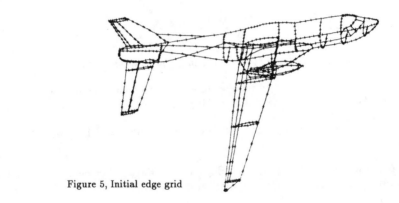

Figure 5, Initial edge grid

Figure 6, Projected sub-faces

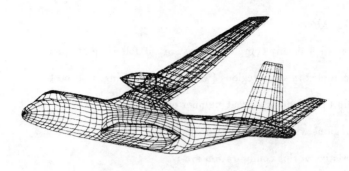

Figure 7, Final grid

6.3 Computing Times

If the number of patches used are not too many (f.e. the commuter case), the program can be run in a workstation, and the computing time is about 5 minutes for the body and 20 minutes for the wing (because there are a lot of points and many patches). whole aircraft. For the commuter configuration the computing time in a CRAY-XMP was of 30 minutes.

7. CONCLUSIONS

-The approach chosen is conceptually simple and can be applied to any type of surfaces.

-It is Recommended to use a low number of patches to define the object (if possible). In this case the computing time will be reduced.

-By modifying the edge grid, it is easy to change the final grid (number of points or shape).

-This method provides a CAD-system-independent surface grid generator and a projection algorithm that can be used for optimization or adaption.

8. REFERENCES

1. De la Viuda, Diet, Ranoux. "Patch Independent Structured Multiblock Grids for CFD Computations", in Numerical Grid Generation in CFD and Related Fields. Barcelona, June 1991.

2. Sony B.K. "Two and Three Dimensional Grid Generation for Internal Flow Applications of CFD", AIAA Paper 85-1526, AIAA 7th Computational Fluid Conference, Cincinnati, Ohio, July 1985.

GENERATION OF SURFACE GRIDS USING ELLIPTIC PDEs

Per Weinerfelt

C.E.R.F.A.C.S.

European Centre for Research and Advanced Training in Scientific Computing
42 avenue Gustave Coriolis, F-31057 Toulouse Cedex, France

SUMMARY

Below follows a brief description of a surface grid generator, developed at CERFACS, aimed at the sub-task 1.3.1 in the BRITE/EURAM EuroMesh project. We assume that the given surface is described on parameter form. A system of nonlinear elliptic PDEs, similar to those used for generating planar curvilinear grids, can be deduced for the parameters. These equations are discretized and solved iteratively. This yields a finite set of parameter values from which the corresponding grid points on the surface are computed. We refer to paper (1) for a detailed description.

AN ELLIPTIC SURFACE MESH GENERATOR

There are today several techniques available for generating curvilinear grids on surfaces. We will here distinguish between algebraic methods and and methods based on differential equations. The present work has mainly been concerned with the last approach. A common way of generating 3D volume grids is to solve the system of elliptic PDEs described below:

$$\alpha_1(\bar{r}_{\xi\xi} + P\bar{r}_{\xi}) + \alpha_2(\bar{r}_{\eta\eta} + R\bar{r}_{\eta}) + \alpha_3(\bar{r}_{\zeta\zeta} + Q\bar{r}_{\zeta}) + 2(\beta_1\bar{r}_{\xi\eta} + \beta_2\bar{r}_{\eta\zeta} + \beta_3\bar{r}_{\xi\zeta}) = 0, \quad (1)$$

where $\bar{r} = (x,y,z)$ denote the coordinates in the physical space, ξ, η, ζ the coordinates in the computational space, P, Q, R are control functions and $\alpha_1, \alpha_2, \alpha_3$ and $\beta_1, \beta_2, \beta_3$ nonlinear functions of the derivatives of $\bar{r}$:

$$\begin{cases} \alpha_1 = (|\bar{r}_{\eta}||\bar{r}_{\zeta}|)^2 - (\bar{r}_{\eta} \cdot \bar{r}_{\zeta})^2, & \beta_1 = (\bar{r}_{\eta} \cdot \bar{r}_{\zeta})(\bar{r}_{\zeta} \cdot \bar{r}_{\xi})^2 - (\bar{r}_{\eta} \cdot \bar{r}_{\xi})|\bar{r}_{\zeta}|^2, \\ \alpha_2 = (|\bar{r}_{\zeta}||\bar{r}_{\xi}|)^2 - (\bar{r}_{\zeta} \cdot \bar{r}_{\xi})^2, & \beta_2 = (\bar{r}_{\zeta} \cdot \bar{r}_{\xi})(\bar{r}_{\xi} \cdot \bar{r}_{\eta})^2 - (\bar{r}_{\zeta} \cdot \bar{r}_{\eta})|\bar{r}_{\xi}|^2, \quad (2) \\ \alpha_3 = (|\bar{r}_{\xi}||\bar{r}_{\eta}|)^2 - (\bar{r}_{\xi} \cdot \bar{r}_{\eta})^2, & \beta_3 = (\bar{r}_{\xi} \cdot \bar{r}_{\eta})(\bar{r}_{\eta} \cdot \bar{r}_{\zeta})^2 - (\bar{r}_{\xi} \cdot \bar{r}_{\zeta})|\bar{r}_{\eta}|^2. \end{cases}$$

These equations can also be used to generate surfaces grids by restricting the equations to the surface. Let us first assume that the surface is expressed on parameter form. This is not a severe limitation since parameter representations of surfaces are used in many practical applications. A complex surface can e.g. be subdivided into simpler patched sub surfaces described by piecewise polynomials. The parametrization is now used to transform the equations in (1) to the following form:

$$\begin{cases} |\bar{r}_{\eta}|^2(s_{\xi\xi} + Ps_{\xi}) + |\bar{r}_{\xi}|^2(s_{\eta\eta} + Rs_{\eta}) - 2(\bar{r}_{\eta} \cdot \bar{r}_{\xi}) s_{\xi\eta} + (s_{\xi}t_{\eta} - s_{\eta}t_{\xi})\phi = 0, \\ |\bar{r}_{\eta}|^2(t_{\xi\xi} + Pt_{\xi}) + |\bar{r}_{\xi}|^2(t_{\eta\eta} + Rt_{\eta}) - 2(\bar{r}_{\eta} \cdot \bar{r}_{\xi}) t_{\xi\eta} + (s_{\xi}t_{\eta} - s_{\eta}t_{\xi})\psi = 0, \end{cases} \quad (3)$$

where ϕ and ψ are terms related to the curvature of the surface. We have hence reduced the original system to a 2×2 system of nonlinear PDEs with the parameters s and t as the unknowns. The solution to these equations yield a set of points, in the parameter space, from which the corresponding grid points on the original surface can be computed. We will now briefly describe how to obtain an approximative solution to (3).

The equations are first discretized using central difference approximations of the derivatives. The resulting nonlinear system of equations are then solved iteratively by the Jacobi method. Alternatively the Gauss-Seidel method can be applied. A multi grid technique has also been introduced in order to speed up the convergence of the iterative procedure.

A start solution is obtained from a linear transfinite interpolation between the four boundaries. It is in many cases necessary to do only a few iterations in order to obtain a fairly good mesh. The iterative method can in this situation be regarded as a smoother of the starting mesh.

Two test examples, a saddle formed surface and a double ellipsoid surface, are shown in figure 1a and 2a. The final surface grids, obtained from the method above, are displayed in figure 1b and 2b.

A computer implementation of this mesh generation technique has also been done. The code, called SURFMESH, is written in FORTRAN 77. It is constructed as a tool box containing different separate subroutines such as grid point generation on the surface edges, generation of the initial mesh, grid smoothing and so forth.

The input to the computer program is the surface parametrization. This can e.g. be given by a tensor product of B-spline functions. In that case the spline coefficients are the input data from which the derivatives along the surface can be formed analytically. The final output from the program is the grid point distribution on the surface.

REFERENCES

[1] Toshiyuki Takagi, Kazuyoshi Miki, Brian C.J. Chen and William T. Sha:
"Numerical Generation of Boundary-Fitted Curvilinear Coordinate Systems for Arbitrary Curved Surfaces", Journal of Computational Physics 58, 67-79 (1985).

FIGURES

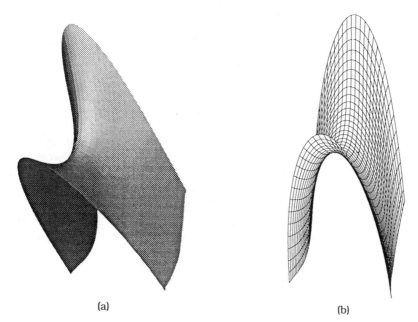

Figure 1: a) Saddle formed surface. b) Curvilinear grid.

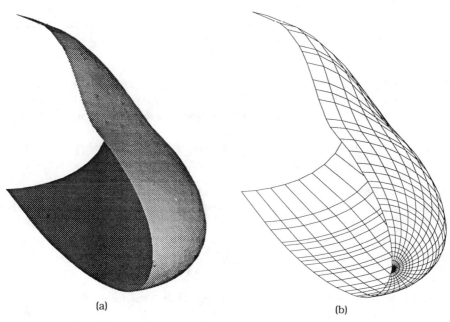

Figure 2: a) A double ellipsoid. b) Curvilinear grid.

47

GENERATION OF STRUCTURED MESHES OVER COMPLEX SURFACES

B. Morin, V. Tréguer-Katossky

AEROSPATIALE DIVISION AVIONS
316, Route de Bayonne
31060 Toulouse Cedex 03, FRANCE.

SUMMARY

A new method for generating structured 2D meshes over assemblies of geometrical patches is presented. The emphasis is on the storage of topological informations into the "Complex Surface" data structure, allowing gaps and overlaps between the patches. The structured grid is generated over a coarse unstructured mesh, in such a way that every grid point lay on the true surface. The steps in the generation of a structured mesh covering a part of a fuselage are shown as an example.

INTRODUCTION

Boundary fitted multi-block structured meshes for 3-dimensionnal complex flow configurations need first generating surface meshes over the body. The geometry data for the surface definition are produced within CAD systems. They generally compound several parametric curved surfaces (geometrical patches) which may be ill-parametrized when mesh generation is concerned. Moreover intersections, gaps and overlaps between patches are often met.

Surface mesh generation ordinarily copes with interpolation problems: in most techniques each structured grid is over one single geometrical patch. Since the blocking structure of the mesh does not necessarily coincide with the geometrical patches, the geometry data needs generally to be prepared: the difficulty is then to control the accuracy when defining new patches that fit the blocking structure.

Although the surface mesh generators are effective, the pre-processing of geometry data is recognized as consuming a lot of man power. It also makes difficult the development of adaptive mesh facilities. Thus, Task 8 "Geometric Surface Modelling" of the Brite/Euram "Euromesh" project aimed at studying mathematical representations of curved surfaces, for coupling with mesh generators.

One partner has investigated how to redefine the geometry data with Coon's multipatches or NURBS before mesh generation [1], another partner has studied the projection onto Bezier Surfaces[2]. These two approaches are worth combining in a mesh generator: first generate a mesh on a Coon's multipatch and then project it onto the true surface.

The other partners have taken a different approach, focusing on the parametrization of unstructured sets of patches[3,4]. No approximation of the patches is to be undertaken. Parameters are computed so that the true surface appears to the mesh generator as a single patch regularly parametrized. This approach was based on Suzuki's work [5,6], who first introduced an unstructured mesh in the process of structured mesh generation.

The main feature added to Suzuki's work is the management of geometry data associated to every elements of the unstructured mesh, that allows to project the structured grid back onto the true surface.

An example illustrates the method, step by step: it consists in generating a structured mesh over the square area ABCD shown on Figure 1.

1. COMPLEX CURVES AND COMPLEX SURFACES

"Complex Curve" and "Complex Surface" denote special data structures used for specifying the topology of a surface area, delimited by a closed contour, over an unstructured set of geometrical patches.

Some CAD systems support very similar notions as "Topological Surface" or "Trimmed Surface", but most of them do not. The Complex Curve and Complex Surface data structures are intended to be implemented on top of CAD systems that support only conventional surfaces.

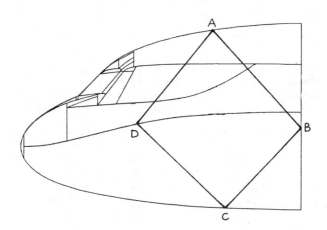

Figure 1: Front part of the fuselage of Airbus A320

1.1. Basic entities

Basic entities should be the most compound entities supported by the CAD package: for example, composite curves should be preferred to curves and regular arrays of patches should be prefered to single patches.

Basic Curves are supposed to admit a global parametrization, that may be piecewise polynomial, B-spline or NURBS:

$$x=x(u) \quad y=y(u) \quad z=z(u), \qquad umin <u< umax, \qquad (1)$$

A similar assumption is made for Basic Surfaces:

$$x=x(u,v) \quad y=y(u,v) \quad z=z(u,v), \qquad \begin{array}{l} umin < u <umax, \\ vmin < v <vmax. \end{array} \qquad (2)$$

Derivatives are also supposed to be available from the CAD package at any point of a basic entity.

1.2. Complex Curves

A Complex Curve is made of one or several "segments". It can be Closed or Open.

A segment of a Complex Curve is associated to a portion of a basic curve limited by 2 endpoints. The segment is oriented from endpoint Origin to endpoint Destination. It may or not have a "Next" segment belonging to the same Complex Curve. It may or not have a "Symmetric" segment belonging to a Complex Curve linked to this one in a Complex Surface.

The orientation of the segments and Next relationship determine the direction of the path along the Complex Curve.

1.3. Complex Surface

A Complex surface is made of one or several "trimmed surfaces" linked to each others through segments of their contours. Its outer contour is explicitely described as a Complex Curve.

A trimmed surface associates a closed Complex Curve (contour) to a Basic Surface. The trimmed surface lays on the left-hand side of the contour, as defined by the orientation of its segments.

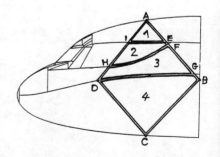

Figure 2: Complex Surface

The following data define the complex surface ABCD shown on Figure 2.

Basic Surfaces	Basic Curves
1. Top 2. Front 3. Side 4. Bottom	1. TopBoundary 2. FrontBoundary 3. SideBoundary 4. BottomBoundary 5 to 8. FuselageCurves

Complex Curves

1. AEI:
   1. FuselageCurve5 from A to E, Next=1.2, Sym=5.1
   2. TopBoundary from E to I, Next=1.3, Sym=2.1
   3. FuselageCurve8 from I to A, Next=1.1, Sym=5.9

2. IEFH:
   1. FrontBoundary from I to E, Next=2.2, Sym=1.2
   2. FuselageCurve5 from E to F, Next=2.3, Sym=5.2
   3. FrontBoundary from F to H, Next=2.4, Sym=3.1
   4. FuselageCurve8 from H to I, Next=2.1, Sym=5.8

3. HFGD:
   1. SideBoundary from H to F, Next=3.2, Sym=2.3
   2. FuselageCurve5 from F to G, Next=3.3, Sym=5.3
   3. SideBoundary from G to D, Next=3.4, Sym=4.1
   4. FuselageCurve8 from D to H, Next=3.1, Sym=5.7

4. DGBC:
   1. BottomBoundary from D to G, Next=4.2, Sym=3.3
   2. FuselageCurve5 from G to B, Next=4.3, Sym=5.4
   3. FuselageCurve6 from B to C, Next=3.4, Sym=5.5
   4. FuselageCurve7 from C to D, Next=4.1, Sym=5.6

5. ABCD:
   1. FuselageCurve5 from A to E, Next=5.2, Sym=1.1
   2. FuselageCurve5 from E to F, Next=5.3, Sym=2.2
   3. FuselageCurve5 from F to G, Next=5.4, Sym=3.2
   4. FuselageCurve5 from G to B, Next=5.5, Sym=4.2
   5. FuselageCurve6 from B to C, Next=5.6, Sym=4.3
   6. FuselageCurve7 from C to D, Next=5.7, Sym=4.4
   7. FuselageCurve8 from D to H, Next=5.8, Sym=3.4
   8. FuselageCurve8 from H to I, Next=5.9, Sym=2.4
   9. FuselageCurve8 from I to A, Next=5.1, Sym=1.3

Complex Surfaces

Trimmed Surfaces:
   1. Top surface restricted to AEI
   2. Front surface restricted to IEFH
   3. Side surface restricted to HFGD
   4. Bottom surface restricted to DGBC

OuterContour: ABCD

Links:
   1. IAE 1.2 to IEFH 2.1
   2. IEFH 2.3 to HFGD 3.1
   3. HFGD 2.3 to DGBC 4.1
   4. ABCD 5.1 to IAE 1.1
   ... (see Sym values in Complex Curve ABCD above)
   12. ABCD 5.9 to AEI 1.3

2. MESH GENERATION OVER A COMPLEX SURFACE

Once several Basic Surfaces have been organized into a 4 sided Complex Surface, it is possible to mesh them in a single operation. Following Suzuki [5,6], the Complex Surface is first approximated with a coarse unstructured mesh. Then the intersections of a regular network of piecewise linear curves are computed. Finally, the structured mesh is obtained by mapping every intersection points to the underlying true surface.

2.1. Triangulation of a Complex Surface

The Triangulation data structure looks like an ordinary finite elements data structure where each triangle K is associated to a Basic Surface BS(K), and to the locations (u_i,v_i), i=1...3 of its vertices relative to BS(K).

The triangulation of a Complex Surface proceeds one Trimmed Surface after the other. The contour of this Trimmed Surface is first discretized, then an Advancing Front algorithm [4] is used for generating mesh points in its interior. Each time a segment of contour is discretized, the information is passed to the symmetric segment (if any) that belongs to the adjacent Trimmed Surface.

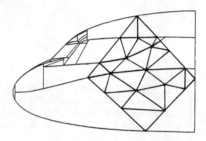

Figure 3: Triangulation

2.2. System of Parameters

The problem at the hand is to construct a regular parametrization (r,s) of a 4 sided triangulated domain Ω. First, curvilinear abcissa σ_i, i=1...4 are built on the 4 sides Γ_i, i=1...4 of the boundary. Then the following system of partial differential equations is discretized:

$$\nabla(\lambda \nabla g)=0, \text{ in } \Omega \qquad \text{where } g = (r,s) \qquad (3)$$

with the following boundary conditions:

$$\begin{aligned} g &= (\sigma_1,0), \text{ on } \Gamma_1 & g &= (\sigma_3,1), \text{ on } \Gamma_3 \\ g &= (1,\sigma_2), \text{ on } \Gamma_2 & g &= (0,\sigma_4), \text{ on } \Gamma_4 \end{aligned} \qquad (4)$$

The discrete solution of this problem gives the values of r and s at each vertices of the triangulation. These values can be interpolated for obtaining piecewise linear functions. Parameter λ controls the grid spacing: if the radius of curvature of the surface is used, the grid will automatically adapt to the curvature.

2.3. Mesh Generation

A regular mesh is first defined in the parametric domain (r,s):

$$g_{i,j} = (r_i, s_j), \quad i=1..N_i, \; j=1..N_j, \quad (5)$$

Each mesh point is then mapped to a triangle $K = \{V_1, V_2, V_3\}$ where:

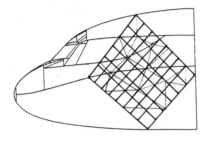

Figure 3: Structured Mesh

$$g_{i,j} = \alpha g(V_1) + \beta g(V_2) + \gamma g(V_3),$$
$$\text{with } \alpha+\beta+\gamma = 1 \quad (6)$$

and then mapped again to the parametric domain (u,v) of the underlying Basic Surface BS(K):

$$u_{i,j} = \alpha u(V_1) + \beta u(V_2) + \gamma u(V_3) \quad (7)$$
$$v_{i,j} = \alpha v(V_1) + \beta v(V_2) + \gamma v(V_3).$$

Finally, mesh points on the true surface are obtained with mapping (2).

CONCLUSION

The method can be coupled with patch dependant mesh generators, for providing patch independant facilities.

In principle, even when the geometry data consists of one single patch, Susuki's method is known to be robust where other methods fail due to singularities in the parametrization. Unfortunately, since method AFT is used in the parametric domain of the basic surfaces, our method is not able to generate the triangulation where singularities are found.

In order to take advantage of this property of Suzuki's method, the triangulation should be build in 3D space.

REFERENCES

[1] E. Chaput, "Brite Euram Euromesh Sub-Task 8.1: Geometric surface

modelling using Coon's multipatches and NURBS", Aerospatiale Technical Report N°, 1992.

[2] J. de la Viuda, "Euromesh Task 8.2, Surface Modeling, Technical Report", CASA TN AVA-AB-GEN-NT-91164, 1991.

[3] B. Morin, V. Tréguer-Katossky, "Brite/Euram Euromesh Sub-Task 8.1: Generation of structured mesh over complex surfaces", Aerospatiale Technical Report, N° 443.512/92, 1992.

[4] S. Farestam, "Brite/Euram Euromesh Sub-Task 8.1: Reparametrization of block boundary surface grids3, CERFACS Internal Report N°.,1991.

[5] M. Suzuki, "Surface grid generation based on unstructured grid", AIAA-90-3052-CP, pp. 948-952, 1990.

[6] M. Suzuki, "An approach to the surface grid generation based on unstructured grid", in Numerical Grid Generation in Computationnal Fluid Dynamics and Related Fields, A.S. Arcilla, J. Haüser, P.R.Eiseman, J.F. Thompson (eds), North-Holland, pp. 947-953, 1991.

SURFACE MODELLING USING COONS MULTIPATCH AND NON-UNIFORM RATIONAL SURFACE

E. Chaput

Aerodynamics Department
AEROSPATIALE Espace & Defense
BP 96 - F-78133 Les Mureaux Cedex - FRANCE

ABSTRACT

The purpose of this paper is to define the mathematical models to be used by any out of CAD system multiblock mesh generator in order to define any surface which supports a face of blocks. Two levels of accuracy are expected to take into account, the fluid interface between two blocks, given here by a Coons multipatch model, and in the other hand the faces of blocks which lie onto the body wall provided by a non-uniform rational surface model. The inverse problem getting parameters of any given point onto the surface is solved by a projection algorithm. Some tools are delivered to check the accuracy of the surface models by comparison with the analytical definition of the surface if available.

INTRODUCTION

The multiblock mesh generation for CFD around a complex body requires an effective interactive tool to create all the curved surfaces involved in block faces geometrical support. In industrial environment these surfaces are defined inside CAD/CAM system according to a specified multiblock topology. The underlying mathematical model is then available to create a mesh and to handle nodes onto faces in order to optimize it.

However most of the applications onto realistic configurations has pointed out that an adaption of such a 3D mesh to the flow solver solution is often needed to improve the quality of the result. The most effective way is then to perform this step without going back to CAD system in order to be able to run an iterative procedure very close to the flow solver. This adaption cycle only makes use of the existing surfaces. However it requires the knowledge of their mathematical models in order to allow mesh nodes displacement onto surfaces according to the constraint defined by the flow solver sensors.

The work presented here gives some mathematical tools providing to any out of CAD system mesh generator the capability to move nodes lying onto 3-D curved surfaces defined only by a set of m x n points as reference grid. A Coons multipatch model [1] using Hermite blending functions is proposed to describe interblock boundaries within the flow field which require a lower level of accuracy than body surfaces. The tangent vectors and twist vectors required for Coons model are provided by a quasi-intrinsic parametrization along the curves defined by reference grid [2,3]. For a very uneven reference grid a mapping onto the parameter patch is introduced [4].·A more accurate model is obtained using a non-uniform rational Bezier surface according to Piegl's method [5].

For the two models previously defined, the coordinates of points on the surface are provided using a routine Get_xyz knowing the desired parameter set (u,v). An inverse algorithm has been developed to get the parameter u and v of any given point x, y, z lying or not on the surface. Some analysis tools are available to check the quality of the resulting interpolated surface if the analytical definition is known. An application of the program to a double-ellipsoide shape is presented.

1 - COONS MULTIPATCH

We assume that all the faces of a 3D multiblock mesh are available into a file which describes these faces by the mean of a set of points placed onto a reference grid. The only available data are: *NS* the number of reference grid in the file, (*nk, nl*) the number of points in each direction of the grid, and $x_{k,l}$, $y_{k,l}$, $z_{k,l}$ for $1 < k < nk$ and $1 < l < nl$, the coordinates of the points. The boundary conditions at the border of each faces are specified using a code *kb* equal to zero if an extrapolation from inner points is required, lower than zero for a symmetry (normal vector to the symmetry plane is given), and greater than zero if additional points are available (*kb* points are given).

A bi-cubic patch is defined in each individual cell of the grid by the interpolation formula of Coons' model :

$$X(\xi,\eta) = F(\eta) \, A \, F(\xi)^T \qquad 0 < \xi < 1 \text{ and } 0 < \eta < 1 \qquad (1)$$

using Hermite blending functions: $F_1(t) = (1+2t)(1-t)$, $F_2(t) = (3-2t)$, $F_3(t) = t(1-t)^2$, $F_4(t) = t^2(t-1)$ and the matrix *A* of the sixteen degrees of freedom at the corner of the patch.

In order to obtain a $C^{1,1}$ continuity at the interface between two consecutive patches, the tangent vectors and the cross derivative vector required at the corner of each Coons' patch are computed using the coordinates of points along the two set of curves of the reference grid. A quasi-intrinsic parametrization [2,3] is used to build a parametric piecewise cubic spline passing through the given points of a curve. The parameters obtained in such a way along the two families of curves are then used as *u* and *v* parameters of the whole surface, according to *k* and *l*-indice variation respectively. The degrees of freedom at the corner of a patch are the position vector X_{kl}, the ξ and η-direction partial derivative vectors $X_{\xi\,kl}$, $X_{\eta kl}$ and $X_{\xi\eta kl}$ which are related to the partial derivative vectors X_u, X_v and X_{uv} with respect to *u*, *v* and *uv* respectively by the following relations:

$$X_\xi(u_{k,l},v_{k,l}) = s_{k,l}\frac{\partial X}{\partial \xi}\bigg|_{k,l} = s_{k,l}X_u(u_{k,l},v_{k,l}) \; ; \; X_\eta(u_{k,l},v_{k,l}) = t_{k,l}X_v(u_{k,l},v_{k,l}) \qquad (2)$$

$$X_{\xi\eta}(u_{k,l},v_{k,l}) = s_{k,l}t_{k,l}\frac{\partial^2 X}{\partial \xi \partial \eta}\bigg|_{k,l} = st_{k,l}X_{uv}(u_{k,l},v_{k,l}) \qquad (3)$$

where $s_{k,l}$ and $t_{k,l}$ are the arc length of the curvilinear borders of the patch respectively in u and v-direction with respect to X_{kl} position. If the reference grid has been built using iso-parametric lines of the surface with an even distribution of points, the relationship between the local parameters ξ, η and the global parameters *u, v* are:

$$\xi = \frac{(u - u_{k,l})}{(u_{k+1,l} - u_{k,l})}, \quad \eta = \frac{(v - v_{k,l})}{(v_{k,l+1} - v_{k,l})} \, . \qquad (4)$$

Otherwise the local parameters must be computed using a Newton algorithm in order to inverse the bi-cubic interpolation expressions of $u(\xi,\eta)$ and $v(\xi,\eta)$ which define a mapping onto the curvilinear parameter patch[4]

The partial derivatives with respect to *u* and *v* parameters are carried out using a 5-point scheme[5] which allows to avoid any oscillation when mixing linear and curved patches. The expression used is the following:

$$X_{u_{k,l}} = (1-\alpha) \Delta X_{k,l} + \alpha \Delta X_{k+1,l}, \text{ with } \alpha = \frac{|\Delta X_{k-1,l} \times \Delta X_{k,l}|}{|\Delta X_{k-1,l} \times \Delta X_{k,l}| + |\Delta X_{k+1,l} \times \Delta X_{k+2,l}|} \quad (5)$$

The figure 1 shows the definition of the linear segment $\Delta X_{k,l}$ used in the above formula.

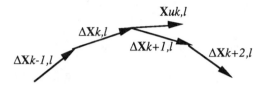

Fig. 1 - 5-point scheme for partial derivatives

The whole surface modelling is then well defined by the Coons multipatch with irregular patches shown in figure 2. Its general expression is:

$$X(u,v) = F(\eta) \, B \, F(\xi)^T \quad (6)$$

where F is the vector of Hermite blending functions.and B is the matrix of coordinates $X(u_{k,l}, v_{k,l})$, tangent vectors and twist vectors at the nodes of the reference grid:

$$B = \begin{bmatrix} X(u_{k,l}, v_{k,l}) & X(u_{k+1,l}, v_{k+1,l}) & s_{k,l} X_u(u_{k,l}, v_{k,l}) & s_{k+1,l} X_u(u_{k+1,l}, v_{k+1,l}) \\ X(u_{k,l+1}, v_{k,l+1}) & X(u_{k+1,l+1}, v_{k+1,l+1}) & s_{k,l+1} X_u(u_{k,l+1}, v_{k,l+1}) & s_{k+1,l+1} X_u(u_{k+1,l+1}, v_{k+1,l+1}) \\ t_{k,l} X_v(u_{k,l}, v_{k,l}) & t_{k+1,l} X_v(u_{k+1,l}, v_{k+1,l}) & st_{k,l} X_{uv}(u_{k,l}, v_{k,l}) & st_{k+1,l} X_{uv}(u_{k+1,l}, v_{k+1,l}) \\ t_{k,l+1} X_v(u_{k,l+1}, v_{k,l+1}) & t_{k+1,l+1} X_v(u_{k+1,l+1}, v_{k+1,l+1}) & st_{k,l+1} X_{uv}(u_{k,l+1}, v_{k,l+1}) & st_{k+1,l+1} X_{uv}(u_{k+1,l+1}, v_{k+1,l+1}) \end{bmatrix} \quad (7)$$

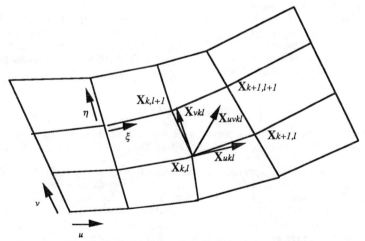

Fig. 2 Coons multipatch with irregular patches

2 - NON-UNIFORM RATIONAL SURFACE

An accurate representation of the body wall surfaces is achieved by using rational polynomials as blending functions in the previous Coons multipatch model. According to Piegl's studies [5], the introduction of a rational cubic patch modelling instead of Coons patch modelling requires to replace the Hermite cubic blending functions by the following rational blending functions:

$G_1(t) = (1-t)^2 D(t)^{-1}$, $G_2(t) = t^2 D(t)^{-1}$, $G_3(t) = 2t(1-t)^2 D(t)^{-1}$, $G_4(t) = 2t^2(1-t) D(t)^{-1}$; $D(t) = (1-t)^2 + t^2$.

The general expression of the rational surface modelling is then:

$$X(\xi,\eta) = G(\eta) \, P \, G(\xi)^T \qquad 0 < \xi < 1 \text{ and } 0 < \eta < 1 \, . \tag{8}$$

P is the matrix of the control points where the infinite control points at the border of the patch will be used to replace the tangent vectors in u and v-direction and the inner infinite control point will be used to replace the twist vector in the A matrix. The relationship between the elements of the two matrix are:

$P_{11} = X(0,0);$ $P_{12} = X(1,0);$ $P_{13} = X_\xi(0,0)/4;$ $P_{14} = -X_\xi(1,0)/4$
$P_{21} = X(0,1);$ $P_{22} = X(1,1);$ $P_{23} = X_\xi(0,1)/4;$ $P_{24} = -X_\xi(1,1)/4$
$P_{31} = X_\eta(0,0)/4;$ $P_{32} = -X_\eta(1,0)/4;$ $P_{33} = X_{\xi\eta}(0,0)/16;$ $P_{34} = X_{\xi\eta}(1,0)/16$
$P_{41} = X_\eta(0,1)/4;$ $P_{42} = -X_\eta(1,1)/4;$ $P_{43} = X_{\xi\eta}(0,1)/16;$ $P_{44} = X_{\xi\eta}(1,1)/16 \, (9)$.

The two patches then require the same input data which can be created from the reference grid previously defined.

The advantage of the rational patch model is its ability to provide an exact definition of cubic spline surfaces, general conics surfaces and planes which are typical surfaces used in CAD system.

Introducing the new blending functions and the correct coefficients in front of the partial derivatives into the matrix B, we obtain a general composite surface which is a uniform rational surface for even distribution of points onto the reference grid and non-uniform rational surface for uneven distribution.

The extension of the model to a general non-uniform rational B-spline surface (NURBS) is obtained when replacing the previous rational blending functions by the non-uniform cubic B-splines and related vertices. However the relationship between vertices and tangent and twist vectors becomes more difficult and requires to solve a system using an inverse algorithm not available in the program.

3 - USER INTERFACE

The two models presented in the previous sections can be implemented in any multiblock mesh generator which requires surface modelling facilities by the mean of two routine packages.

The first one, named Get_xyz, provides the three cartesian coordinates of any point of the surface for a given set of the parameters u and v, using the algorithm presented in (1).

The second one named Get_uv, allows to find the parameters u and v of any point P in the vicinity of the surface. A first step is required to find the nearest patch from the point P before to project it onto the surface. The direction of the search [6] is given by the direction of the vector joining the corner of the patch (position X_{kl}) to the point P. The direction of the displacement is indicated by the sign of the scalar products of this vector with the tangent vectors at the corner of the patch (X_{ukl} and X_{vkl}) as shown in figure 3. A second pass to the

the same corner indicates the convergence. The projection step is achieved by the inversion of the following system using a Newton algorithm:

$$\text{Find } u,v \text{ / } \begin{array}{l} (\mathbf{P} - \mathbf{X}(u,v)) \cdot \mathbf{X}_u(u,v) = 0 \\ \text{and } (\mathbf{P} - \mathbf{X}(u,v)) \cdot \mathbf{X}_v(u,v) = 0 \end{array} \quad (10)$$

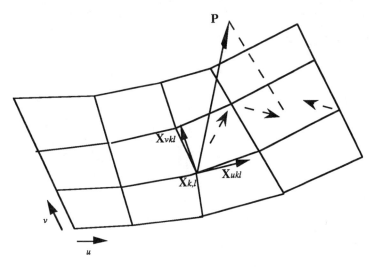

Fig. 3 - Sketch for projection algorithm

4 - SURFACE QUALITY

Some tools are available in the program in order to check the quality of the surface models when an analytical definition allows to compute the exact solution as reference. Three criteria have been defined:

a) the distance to the true surface: the position vector $\mathbf{X}(u,v)$ of any interpolated point given by the model for a couple of parameter (u, v), are compared with the exact position $\mathbf{S}(u,v)$ provided by the analytical definition of the surface. The distance d from the point to the surface is:

$$d = |\mathbf{X}(u,v) - \mathbf{S}(u,v)| \quad (11)$$

b) the normal deviation : the normal vector to the surface at a given point is defined using the cross product of the two tangents at this point:

$$\mathbf{X}_u(u,v) = 1/s \cdot \mathbf{F}(\eta) \; \mathbf{B} \; \mathbf{F'}(\xi)^T \qquad s = (1- \eta) \; s_{k,l} + \eta \; s_{k+1,l} \quad (12)$$
$$\mathbf{X}_v(u,v) = 1/t \cdot \mathbf{F'}(\eta) \; \mathbf{B} \; \mathbf{F}(\xi)^T \qquad t = (1- \xi) \; t_{k,l} + \xi \; t_{k+1,l} \quad (13)$$

$$\mathbf{N}(u,v) = \mathbf{X}_u \times \mathbf{X}_v \quad (14)$$

The angle of deviation with the exact normal direction is then computed.

c) the curvature default : the Gaussian curvature κ of the surface [7] is defined by:

$$\kappa = \begin{vmatrix} \mathbf{N}.\mathbf{X}_{uu} & \mathbf{N}.\mathbf{X}_{uv} \\ \mathbf{N}.\mathbf{X}_{uv} & \mathbf{N}.\mathbf{X}_{vv} \end{vmatrix} / \begin{vmatrix} \mathbf{X}_u.\mathbf{X}_u & \mathbf{X}_u.\mathbf{X}_v \\ \mathbf{X}_u.\mathbf{X}_v & \mathbf{X}_v.\mathbf{X}_v \end{vmatrix} \quad (15)$$

and compared with exact curvature of the surface.

The figure 4 shows the result of the distance checking for a part of a sphere defined by a very coarse reference grid with only 3 points in each direction (fig. 4b). The tangent vector evaluation at the border of the grid requires two extra points (fig. 4a). The result of the Coons' model interpolation (fig. 4c) is compared with the result of the rational model (fig. 4d). The iso-contours of the distance d to the surface are plotted respectively on fig. 4e and fig. 4f, with an increment of 0.1% of the radius of the sphere from 0 to 0.7%. The greater distance is reached by Coons' model at the middle of the patches. The remaining error on rational model is related to the approximation of the length of the curves passing through the points of the reference grid.

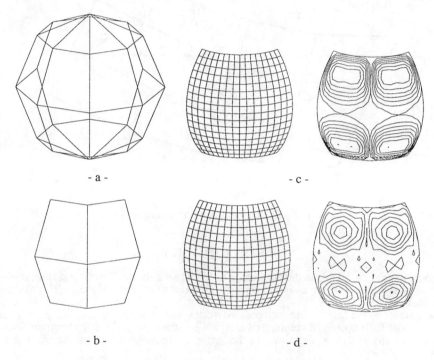

Fig. 4 - Distance checking on a sphere a) reference grid with 2 extra points
b) reference grid; c) Coons' model interpolation with Iso-d contours;
d) Rational Bezier interpolation with Iso-d contours

5 - APPLICATIONS

The comparison of the Coons' multipatch model and the non-uniform rational Bezier surface model has been performed on some analytically defined 3D shapes. The example presented here involves a 3D double-ellipsoide on which is applied a parametrization based upon iso-curvature lines according to the wall mesh topology defined by ONERA [4]. The figure 5a shows the topology of the reference grids (9 x 9 points). The symmetry condition is used at the borders except for the intersection curve between the two ellipsoides where an extrapolation is applied.

A 25 x 25 mesh is then generated upon each patch of the surface using Coons' model (fig. 5b) and rational Bezier model (fig. 5c). The same size of the mesh cell is prescribed at the both side of an interface between two patches. The mesh generated using Coons' model is presented on figure 5b. It can be compared with the resulting mesh of rational Bezier model on figure 5c. There is no significant differences between the two models with such a fine reference grid.

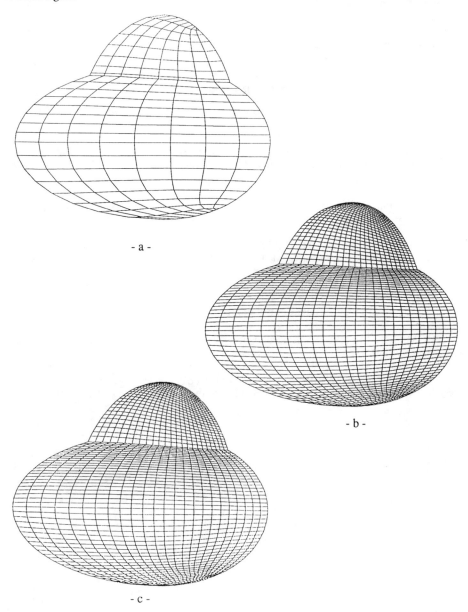

Fig. 5 - Double-ellipsoide test case : a) reference grids (9x9 points) - b) Coons' model - c) Rational Bezier model

CONCLUSION

A program has been developed to provide to any multiblock mesh generator the capabilities to move the nodes of a mesh onto a given curved surface only defined by a reference grid. Two mathematical models have been implemented and checked using analytically defined shapes in order to compare the errors of the two interpolation formulae. An application to a double-ellipsoide with a severe topology has prouved the robustness of the method.

A first version of the code with Coons model, has been delivered at ONERA and introduced in a general optimization code[4].

REFERENCES

[1] - COONS S.A. (1967) - Surfaces for computer. Aided Design of Space Forms MAC-TR-41 M.I.T.

[2] - Mc CONALOGUE D.J. (1970) - A Quasi-Intrinsic Scheme for Passing a Smooth Curve throught a Discrete Set of Points.- The Computer Journal Vol. 13 n° 4, November 1970.

[3] - MORICE Ph. (1983) - Numerical generation of boundary-fitted coordinate systems with optimal control of orthogonality.- ASME Fluids Engineering Conference, Houston.

[4] - DESBOIS F. and JACQUOTTE O.-P. (1991) - Surface Mesh Generation and Optimization - Third International Conference on Numerical Grid Generation in Computational Fluids Dynamics, Barcelona, June 1991.

[5] - PIEGL L. (1988) - Hermite and Coons like Interpolants using Rational Bezier Approximation Form with Infinite Control Points. Computer-Aided Design Vol. 20 n° 1.

[6] - THOMPSON J.F. (1987) - A composite grid generation code for general 3D regions - AIAA paper 87-0275.

[7] - DING QIULIN and DAVIES B.J. (1987) - Surface engineering geometry for computed-aided design and manufacture, Ellis Horwood Series in Mechanical Engineering, 1987.

REPARAMETRIZATION OF BLOCK BOUNDARY SURFACE GRIDS

Stefan Farestam

C.E.R.F.A.C.S.

European Centre for Research and Advanced Training in Scientific Computing
42 avenue Gustave Coriolis, F-31057 Toulouse Cedex, France

SUMMARY

This article describes the work completed at CERFACS in task 8 of the BRITE/EURAM EuroMesh project, concerning reparametrization of block boundary surface grids. A new surface patch is arbitrarily positioned on a set of existing, and interconnected, surface patches. The parametrization is found by representing the patch in the parameter space of the original patches using an unstructured mesh, and then solving a set of partial differential equations in the physical space of the new patch, using the finite element approach.

1 INTRODUCTION

The generation of multi block meshes for complex 3d geometries often means creating blocks that are not aligned with the surface grid. This problem is difficult to overcome, since the surface grid generation is dictated by the constraints of the geometry of the bounding object, whereas the topology of the multi block grid is linked with the flow model. The need to arbitrarily position a new patch on the surface, with respect to existing patches, is therefore considerable. There is also the problem of accuracy. If the new patch has a new physical description, then it will by necessity differ somewhat from the set of original patches, since its description is based on a discrete representation. This is a highly undesirable property, but it is hard to avoid using conventional methods. Recently Suzuki, [1, 2], proposed an elegant solution to this problem, well suited for industrial problems. He suggests representing the new patch in the parameter space of the existing patches, employing an unstructured grid. The new parametrization along the sides of the patch may then be found easily, using a chord length parametrization. The parametrization of the interior of the patch is obtained by solving a set of partial differential equations in the physical space of the patch, using the finite element approach. In the simplest case, this means solving for Laplace's equation. The principal steps are:

- Discretizing the patch boundary.
- Subdividing the patch into subpatches.
- Triangulating the subpatches.
- Combining the triangulations, and parametrizing the patch.

To access a point on the new patch, the triangle containing the new point is located, and the coordinates in terms of the old patch, to which the triangle belongs, are obtained through linear interpolation, and from this, all physical data for the point is obtained. The triangulated patch is thus only an interface to the parameter space of the set of old patches, consequently all properties of the original surface are preserved, which is one of the principal advantages of this method. Other advantages are speed and robustness, especially when dealing with surfaces containing discontinuities.

2 THE BASIC ALGORITHM

Each patch, $\bar{p}_i$, of the original set of n patches maps a rectangle in the (ξ, η) parameter space

$$(\xi_{i_{min}} < \xi < \xi_{i_{max}};\ \eta_{i_{min}} < \eta < \eta_{i_{max}}) \tag{1}$$

onto the physical space, by the function:

$$x = x_i(\xi, \eta). \tag{2}$$

The connectivity of the patches is assumed to be known. The new patch, $\bar{q}$, is initially defined by the sequence of nodes $r_{i=1,m}$ which specifies its boundary. These nodes are stored in a linked list, L, and carry information about which original patch, $\bar{p}_i$, they belong to, as well as the local coordinates of the node in the parametrization of $\bar{p}_i$. An example of a typical configuration is shown in figure 1a, where the new patch partially overlaps four original patches.

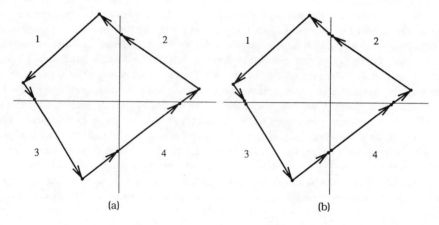

Figure 1: a) The initial configuration. b) Extra boundary nodes added.

2.1 Boundary Crossings

Consider a point located on a boundary shared by two patches $\bar{p}_i$ and $\bar{p}_j$, and thus described by (ξ_i, η_i) as well as (ξ_j, η_j). Such points are stored as two consecutive nodes in L, and it is ascertained that they represent the same physical point by fixing one of the nodes, and moving the other along the common boundary of $\bar{p}_i$ and $\bar{p}_j$ until the physical distance between the nodes reaches a minimum. Ideally the minimal distance should be 0, but this will be the case only if the patches are perfectly lined up.

If $\bar{q}$ extends over more than one of $\bar{p}_{i=1,n}$, all the boundary crossings are identified, and extra nodes are inserted if necessary, as outlined above. This is a necessary step to prepare for the subdivision of $\bar{q}$.

2.2 Subdivision of the Patch

Each $\bar{p}_i$ is assumed to be continuously differentiable in the interior. Discontinuities in the original surface are therefore always located along boundaries of the $\bar{p}_i$. For this reason, it is important that the triangulation of $\bar{q}$ has no elements that traverses these boundaries. This is accomplished by subdividing $\bar{q}$ into $\bar{q}_{i=1,k}$, where each $\bar{q}_i$ is uniquely contained in only one of $\bar{p}_i$. The algorithm accomplishing this walks along the current boundary of $\bar{q}$ and gradually identifies subpatches, $\bar{q}_i$. For each subpatch $\bar{q}_i$ that is identified, the current boundary is led around $\bar{q}_i$, as shown in figure 2a. Eventually $\bar{q}$ is entirely subdivided (figure 2b).

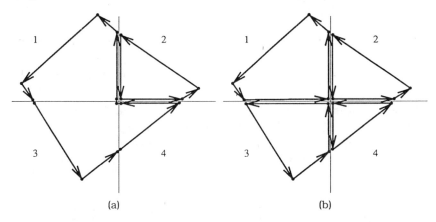

Figure 2: a) The first subpatch has been detached. b) The patch is fully subdivided.

2.3 Redistribution of Boundary Points

The triangulation of a patch can only proceed properly if the discretization of its boundaries is in accordance with the element size prescribed in the interior. All the patch boundaries of $\bar{q}_{i=1,k}$ are therefore reparametrized, i.e. the original boundary nodes are interpolated by a spline in parameter space, and new nodes are inserted, if necessary. For the external boundaries of $\bar{q}_{i=1,k}$, this process is straightforward. The internal boundaries, however, may have two conflicting discretizations, since an internal boundary may occur at a discontinuity in the surface represented by $\bar{p}_{i=1,n}$. In this case, the side of the boundary that is most densely discretized is chosen, and the points along the other side are found by minimizing the physical distance, as described in section 2.1.

2.4 Triangulation of Subpatches

Each subpatch $\bar{q}_i$ is triangulated using the Advancing Front Technique [3], commonly referred to as AFT. However, instead of using a background grid to determine the element size and

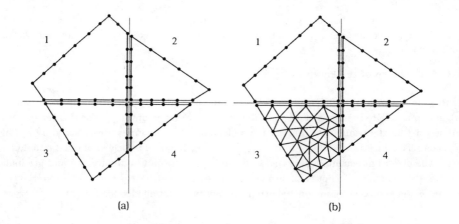

Figure 3: a) The boundaries are reparametrized. b) One subpatch is triangulated.

the stretching in the subpatch $\bar{q}_i$ to be triangulated, the curvature of the surface is used to determine these parameters.

Any continuously differentiable surface has two principal curvatures, as well as two principal directions of curvature, at any point. These are defined for each patch $\bar{p}_i$, [4, 5], as the two generalized eigenvalues, $\kappa_{1,2}$, and unit eigenvectors, $\epsilon_{1,2}$, of the system:

$$D\epsilon = \kappa G\epsilon \tag{3}$$

where

$$G = \begin{bmatrix} \frac{\partial x}{\partial \xi} \cdot \frac{\partial x}{\partial \xi} & \frac{\partial x}{\partial \xi} \cdot \frac{\partial x}{\partial \eta} \\ \frac{\partial x}{\partial \eta} \cdot \frac{\partial x}{\partial \xi} & \frac{\partial x}{\partial \xi} \cdot \frac{\partial x}{\partial \xi} \end{bmatrix} \qquad D = \begin{bmatrix} n \cdot \frac{\partial^2 x}{\partial \xi^2} & n \cdot \frac{\partial^2 x}{\partial \xi \partial \eta} \\ n \cdot \frac{\partial^2 x}{\partial \eta \partial \xi} & n \cdot \frac{\partial^2 x}{\partial \eta^2} \end{bmatrix} \tag{4}$$

and

$$n = \frac{\partial x}{\partial \xi} \times \frac{\partial x}{\partial \eta} \bigg/ \left\| \frac{\partial x}{\partial \xi} \times \frac{\partial x}{\partial \eta} \right\| . \tag{5}$$

G is called the first fundamental matrix of the surface, and D the second. $\epsilon_{1,2}$ define two orthogonal directions in the physical space of the surface, but not necessarily in its parametrized description. We want the generated triangles to be aligned along $\epsilon_{1,2}$, and with a size in each direction inversely proportional to the curvatures $\kappa_{1,2}$. We refer to the constant of proportionality, δ, as the geometry resolution. A minimum- and maximum element size, $S_{min,max}$, is also defined. To achieve the triangulation, we define a control space $\mathcal{R}^{2*}$, in which we strive to generate equilateral triangles of unit size. The mapping from the local parameter space of the patch to $\mathcal{R}^{2*}$, is governed by the tensor T, defined by:

$$T = \begin{bmatrix} \epsilon_{2_\eta}/C/M_1 & -\epsilon_{1_\eta}/C/M_2 \\ -\epsilon_{2_\xi}/C/M_1 & \epsilon_{1_\xi}/C/M_2 \end{bmatrix} \tag{6}$$

where

$$C = \|\epsilon_1 \times \epsilon_2\| \tag{7}$$

and

$$M_{1,2} = S_{1,2} / \epsilon_{1,2} G \epsilon_{1,2} \qquad S_{1,2} = \delta / \kappa_{1,2} . \tag{8}$$

$S_{1,2}$ are constrained to the interval $[S_{min}, S_{max}]$.

The use of T makes the triangulation independent of the way the surface is described, and this technique therefore has the important property of producing a triangulation that is uniquely determined by the intrinsic geometry of the surface. The whole process is thereby rendered very stable, and the generation of high quality triangulations is greatly facilitated.

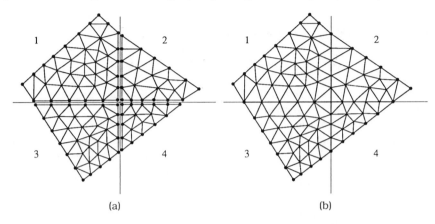

Figure 4: a) All subpatches have been triangulated. b) The final patch.

2.5 Assembly and Parametrization of $\overline{q}$

The subpatches $\overline{q}_i$ are now reassembled into one patch. The accumulated chord length is used as a basis to generate an initial parametrization of the boundaries of $\overline{q}$. The parametrization, (s, t), of the interior of $\overline{q}$ is found by using the algorithm due to Suzuki, [1, 2], which solves a set of partial differential equations in the physical space of the new patch, by using the finite element approach. We have:

$$\nabla(\mu \nabla \xi) = 0 \quad and \quad \nabla(\mu \nabla \eta) = 0 \qquad (9)$$

where μ is a weight function. We use the Gaussian curvature of the surface, $\kappa_1 \kappa_2$, normalized, and constrained, to the interval $[\delta/S_{max}, \delta/S_{min}]$, as μ. The resulting linear system of equations is solved by a relaxation method.

2.6 Physical Data Access

We now discuss how the coordinates, and other data, of a physical point, p, corresponding to a given parameter pair (s_p, t_p) in $\overline{q}$, is obtained. First the triangular element, T, containing (s_p, t_p), is located. This is accomplished by a directed search, starting in a corner of $\overline{q}$. (s_p, t_p) may now be expressed as a barycentric combination of the nodes of T. We have:

$$s_p = c_0 s_0 + c_1 s_1 + c_2 s_2 \quad and \quad t_p = c_0 t_0 + c_1 t_1 + c_2 t_2 \qquad (10)$$

where

$$c_0 + c_1 + c_2 = 1 \ . \qquad (11)$$

The coordinates of (s_p, t_p) in the appropriate original patch are given by:

$$\xi = c_0\xi_0 + c_1\xi_1 + c_2\xi_2 \quad and \quad \eta = c_0\eta_0 + c_1\eta_1 + c_2\eta_2 \qquad (12)$$

where (ξ_j, η_j) are the original patch parameters for (s_j, t_j).

It should be stressed that by only being an interface to the original surface, this technique makes it possible to create "floating" patches that preserve all of the properties of the original surface, and yet are capable of managing discontinuities.

3 EXAMPLE

As an example, we apply the algorithm outlined above to the join of a section of an aircraft wing with a section of its fuselage. In figure 5, we show a structured mesh on the configuration, generated from two original patches, with the unstructered mesh of the new patch overlayed. The joint between the two patches is highly discontinuous. Figure 6 shows the unstructured grid in detail. Note how the parameterization of the wing takes precedence along the inter patch boundary, as well as the stretching of the elements along the leading edge of the wing. Figure 7. shows the final mesh.

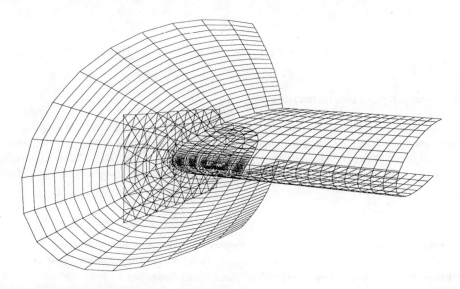

Figure 5: The initial structured mesh, with the new patch superimposed

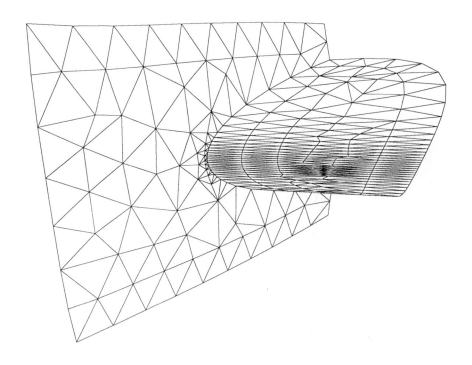

Figure 6: Detail of the triangulation of the new patch

REFERENCES

[1] Masahiro Suzuki. An approach to the surface grid generation based on unstructured grid. In A. S. Arcilla, J. Häuser, P. R. Eiseman, and J. F. Thompson, editors, *Numerical Grid Generation in Computational Fluid Dynamics and Related Fields*, pages 947–953. Elsevier Science Publishers, 1991.

[2] Masahiro Suzuki. Surface grid generation based on unstructured grids. *AIAA Journal*, 29(12):2262–2264, 1991.

[3] Rainald Löhner and Paresh Parikh. Generation of three-dimensional unstructured grids by the advancing front method. In *26th Aerospace Sciences Meeting*. AIAA, January 1988.

[4] D. Qiulin and B. J. Davies. *Surface Engineering Geometry for Computer-Aided Design and Manufacture*. Ellis Horwood, 1987.

[5] Gerald. E. Farin. *Curves and Surfaces for Computer Aided Geometric Design*. Academic Press, 1988.

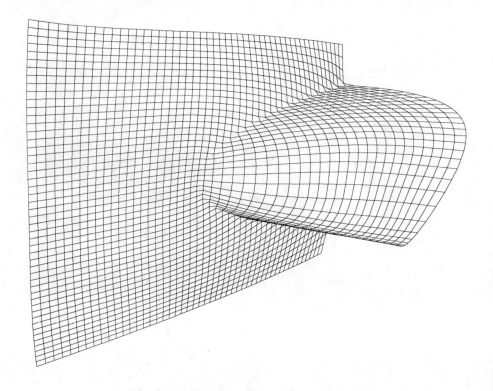

Figure 7: The final mesh

Aircraft Surface Generation

by

Helmut Sobieczky
DLR Inst. Theor. Fluid Mechanics
Bunsenstr. 10
Göttingen, Germany

ABSTRACT

Surface generation software is being created with emphasis on parametric description of aircraft components and their junction. Applications are illustrated and examples are shared with project partners for their development of CFD mesh generation. Industry partners in the meantime and within this collaboration use this approach to develop a powerful tool for transport aircraft component design, establishing practical interaction between a theoretical aerodynamics group and the project design engineers.The goal is shaping geometry generator codes to become fast interactive tools supporting the use of aerodynamic knowledge bases in future design expert systems using modern graphic workstations.

INTRODUCTION

Aerodynamics has become an important component in the knowledge base of the Aerospace Sciences. In this situation of increased usage of the computer in aerodynamics we may ask how these successful developments blend into the practical work of leading an aerospace project from the idea toward production. Among many other necessities to link the project goal with knowledge bases of many disciplines, we realize a simple one, namely to generate surface data for various purposes: an input description for the mathematical modelling of aerodynamic performance as well as for non-aerodynamic investigations, finally creating data for numerically controlled (NC) production. The importance of this interface between the disciplines is commonly acknowledged: CAD/CAM technology has become an integral part in modern industry. In aerodynamics we have a growing need for CAD systems allowing not only for production-oriented purposes but also for two reasons involving analysis and design:

(1) Computational fluid dynamics (CFD) needs mathematically well-defined boundary conditions for the numerical evaluation of differential equations modelling physical phenomena, and (2) applying these numerical methods to practically useful structures requires a careful parameterization of generating these shapes or boundary conditions, to have control over their development in the course of carrying out optimization strategies for achieving optimum aerodynamic performance.

Transonic aerodynamics is an excellent field to demonstrate the value of choosing geometry parameters from the gasdynamic knowledge base of low drag airfoil design. It stimulated the publication of an article [1], and some test case configurations were generated with the first versions of a computer program to define wings, fuselages and hydraulic configurations. Aircraft or propulsion components require a flexibility to fine-tune surface details especially where local supersonic regions occur and, at design conditions, a minimization of shock wave strength is desirable. Airfoil theory with 2D direct and inverse design methods is a mature tool useful for large aspect ratio swept wings. At wing roots, however, the concept of swept wing theory breaks down, the flow field is heavily three-dimensional. Blending efficient wing sections with root fairings is therefore an important work topic for the practical aerodynamicist, computational geometry should be used for flexible shape control.

To this point, a geometry generator is used and extended with some new functions. Testing these tools in transonic and supersonic configurations is the goal, here a brief illustration of recent work will be given.

GEOMETRY TOOLS FOR CURVE AND SURFACE GENERATION

Curve and surface theory for application in Computer Aided Geometry Design (CAGD) is represented by the names of P. Bezier and S. Coons, and with B-spline methods. Mature and versatile methods for surface generation have been developed for the automotive industry. With a strong background like CAGD systems derived from these theories, the ideas given here can be seen as an attempt to add a few options to existing systems, with emphasis to problems specifically occuring in aerodynamic design. Here the use of many different kinds of analytic functions is stressed, not only polynomials, for curve generation. Some of the functions used allow inversion of dependent and independent variables which efficiently supports the juncture of components and the establishment of surface metrics. Surfaces here are always constructed from curves, not patches, allowing for easier explicit description of one coordinate as function of the two others in 3D space. Flexibility in curve generation is therefore the backbone of this method to arrive at a multiplicity of useful shapes. In the following a general outline is given how curves are defined and subsequently used for surface definition.

Curves:
Non-dimensionalization of plane curve arcs allows for their representation in a square, (Figure 1). Depending on the structure of the shape within this interval we may choos e from a variety of functions $Y(X)$ having in common only the start $Y(0) = 0$ and the end $Y(1) = 1$. Besides naming the chosen function by an identifier g, we may have some free parameters to control slopes, curvatures, powers or exponential behavior of the curve at the interval ends. Restricting the number of parameters to 4 we already may describe nearly every kind of shape within the interval by a function family

$$Y = Y_g(a, b, e, f; X) .$$

In contrast to using a set of given support points $P_n(x,y)$ for spline interpolation, we non-dimensionalize the resulting intervals, select a function (g) and the parameters (a,b,e,f) for each interval. This technique, in principle, has been explained [1], here we stress the advantage of such a concept with a variety of functions especially whith interactive control using the graphic tools of workstations.

Surfaces:
The present surface generator is a specialized tool for aerospace and hydraulic applications with emphasis to creating input for analysis with aero- or hydrodynamic CFD codes. In Design Aerodynamics, we have a need to define generic shapes for code development, but with an eye on the requirements of practical configuration design. In contrast to conventional aircraft with

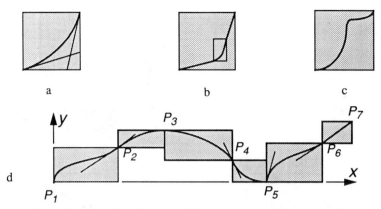

Figure 1: Curves in normalized interval: (a) parameters used for slope and curvature control at interval ends; (b) embedding a standard function between straight ends, (c) using one function to redistribute another function, (d) scaling the intervals and composing curves of arbitrary complexity but known local analytic structure

medium or large aspect ratio wings, supersonic flight vehicles with high lift/drag ratio requirements are more suitably described by cross sections and a full integration of wing and body must be achieved. Fairing a conventional wing root was explained earlier [1], here we extend the concept: successive use of functions, as illustrated for curves in Fig. 1, allows to deform one part of a function while the rest keeps its shape. All curve parameters are now functions themselves of, for example, the axial direction, this way allowing for arbitrary changes of cross sections or other local properties.

Grids:
Algebraic grids of CH- or CO-type are optionally available from the original [1] code. Extensions for many practical applications have been introduced by the industry partners [2]. Many CFD codes seem robust enough to perfectly work with algebraic grids, others require solutions to partial differential equations resulting in elliptic or hyperbolic grids. Because of the various requirements by different CFD analysis codes, only the needed flexibility in creating the solid surfaces of aircraft and farfield boundaries is the topic of this contribution. However, test cases for CFD codes validation have been published and examples will briefly be illustrated in the following.

AIRCRAFT COMPONENT JUNCTURES

Some recent extensions to the existing geometry generator are presently applied to new generic configurations, they include refinements of options available for the wing/body junction. Previous case studies were carried out with the body given by a generalisation of superelliptic quarter cross-sections, now we have the option to add bumps to this fuselage with key parameters in the same flexible way as the original fuselage is created. Wing fillets, canopy and wheel gear boxes may suitably be modeled this way. Figure 2 shows the mid-portion of a fuselage, with the root portion of a wing, both components joining with a controlled smoothness at both the wing root sections and the body fillet bump.

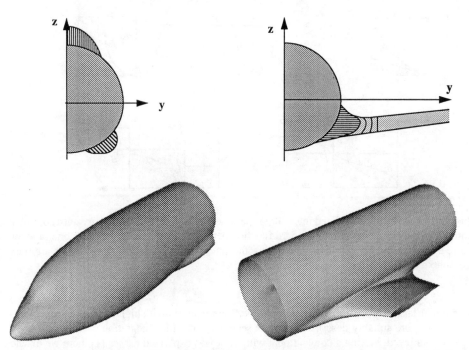

Figure 2: Fuselage cross sections (a, b), with bump additions (hatching marks addition direction) and blended projection of the wing toward the fuselage. Shaded surface graphics of generic transport aircraft forebody and midsection (c, d)

Further extensions of the technique have been under development and are currently refined, mainly for a generation of aerodynamically optimized pylons and flap gear fairings. In another research project, a similar approach has been used to model supersonic and hypersonic configurations, with a strong integration of airframe and propulsion [3]. The use of geometry generators based on these techniques is greatly accelerated and made comfortable by the use of new computer science tools, a new interactive version is currently under development to be combined with fast design and analysis codes to become an aerodynamic expert system.

EXAMPLE 1: DLR - F5 TEST WING CONFIGURATION

Within the frame of collaboration between European partners, an example test case has been made available for CFD mesh generation purposes: the DLR - F5 test wing has been generated with this tool and was produced as a 700 mm span wind tunnel model. Wind tunnel tests were carried out earlier [4], the welldefined half - model experiment proved to be a difficult test case for Navier / Stokes CFD code validation. A downgraded version of the generator code [5] was distributed on diskette to the partners of the Euromesh project. A data set of an algebraic grid around the wing in the windtunnel, as it was used for first Navier / Stokes calculations, was distributed also, as a sample 3D grid for further optimization. Figure 3 shows the wing with large fillet, mounted onto a splitter plate in the wind tunnel and a detail of the surface grid including the wing root, with a graphic visualization of the surface pressure.

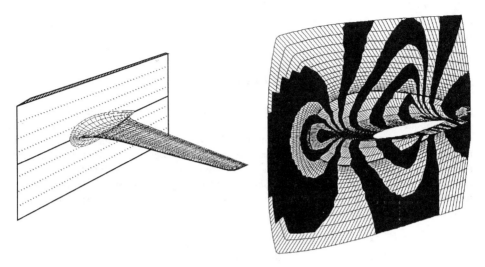

Figure 3: DLR - F5 Test Wing for computational code validation. Algebraic grid detail in wing root area with surface pressure isofringes illustrating result from N/S code [6].

EXAMPLE 2: A310 TRANSPORT AIRCRAFT CONFIGURATION

An example reflecting collaboration within European aerospace establishments and industry is illustrated next. Project study groups should use the same software for preliminary design studies as those verifying aerodynamic performance with numerical fluid mechanics simulation codes. A first attempt to convince the partners in industry, that the present geometry generator might be useful for such common software, was the geometric modelling of the A310 aircraft wing-fuselage configuration. For this purpose, wing planform and body shape data were transformed into input parameters for the generator. Wing section geometry, with twist and dihedral extracted, was non-dimensionalized to form a set of airfoils. All data created are analytically smooth or with controlled curvature and slope discontinuities, interpolation with splines is used only for airfoil input data which were given as a prescribed set of data.

With this effort and a first task to provide input data for Euler codes validation, the resulting geometry (Figure 4) for wing and fuselage was found sufficiently close to the actually given A310 data. As a first application an algebraic grid of nearly 200 000 points was generated around the wing-body configuration to allow for comparison of two Euler analysis codes for inviscid transonic flow past the A310. After this, new configurations were designed and analysed with CFD codes[7], but the real reward for the effort to find geometry input parameters of an already existing shape, (which originally was generated of course with different tools), is the interest of the project groups to consider using the code for their various conceptual design studies. This in turn allows now for a rapid interchange of data between project and CFD analysis groups, requiring only common input parameters exchange.

Future work at DLR is aimed at shaping geometry generator codes to become fast interactive tools for the aerodynamic design expert on modern graphic workstations. The use of a new interactive version in the X-Windows environment will allow to combine components towards full aircraft configurations, (Fig. 4), in a very rapid and practical way.

CONCLUDING REMARK

Surface generation software has been created with emphasis on parametric description of aircraft components and their junction. Applications have been illustrated and examples are shared with project partners for their development of CFD mesh generation. Industry partners in the meantime and within this collaboration have developed a powerful tool for transport aircraft component design to establish practical interaction between a theoretical aerodynamics group and project design engineers. The goal is shaping geometry generator codes to become fast interactive tools supporting the use of aerodynamic knowledge bases in future design expert systems using modern graphic workstations.

REFERENCES

1. Sobieczky, H., 'Geometry Generation for Transonic Design', Recent Advances in Numerical Methods in Fluids, Vol. 4, Ed. W. G. Habashi, Swansea: Pineridge Press, pp. 163 - 182. (1985).

2. Barnewitz, H., Becker, K., 'Interactive Netzgenerierung mit dem algebraischen Netzgenerator INGRID - Hinweise zur Benutzung', MBB - Bericht No. 1666, (1988).

3 Sobieczky, H., Stroeve, J. C.: 'Generic Supersonic and Hypersonic Configurations', AIAA-91-3301, (1991)

4 Sobieczky, H., Hefer, G., Tusche, S.:DFVLR-F5 Test Wing Experiment for Computational Aerodynamics. Notes on Numerical Fluid Mechanics, Vol. 22, pp.4 - 22, ed. W. Kordulla (1988)

5 Sobieczky, H.: 'DLR-F5 Test Wing Configuration for Computational and Experimental Aerodynamics', Wing Surface Generator Code, Control Surface and Boundary Conditions. Report prepared for B/E Area 5 Aeronautics Project Aero 0018 'Euromesh' (1990).

6 Kordulla, W., Schwamborn, D., Sobieczky, H.: 'The DFVLR-F5 Wing Experiment - Towards the Validation of the Numerical Simulation of Transonic Viscous Wing Flows', AGARD CP 437 "Validation of Computational Flud Dynamics", (1988)

7 Rill, S., Becker, K.: 'Simulation of Transonic Flow over a Twin Jet Transport Aircraft', AIAA-91-0025, (1991)

IV. VOLUME GRID GENERATION

USE OF ONERA GRID OPTIMIZATION METHOD AT CASA

José M. de la Viuda, Juan J. Guerra and A. Abbas

Aerodynamics Dept., Dirección de Proyectos

CASA, Avda John Lennon s/n, GETAFE (SPAIN)

SUMMARY

This article describes the use of ONERA grid optimization method at CASA. The method has been used to improve the quality of 3-D mutliblock grids generated using the ICEM-CFD tool.Four configurations have been optimized and different levels of improvement from the initial grids have been obtained. This work is part of the EUROMESH project, a project sponsored by the Comission of the European Communities.

1. INTRODUCTION

In all CFD applications the grid modelling the physical space plays a key role. The solutions are highly grid-dependent so, a good grid is the basis for a good result. The current available grid generation tools used at CASA provide grids of "poor" quality but they are generated in short time and easy to change. A grid optimization method is the obvious complement required to produce "good" initial grids. The criteria for quoting a grid as "poor" or "good" will be described in this article, as well as the cases tested and the final conclusions. The grids that have been optimized are :

- DLR F5 wing, free stream (4 blocks)
- DLR F5 wing, in tunnel, O-H (4 blocks)
- CN-235 vertical tailplane (8 blocks)
- Fighter aircraft (forebody+canard) (11 blocks).

2. INITIAL GRID GENERATION

The initial grids are created using the ICEM-CFD tool, that CASA has developped in colaboration with Aerospatiale and Control Data (see ref 1). The topology of the configuration as well as the grid point distribution are defined within the CAD system, and a command file is automatically generated for the grid generator. The grid generator computes first the grid in the block faces, and after, generates the inner points using algebraic interpolation. This algebraic interpolation can be performed at two levels, first, the basic one and the second level, that yields grids of better quality but is more time-consuming, and uses transfinite interpolation (ref 2).

This method allows to generate multiblock grids for CFD in short time, and it is extremely flexible. But, on the other hand, to have grids of good quality (see below for quality criteria), it would be necessary to generate a great number of blocks. The approach chosen has been to generate initial grids with small number of blocks and to optimize the algebraic grid, using ONERA's method.

3. QUALITY CRITERIA

Four quality criteria have been followed to compare the initial and the optimized grid, and between two optimized grids.

<u>Negative Volumes</u>: The existence of negative volumes is not suitable for computations, and the grid is considered unacceptable.

<u>Visual Observation</u>: The detailed observation of some grid planes can show grid distorsions that can not be accepted by the solver, the grid is also considered unacceptable.

<u>Solver Convergence</u>: It may occur that, after several iterations, the solution does not converge. In this case the grid is unacceptable,too.

<u>Results Comparison</u>: If the solver has converged for both, initial and optimized grid, the comparison of the results (pressure distribution and others), are used to determine the quality of the grid.

4. OPTIMIZATION METHOD

The optimization method is widely described in refs. 3 and 4. The philosophy of the method is the possibility of defining a measure of grid cell deformation with respect to a reference cell; the assembly of elementary contributions leads to a global measure of mesh quality that can be used as an optimisation criterion. The optimization process follows then an analogy with the finite-elemt methods and, using variational methods, optimizes the grid quality.

5. TEST CASES

5.1 DLR F5 wing (free stream)

An O-O grid made of four blocks has been arround the wing. It has 91 points around the airfoil (I), 28 spanwise (K) and 20 from the wing to the infinity (J).

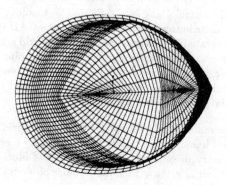
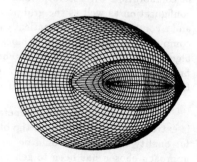

Figure 1, Initial grid, J=15 Plane   Figure 2, Optimized grid, J=15 Plane

The initial grid had cells with negative volume, so , it was necessary to use optimization method. After the first optimization, some distorsions appeared in the I-constant plane placed at the leading edge, so the optimization process was repeated imposing this surface as fixed.

As Figures 1 and 2 show, the optimization improves the grid quality.

5.2 DLR F5 wing (in tunnel, O-H)

In the Second Intermediate Report the results describing the optimization of an O-O mesh around the DLR F5 wing in a tunnel were shown. First computations failed because of the high variations in cells aspect ratio, so, it was decided to generate an O-H grid. This grid has 89 points around the airfoil (I), 20 from the airfoil to the wind tunnel inflow and outflow sections (J), 28 spanwise in the wing and 12 from the wing tip to the side wall (both K direction).

The initial grid had negative volumes, therefore no computations were done. In the first optimization process there was a mistake in the automatic generation of topology data and intermediate K plane (K=20) was imposed as fixed surface, letting the wind tunnel wall (K=40) as free surface. The computational results showed an asymmetric pressure distribution for upper and lower side.

Once that the error was found, the optimization process was repeated, starting with the optimized grid, but now with the proper boundary conditions.

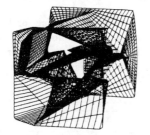

Figure 3, Initial grid, J=15 Plane

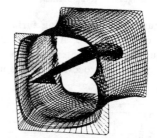

Figure 4, Optimized grid, J=15 Plane

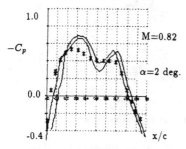

Figure 5, Initial grid, Cp distrib.

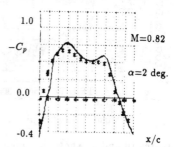

Figure 6, Optimized grid, Cp distrib.

Figures 5 and 6 show Cp distribution (M=0.82 alpha=0) for the section y/b=0.95, for the 1st optimized and for the final grid.

5.3 CN-235 vertical tailplane

The grid is made of 8 blocks, in this case the block definition has been done more carefully, and the result is a better initial grid. The 8 blocks are placed in a concentric way, the 4 inner blocks, model the region near the wing, in a O-O mesh. The 4 outer blocks cover the far field region. The mesh has 91 points around the airfoil (I), 28 spanwise (K) and 11 points in K direction for inner and outer domains. The optimization process converge after an small number of iterations (300).

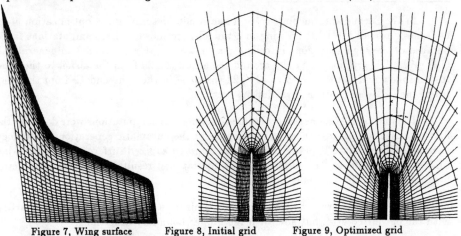

Figure 7, Wing surface   Figure 8, Initial grid   Figure 9, Optimized grid

Figure 7 shows the wing surface. Figures 8 and 9 show I=23 plane for the initial and the optimized grid. As the initial grid has a greater number of blocks and was generated using transfinite interpolations, the differences between the two grids are small. The optimization process converges until the fixed residue value in few iterations (276).

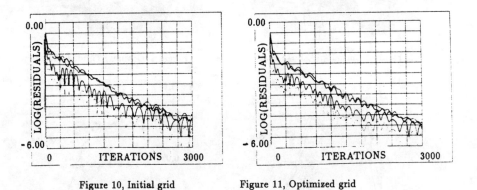

Figure 10, Initial grid   Figure 11, Optimized grid

Figures 10 and 11 show the small differences in the convergence of the solution.

5.4 fighter aircraft

This grid is made of 11 blocks, it is an O-H type, although the canard introduces some topology modifications. At the present moment there has been no computation using this grid. The grid has 98 points in longitudinal direction (J), 28 points from the body surface to the infinity (I) direction and 26 points in the meridians (K direction).

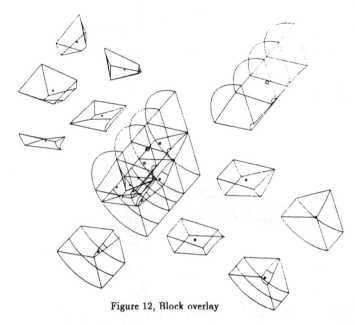

Figure 12, Block overlay

Figure 12 shows the block overlay. Figures 13 and 14 show the I=18 plane. There are no great differences between the initial and the optimized grids, specially in the K direction. The benefits of the optimization are more importants near the canard.

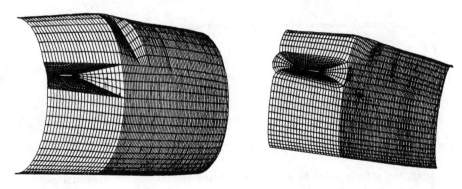

Figure 13, initial grid   Figure 14 Optimized grid

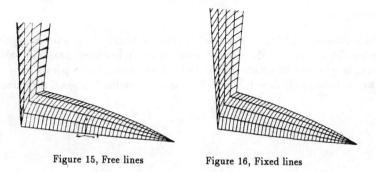

Figure 15, Free lines   Figure 16, Fixed lines

Figures 15 and 16 show the effect of imposing lines as fixed ones.

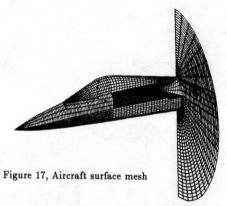

Figure 17, Aircraft surface mesh

6. COMPUTING TIME

The following table shows the number of blocks, number of points, number of iterations and CPU time for the configurations optimized :

conf name	blocks	Nb points	Iter	CPU time
F5 free	4	50960	500	30
F5 tunnel (1)	4	71200	800	50
F5 tunnel (2)	4	71200	1000	77
CN235	8	52160	276	15
Fighter	11	98178	800	60

All CPU times in minutes for CRAY-XMP

For F5 wing, free stream, after 50 iterations a steady residue of 0.15 was obtained

For F5 wing in tunnel (2), after 800 iterations a steady residue of 0.25 was obtained

For fighter configuration, after 60 iterations a steady residue of 0.90 was obtained

7. CONCLUSIONS

After testing ONERA grid optimization method with different configurations, some conclusions can be extracted :

-For complex configurations, the definition of block interfaces becomes a tedious work. Mainly because of the definition of curve interfaces.

-Wall type surfaces have been set as fixed ones for the optimization process, and the remaining are left free. However, in some cases, additional surfaces (f.e., block interfaces) must be fixed.

-Increasing the number of blocks, improves grid quality and reduces the number of iterations required.

-An initial grid of good quality (f.e. an elliptic grid) reduces the required number of iterations and reduces the differences between initial and optimized grids.

-The use of this optimization technique improves the mesh quality and enables the use of "poor" initial meshes, which can be created easily. The combination of both methods provides good grids in short time.

8. REFERENCES

1. De la Viuda, Diet, Ranoux. "Patch Independent Structured Multiblock Grids for CFD Computations", in Numerical Grid Generation in CFD and Related Fields. Barcelona, June 1991.

2. Sony B.K. "Two and Three Dimensional Grid Generation for Internal Flow Applications of CFD", AIAA Paper 85-1526, AIAA 7th Computational Fluid Conference, Cincinnati, Ohio, July 1985.

3. Jacquotte, O., and Gaillet C. " Three-Dimensional Multi-block Mesh Optimization Using a Variational Method".

4. Jacquotte, O., " Recent Progress on Mesh Optimization" in Numerical Grid Generation in CFD and Related Fields. Barcelona, June 1991.

MULTI-BLOCK MESH GENERATION

FOR COMPLETE AIRCRAFT CONFIGURATIONS

Klaus Becker, Stefan Rill

Department EF 10, Deutsche Airbus GmbH

P.O. Box 10 78 45, D-2800 Bremen 1, Germany

SUMMARY

Mesh generation is a basic part of computational fluid dynamcis (CFD). However, as it is one of the major bottlenecks of CFD application in industry, increasing attention has to be payed to the development of fast and user-friendly, good quality mesh generation systems.

In this paper, the fundamental elements of the Deutsche Airbus GmbH's (DA) interactive mesh generation system INGRID are described. Current enhancements deal with airplane component construction, surface mesh generation, 3D topological considerations, basic 3D block mesh generation algorithms, block mesh implantation and mesh smoothing experiences. Special emphasis is layed on a local sub-block embedding technique which directly fits into the context of multigrid flow solving.

The flow field is determined with DA's MELINA code, a cell vertex, multigrid, multi-block Euler solver which is based on central differences, artificial viscosity and explicit 5-stage Runge-Kutta time stepping. MELINA already has a considerable impact on the aerodynamic design of complex 3D flow problems at DA such as laminar glove design, integration of propulsion systems or simulation of flap track fairings on the wing.

The validation experiments for INGRID/MELINA with wing/body/pylon/engine/flap-track fairing configurations have shown that the reliability of CFD results is very high even for that complicated geometries. This offers great chances for future developments of efficient, safe and environmentally clean airplanes.

INTRODUCTION

Considerable advances have been recognized in the area of numerical flow simulation for complex configurations during the last years which have reached a stage of real applicability in industrial context. Future developments are now aiming at the full viscous simulation for realistical aircraft models. However, mesh generation for those complex configurations, which is the basis for CFD work, is still the most time-consuming part and thus the clear bottleneck for industrial application. For sure, a major effort in this game is the development of algorithms and systems that generate suitable meshes around these complicated objects either on the surface for low speed panel method flow solving or for transonic inviscid and even viscous 3D simulations. This paper will report on the way DA is attempting to attack the problem of increasing the overall efficiency of CFD simulations. This means not only reduction of costs and time but also an increase in accuracy of the numerical solution. At DA, the basic package for 3D compressible flow simulation consists of the INteractive GRID generation system INGRID and the Multi-block EuLer INtegration Algorithm MELINA. In this paper we concentrate on the mesh generation part and only touch the flow solver as it is directly involved in the conceptual considerations.

Within INGRID, the first section is dealing with surface constructions. This part can directly be used as surface mesh generation part for panel methods. The generation of 3D grids around wing/fuselage or more complicated aerodynamic configurations which is the main issue of this paper, follows a step by step approach (Fig. 1).

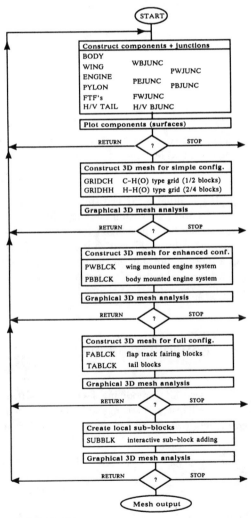

Fig. 1 Basic flow chart of INGRID.

- Firstly, a multi-block mesh is constructed for a simple wing/fuselage combination. Thereby, a basic topological structure is chosen.

- Secondly, main components like pylon and engine (nacelle) are added to the configuration. Suitable block frames are cut out of the prior blocks and filled with the surface of those components and the respective 3D mesh.

- Thirdly, smaller components like flap track fairings are put into place. One or more of the prior blocks are split at a suitable mesh plane and free space is generated by pushing the split plane aside using spline techniques. The space is then filled with new mesh blocks which are called implanted blocks.

- Additionally, the mesh can be refined by overlaying so-called sub-blocks onto the mesh generated so far. These sub-blocks are included in a multigrid sequence of mesh levels and treated in the multigrid manner of local refinement.

At each of these stages, the multi-block mesh topology is updated automatically. If one or more of the above sections are missing it is possible to skip the respective section in the flow diagram (Fig. 1) and proceed with the remaining steps. Thus local sub-blocks can be added at all stages of configuration complexity, however, only as the very last step in the mesh generation process.

INGRID uses only explicit mesh point calculation techniques, no system of equations has to be solved. Therefore it is very fast and allows the efficient construction of 3D aerodynamic meshes on workstations. There are two basic techniques: The first one is a simple line construction algorithm which connects two points via a line with prescribed start and end directions and bending and a prescribed distribution of points. The second one is a block filling technique based on a sophisticated arc length blending of the given boundary of a 2D field of 3D coordinates. Recent improvements of both techniques led to more smooth mesh point distributions within the mesh blocks.

The block filling technique has been developed for easy 3D block mesh generation. A first version has been enhanced by making it independent of user given parameters. Along with this the use of a mesh smoothing on the basis of optimization techniques has been investigated. Modifications of the driving functional have been derived and tested in order to improve the mesh in the vicinity of the boundary.

COMPONENT MODELING

The first step of INGRID consists of a modeling facility for airframe components like wing, fuselage, engine, pylon or flap-track fairings and their relation to each other. The single objects are defined in their own coordinate systems and then manipulated to build the total configuration.

Surface Definition

Different approaches are used so far for the definition of airframe components currently under consideration. They are split into two classes: wing-type objects and body-type objects.

- All <u>wing-type objects</u> are defined by normalized airfoils at representative span stations and some characteristic functions along the plane wrap of the wing. From these functions the geometrical twist angle, expected load twist angle, thickness, interpolative blending of airfoils, etc. can be evaluated. The airfoils are given by sets of points, and they are splined along their arc length to redistribute points according to the requirements from the flow solver that will be applied on the final mesh.

- <u>Body-type objects</u> are formed from simple quasi-double-superelliptic cross-sections which can vary along the length of the body. The center line of a fuselage is assumed to be straight so that the cross-sections are all parallel. In an enhanced form, the engine is treated as a "full" body which contains a second inner "body" to form the inner engine tube. In contrast to the airplane fuselage, both parts of this object follow a center line that can be bended. Thus drooped nacelles can be constructed. Flap-track fairings are considered as boat-shaped half-bodies which are mounted to the wing.

The major advantage of this included mini design system for airframe components appears in connection with airplane design. Prior to any time consuming constructions or CAD descriptions of a new configuration, CFD results can be obtained using INGRID and thus an optimization in the design process is possible at a very early stage.

Interfaces for input from CAD systems and output to other flow simulation software are under development to ease the use of this system for configurations that are already existing. Special problems arise with the transfer of surface and component descriptions from one software system to the other. Standards like IGES [5] that have been defined for this purpose can be used, but these are usually not able to transfer the identical information. Additional problems arise with the typically different style of working with CAD or CFD. Though CAD systems are able to create meshes on surfaces, these meshes do not fit to the essential requirements of CFD. Additionally, non-uniform surface definitions which are often used with CAD, are to be unified before usable with CFD. Thus, a lot of difficulties are to be overcome just before the real object is defined, and surface mesh generation suffers from these difficulties as well as 3D mesh generation.

Surface mesh generation is the first item for both panel surface meshes and 3D meshes. However, the needs of the different types of flow solvers to be applied are quite different and thus influence the typical requirements for the mesh generation system. Panel mesh generation is very much aimed at reasonable surface descriptions and often based on an unstructured type of mesh. Nevertheless, certain rules of continuity of mesh lines and closure of surface patches must be taken care of. A main effort is in the construction of aerodynamic meshes that is keeping track of the essential flow features with the help of a pre-adapted mesh refinement. A second requirement is to keep the total number of panels as low as possible. For 3D grids, the surface meshes are even more restrictive in their topology if they refer to a structured form like in our case. The fineness of the mesh at one edge of the configuration can directly influence that on an other edge and can thus not be arbitrarily defined - compared to the unstructured approach.

Various geometrical surfaces can be generated with the INGRID system. Figures 2 to 4 show some specific examples.

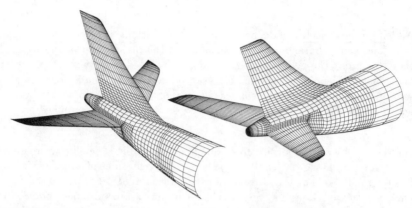

Fig. 2 Tailplane region of a transport aircraft - panel mesh.

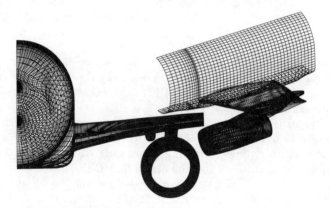

Fig. 3 Body with belly and wing root fairing.

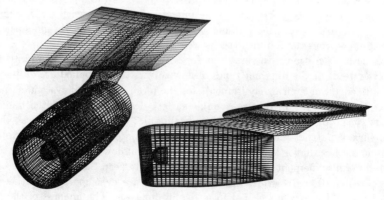

Fig. 4 Pylon/engine/wing configuration.

Component Junction Problems

A general problem with the definition of components and their junction is the determination of intersections. The system must be able to find intersection lines of components as well as to define a wrap or blending of one component to the other.

A typical situation where wrapping is applied is at a wing/body junction. Usually, a wing is defined from a root section to a tip section, and the root section lies outside the body. The fairing region is not specified in advance and has to be created. In that case, the wing has to extrapolated along rays on its surface until it hits the body. Thus the body intersection line is determined from an iterative process that finds the point where the ray runs into the body surface. After that process has been applied to the root profile, the other airfoils within a certain wing attachment region are bended as to follow the body shape (Fig. 5). The amount of bending depends on the relative distance of the airfoils to the root profile.

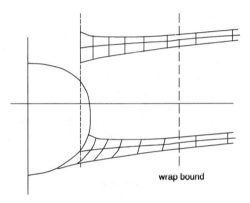

Fig. 5 Wing/fuselage junction.

Real intersection of components can be viewed as interpolation problem where one component lies part inside the other. Rays on it's surface run through the surface of the second component and the intersection point has to be determined.

In the current version of INGRID, the rays are defined as straight lines. Their direction is determined from the nearest available information. However, this process assumes a ray-linear surface definition of the first component in the vicinity of the junction. Therefore heavily nonlinear intersection regions have to be modelled with care.

SPACE MESHES

As outlined above, the philosophy for the construction of 3D multi-block meshes for complex configurations is a step by step one (Fig. 1). Starting from a simple baseline configuration, i.e. a 3D mesh around a wing/fuselage combination more complex configurations are surrounded by a mesh using a cut-out technique. A block is cut out of the global mesh and filled with a new block mesh system that describes the new component.

The outer boundaries of the local system are required to fit to the boundaries of the cut-out block. For further components, 3D space meshes are built into the multi-block grid by an implantation technique. One or more existing blocks are split at an appropriate mesh plane position and the mesh is pushed aside in order to get free space for the new component. This space is then filled with new mesh blocks. As the new blocks usually don't extend to the far field but their faces collapse in singular lines, they are fully enveloped by the original mesh and thus called implanted.

Topology Definition

The basic structure of a multi-block mesh is given by the topology, i.e. the definition of the inter-connection of the blocks and the boundary conditions at their faces. For any new airplane configuration, the topology may be different and thus a lot of variations have to be implemented in the multi-block mesh generation system for complex configurations. Within INGRID, a building box principle is used: At a certain stage, the system is only able to create a restricted number of topologies, but is continuously extended to the treatment of additional configurations. However, this strategy limits the system to special configurations, but fortunately, the number of essentially different configurations is limited within a transport aircraft environment.

The topology of the multi-block mesh around a certain configuration is completely defined within the mesh generation system and cannot be changed by the user. Thus the difficulties with topological definition of blocks and their relation to each other are already solved for a list of configurations and do not require too much insight from a design engineer.

For example, an H-H mesh for a basic wing/fuselage combination consists of at least one lower and one upper block. Usually, the outer region of the mesh is put into separate blocks which can be coarser than the inner ones in all three coordinate directions.

For the wing mounted body/wing/pylon/nacelle case, the grid consists of 8 blocks which can be identified in Fig. 6 as blocks 2 to 9. Blocks 2 to 4 are identical with the blocks of the body/wing configuration. Grid blocks 5,6,7,8 and 9 result from rearranging the former body/wing-block no. 1 under the wing according to the cut-out technique. Grid blocks 7 and 8 form the nacelle inlet and exit tubes which extend to the farfield up- and downstream, respectively. Block 9 is a special block that had to be introduced, because the pylon extends over the nacelle rear end into the jet.

For the body mounted engine case the topology definition follows a similar way. The major difference is two ring blocks around the engine if the simulation of an unducted propfan is required. Similar considerations with a double ring are used for bypass engines, for which a detailed view is also given in Fig. 6.

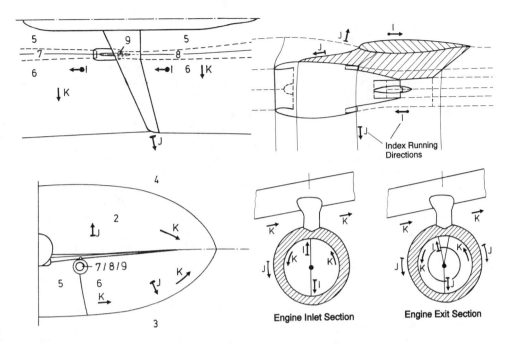

Fig. 6 Mesh block topology for wing mounted engine case - details near a bypass engine.

Block Mesh Generation Strategies

3D automatic mesh block generation is a helpful tool if boundary coordinate values of a block are known. As this is often the case in multi-block mesh generation programs, the development of an automatic point distribution mechanism is desirable which fills a block with points. One of the problems with such a method is to avoid intersection of lines or planes. This can be achieved by the use of sufficiently small perturbed Laplace equation mesh generators, for example. However, these methods are relatively expensive and a sophisticated definition of Poisson driving terms for proper concentration of mesh points in relevant regions is necessary. In this paper, some much simpler approaches are described that fulfill the mesh generator's needs in many cases. Direct calculation of mesh points is used with easily handable point mesh distribution tools. The methods provide smooth and well-distributed meshes within minimum computation time.

Two basic methods are described: The first one is a 3D line shooting algorithm, the second one can be used to fill a logically 2D array in 3 dimensions and thus applied in a successive way will fill a complete 3D block.

Free Line Construction

If only two opposite sides of a 3D mesh block are given, a simple technique to fill the block automatically is to shoot lines from one given side to the opposite one. The shooting algorithm we use has been firstly developed by Sobieczky and Schoen [12],[13] but was subsequently refined to enable more flexible use. The curve starts and ends at prescribed points; it may follow a prescribed start and/or end direction. Additionally, the distribution of points along the line can be prescribed as well as the bending of the curve at start and end point. If, now, all input values like directions and bending factors vary smoothly throughout the known sides of the block, then the lines also vary smoothly. Thus the block can be filled with a reasonable mesh.

The algorithm works quite well but has several disadvantages. Firstly, it is only possible to fit to the boundary points given on the start and end plane. The four other boundary faces of the block are also constructed this way and cannot be matched to any predefined values. Secondly, the distribution of parameters must be selected very carefully because of a nonlinear change in behavior of the curves. The curves vary rapidly on small changes of the bending parameters, for example. Additionally, the total curve variation achieved by the variation of all parameters as directions, distributions and bending parameters can hardly be controlled because of overlaying nonlinear effects. However, this technique is very useful for smoothly curved mesh blocks.

If not to be used for complete block filling, this method can be used for calculating block faces or planes inside a block. These kind of planes are often necessary or at least useful to split up blocks along strongly curved edges to cater for better controlled mesh concentration and avoid cell distortion.

In detail, the free line construction technique steps from one from one point on the line to the next. As depicted in Fig. 7, it runs from point a P_1 to a point P_k using $\vec{n}_1$ as start direction in a first sweep. The second sweep is reverse from P_k to P_1 with the start direction at P_k. A blending of both curves which is based on arc length factors results in the final curve.

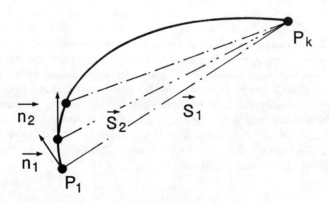

Fig. 7 Free line construction.

The algorithm can be described by the following steps:

- 1. Start at P_1, old direction vector is $\vec{n}_1$.
- 2. Compute the new direction vector as $\vec{t}_1 = f_2^\beta \frac{\vec{s}_1}{\|\vec{s}_1\|} + (1 - f_2^\beta)\frac{\vec{n}_1}{\|\vec{n}_1\|}$.
- 3. Set (the new length of this vector to) $\theta = \|\vec{s}_1\| \frac{f_2 - f_1}{1 - f_1}$
- 4. The new point is $P_2 = P_1 + \theta \vec{t}_1$.
- 5. For next step use $\vec{t}_1$ for $\vec{n}_1$, P_2 for P_1, $\vec{s}_2$ for $\vec{s}_1$, f_2 for f_1 and go to 1 until is P_k reached.

In this process, f_1 is the value of the normalized distribution at point P_1, starting with zero. f_2 is the desired distribution value for point P_2. It's value is between zero and one, by definition. $\vec{s}_1$ is the vector from P_1 to the final target point P_k. f_2^β determines the amount of influence of the vector $\vec{s}_1$ on the next direction vector. Thus, the larger the bending factor β is, the more the line tries to follow the start direction and therefore is more bended. If β tends to zero, the influence of the start direction is reduced so that a straight line is constructed in the limit.

Arc Length Mixed Array Filling

A second technique constructs the block entries plane by plane thus taking care of four block faces in a ring. As the coordinates of a structured mesh within the block shall be distributed such that the variation from one edge of the block to the opposite one is smooth, a blending strategy is used for the construction of the inner nodes. For a logically 2D frame of a single mesh plane, a sophisticated blending of boundary arc length distributions has been developed which is the basis for array filling.

The first idea for such a method can be described by a short algorithm. Following the notation of Fig. 8, an auxiliary straight column s_i is calculated first from one boundary point to the corresponding opposite one. After that the deviation of the given left and right side boundaries from ficticious straight columns (s_1 and s_n) is blended to the position of the ith column and added to the auxiliary straight column s_i.

In detail, any interior point X_i^j is defined as

$$X_i^j = A_i^j + D_i^j. \tag{1}$$

$$A_i^j = p_i^j X_i^m + (1 - p_i^j) X_i^1 \tag{2}$$

is the linear interpolant between the boundary points X_i^1 and X_i^m, i.e. point no. j of a ficticious straight column at the ith column position. The interpolation coefficient p_i^j which represents the ratio of the length of column i up to point j and the total length of column i is calculated from a blending of the normalized arc length distribution of points on the left and right boundaries, i.e. column $i = 1$ and $i = n$ (see eqn (4)).

The second part D_i^j describes the deviation of the final mesh column i at point no. j from the ficticious straight line. It is calculated from the deviations at the left and right boundaries ($i = 1$ and $i = n$) which are additionally scaled by the ratio of the length of column i and column 1 or n, repectively. The left and right deviations are linearly blended using the relative length of line j up to position i as interpolation coefficient q_j^i:

$$D_i^j = q_j^i \frac{\Delta_i}{\Delta_n} D_n^j + (1 - q_j^i) \frac{\Delta_i}{\Delta_1} D_1^j. \tag{3}$$

Δ_i is the length of column no. i.

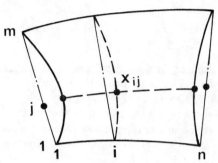

Fig. 8 2D array filling.

Within this general approach, the interpolation coefficients p_i^j and q_j^i are unknown. They depend on the final point positions X_i^j and thus the above settings form a system of nonlinear equations.

In the first version, an approximative solution was defined by

$$q_j^i = \beta q_m^i + (1 - \beta) q_1^i, \tag{4}$$
$$p_i^j = q_j^i p_n^j + (1 - q_j^i) p_1^j \tag{5}$$

with a user-defined value of β between 0 and 1. This type of definition has been used within INGRID for a long time. The specific choice of the variable β did not appear too difficult. Rather complicated meshes can be generated by the use of this algorithm. It is even possible to treat rather distorted arrays in 3D space. Degenerated edges are allowed, where all mesh points collapse to one physical point. In that case the respective distributions are assumed to be equidistant (Fig. 14).

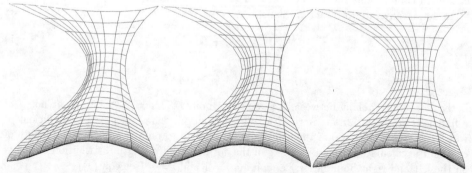

Fig. 9 2D array filling - a)column blend $\beta = 0$ - b) column blend $\beta = 1$ - c) line blend.

If the left column of the array shown in Fig. 9a-c is blended to the right one, $\beta = 0$ gives a reasonable result (Fig. 9a). However, for $\beta = 1$ the lines intersect at the right boundary (Fig. 9b). Fig. 9c shows the result if the bottom line is blended to the top line; it is nearly independent of β.

A more automatic parameter independent definition has been found desirable for the programming of mesh block generation for new configurations. This can easily be achieved by the "vice versa" definition

$$q_j^i := p_i^j q_m^i + (1 - p_i^j) q_1^i, \tag{6}$$
$$p_i^j := q_j^i p_n^j + (1 - q_j^i) p_1^j, \tag{7}$$

which has the solution

$$p_i^j = \frac{p_1^j(1 - q_1^i) + p_n^j q_1^i}{1 - (q_m^i - q_1^i)(p_1^j - p_n^j)}, \tag{8}$$
$$q_j^i := p_i^j q_m^i + (1 - p_i^j) q_1^i. \tag{9}$$

As shown above, a drawback of the former methods is that they are direction-dependent: column $i = 1$ is blended to column $i = n$ and this is different to the line $j = 1$ to line $j = m$ blending (Fig. 9). Therefore a new direction-independent method has been developed where the straight part A_i^j as well as the deviation part D_i^j in (1) are defined by a mixing from the four corners and edges of the 2D array:

$$A_i^j = p_i^j(1 - q_j^i)X_1^m + p_i^j q_j^i X_n^m + (1 - p_i^j)(1 - q_j^i)X_1^1 + (1 - p_i^j)q_j^i X_n^1, \tag{10}$$
$$D_i^j = \frac{\Delta_i}{\Delta_1}D_1^j(1 - q_j^i) + \frac{\Delta_i}{\Delta_n}D_n^j q_j^i + \frac{\Gamma^j}{\Gamma^1}D_i^1(1 - p_i^j) + \frac{\Gamma^j}{\Gamma^m}D_i^m p_i^j. \tag{11}$$

Γ^j is the length of line no. j.

The application of this algorithm is more general and straight forward. For the above example, we end up with a mixing of both line and column blended meshes which looks quite nice (Fig. 10a). However, some difficulties appear if the 2D array is even more squeezed (Fig. 10b). The interpolation of blending distributions is not absolutely save in that case concerning the overlap of mesh lines. Thus further development is necessary to improve the method. This is also true for the extension to 3D. Additionally, the enhanced version shall take care of the boundary distance of the first mesh line inside the array. Some more sophisticated strategies which keep some distances more or less fixed in order to create better cells near a wall, e.g., is highly recommended for good quality meshes in the context of Navier-Stokes calculations.

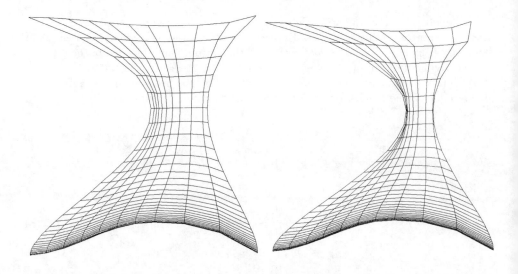

Fig. 10 Automatic 2D array filling - a) normal case - b) squeezed region.

The subsequent plane by plane application of 2D array filling leads to a 3D block filling algorithm. However, this technique does not fully guarantee regular block meshes even if the boundary lines in the third direction do not overlap. Within the block, there is only few information from the boundaries concerning the propagating direction of planes. A generalization of the above 2D filling to 3D is desirable. Nevertheless, the current technique is frequently used in the mesh generation system INGRID [1], [2].

Examples and General Remarks

During the work on 3D block mesh generation we found that it is not worthwhile treating cases where the blocks are too distorted. It seems necessary then to introduce direct control of the mesh spacing and mesh line propagation directions at singular points or points with rapidly varying slopes. These points appear usually on the configuration boundary near component junctions. In this case we prefer a separation of the 3D mesh block and a separated block filling for the sub-regions. The deviding mesh planes are constructed using the above methods adapted to the special situation.

In the vicinity of the nacelle and pylon just underneath the wing there is a zone which needs high quality mesh generation for proper resolution and minimum mesh distortion. The problems arise because of the very flat pylon leading edge and the requirement for high mesh resolution near the wing leading edge. Fig. 11 shows a grid segment in that zone. The left boundary curve is created by line shooting and the segment is filled using array filling. The enlarged views at the wing nose and engine nose demonstrate capabilities of the parameter driven array filling (Figs. 12 and 13).

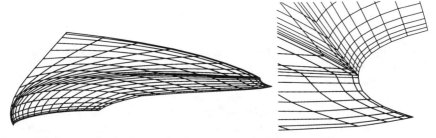

Fig. 11 Mesh segment in front of pylon generated with array filling.

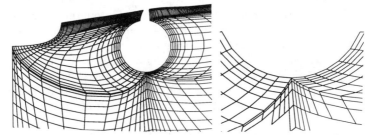

Fig. 12 Vertical cut through configuration - enlarged view below nacelle.

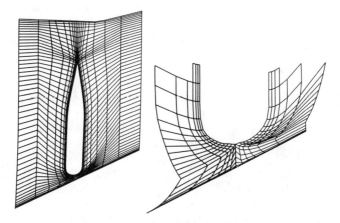

Fig. 13 Mesh on wing surface near pylon - enlarged view of pylon nose.

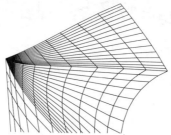

Fig. 14 Singular mesh segment generated with line shooting and array filling.

MESH SMOOTHING

A ususal approach to a posteriori mesh smoothing which may be recommended in the case of the meshes created so far is the minimization of an appropriate cost functional. This definite functional has to describe at least one of the essential properties of a so-called good mesh which are orthogonality, squewness, smoothness, growth of mesh cells, etc. Carcaillet et al. [4] developed a functional that combines a measure for local orthogonality and smoothness. Orthogonality is expressed in terms of the sum of all angles between all neighbor mesh lines that meet at a given mesh point. Smoothness is calculated from the sum over the length of all mesh intervals on lines that meet at the point under consideration. Both terms are linearly combined in order to form a closed cost function. After that, a usual optimization algorithm can be used to drive the mesh so that the costs become minimal. However, because the above conditions cannot both be met - at least at the boundary of the mesh where some corners may appear - it can happen that the mesh will improve in some region but become worse in others. Therefore it is necessary to tune the functional very carefully in order to achieve a minimum overall improvement.

Referring to Fig. 15, Carcaillet et al. give the following cost functional

$$F = \Sigma_i \Sigma_j \Sigma_k [\delta O_{ijk} + (1-\delta) S_{ijk}] \qquad (12)$$

with a linear weighting parameter δ between 0 and 1. The single entries are

$$S_{ijk} = \omega_1 \|\vec{r}_1\|^2 + \cdots + \omega_6 \|\vec{r}_6\|^2, \qquad (13)$$
$$O_{ijk} = (\vec{r}_1 \cdot \vec{r}_2)^2 + \cdots + (\vec{r}_6 \cdot \vec{r}_4)^2 \ (12\,terms). \qquad (14)$$

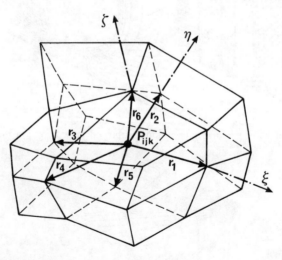

Fig. 15 Mesh cell arrangement.

This basic definition has several disadvantages with respect to three-dimensional grids. Firstly, the scale between smoothness and orthogonality is not balanced because of the different polynomial degrees. Therefore the linear blending between both parts does not correspond to the natural feeling for a certain value of δ. Secondly, the orthogonality does not only measure angles at a node but is reflecting the size of the abutting cells. This means that this part of the functional becomes large for large non-orthogonal cells and thus the measure for small cells near walls is not adequately represented. At least during the first iteration sweeps which are relevant for a posteriori mesh smoothing, the procedure prefers to orthogonalize in the large cells region and not where it is primarily desired. The main difficulty, however, is to avoid mesh movement in a certain direction if this is not desired. This has been tried by special settings for the parameters $\omega_1, \cdots, \omega_6$ but cannot be fully achieved in this way.

For our purposes, the above orthogonality measure has been changed to

$$O_{ijk} = \frac{(\vec{r}_1 \cdot \vec{r}_2)^2}{\|\vec{r}_1\|^2 \|\vec{r}_2\|^2} + \cdots + \frac{(\vec{r}_6 \cdot \vec{r}_4)^2}{\|\vec{r}_1\|^2 \|\vec{r}_2\|^2} \quad (12\, terms), \qquad (15)$$

which means that the cells sizes are eliminated.

For a more or less two-dimensional test section in front of the wing Fig. 16 a) nose we found the following results: The aim was to reduce the mesh slope discontinuity in the H-mesh block near the wing nose. Application of the original method leads to unacceptable cells near the wall (Fig. 16 b). Eliminating the effect of large cells in the orthogonality functional gives better results (Fig. 16 c) though orthogonality is strongly violated near the wall. Points on the wall are not allowed to move, which would have helped in this case, but is undesirable because of general surface approximation problems. Even a drastic enlargement of the boundary entries to the cost functional by additional weighting factors could not help in getting more orthogonal meshes near the wall.

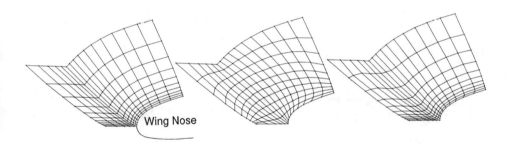

Fig. 16 Mesh smoothing tests - a) initial mesh - b) original method - c) using eqn. (19).

The essentials from these investigations are that mesh smoothing of the described kind is very tricky to use, particularly in 3D. It should not be applied across regions with corners within which a violation of the desired properties of smoothness or orthogonality is unavoidable. A much more sophisticated method is necessary to fulfill the requirements of automatic block mesh smoothing. Based on this experience it has been decided to postpone mesh improvement with repect to geometrical properties for a while. The robustness and accuracy of the cell vertex Euler solver to be applied on those meshes helps very much in this situation.

BLOCK IMPLANTATION TECHNIQUE

As airframe configurations become more and more complex, it has been found comfortable to add additional components through the implantation of new mesh blocks. That is existing mesh blocks which surround the location of the new component are split at an appropriate position. The mesh planes are pushed aside to make some space available for a new local block system which contains the added object. Pushing is done on mesh lines perpendicular to the split plane. All three coordinates of points along such a line are splined along the arc length of the line and then redistributed following a new normalized distribution. In the free space no problems could be observed, however, near a wall boundary which is parallel to the line direction a loss of surface accuracy has to be encountered. This can be repaired by a projection of the first mesh plane on the real surface.

Along with the computation of the coordinates of the new blocks, the topology of the mesh has to be adapted to the new situation. The result of this can be seen from Fig. 17 if this is compared to Fig. 6: Three fairings have been added in sequence from left to right. The successive block numbering is broken because the previous blocks are split into new ones.

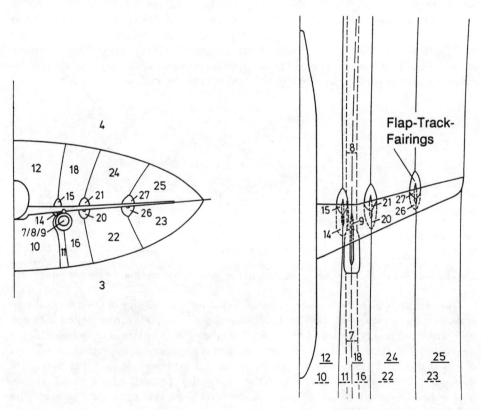

Fig. 17 Block implantation - topological information for flap track fairing case.

The implantation technique is used for 3D meshes around flap track fairings, for example. Figs. 18-20 show some typical views of the mesh.

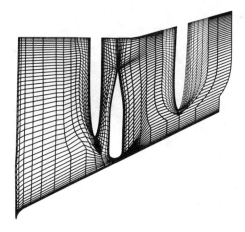

Fig. 18 Flap track block implantation - lower wing surface.

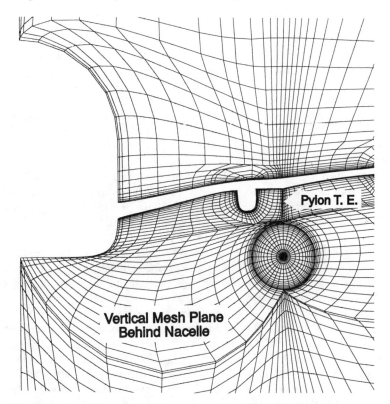

Fig. 19 Block implantation - vertical cut through mesh behind nacelle.

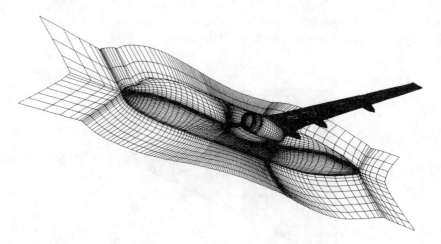

Fig. 20 Complete surface mesh and part of symmetry plane.

GENERATION OF LOCAL SUB-BLOCKS

Computing a 3D flow field around complex configurations it is difficult to resolve all features of the solution to the same accuracy with a uniform mesh. This problem is already somewhat relaxed by using multi-block meshes with the option of enrichment and coarsening of the mesh from one block to the other. The drawback of this feature is the fact that the desirable surfaces of block splitting and the requirements for mesh refinement usually do not coincide. Another pitfall is, that for a posteriori changes of the spatial resolution a complete change of the block topology is required.

This leads to the development of sub-block refinement. The idea of this is to simply "patch" locally refined mesh blocks onto the existing mesh and connect the additional fine sub-blocks with the mesh via a multigrid technique. At the time beeing, sub-blocks are defined a priori or a posteriori to the solution of the flow problem by the grid generator INGRID. Their orientation or extension in the computational domain can be judged and changed interactively with mouse and menu technique. The only limitation to the topology is that a sub-block has to lie completely in a grid block of the existing mesh, which includes touching the block boundary. But a grid block may have various sub-blocks and a sub-block may have several sub-blocks itself. In any case, the topological connections to the existing block system are updated automatically.

The mesh points of the locally refined sub-blocks are constructed with a simple trilinear interpolation of the coarse cells. On any component's surface, a special interpolation of the points based on Coons' local patches is used. Thus yielding a doubling of the grid density in all three coordinate directions. Although flow solutions on meshes obtained with this strategy are very encouraging, as will be shown later, it is a matter of future investigations to refine the method of sub-block construction in such a way that smoother variations in grid stretching are obtained. This is of special importance with respect to the application of the sub-block technique to Navier-Stokes simulations.

The sub-block approach can be viewed as a compromize between structured an unstructured meshes, combining the benefit of high computational efficiency on structured meshes and of clustering grid points in a "quasi unstructured" way by scattering sub-blocks and even further refined blocks in regions of strong gradients. It is envisaged but not yet realized to use this method for solution adaptive mesh refinement if the regions of sub-block refinement are determined automatically during the iteration by suitable sensor functions.

The interior grid points of the sub-blocks are interacting with the coarser block via a multigrid technique that will be described in the next section. For the boundary points of the sub-blocks we use the same boundary condition routines as for the coarser blocks. For the cut conditions along the edges of the fine sub-block we introduced guard cells that are updated through an additional inter-block boundary exchange buffer. This buffer field had to be added, because with a sub-block the coarse block requires not only data exchange with other blocks along its "external" block faces but also with one or more sub-blocks along "internal" block faces.

Examples for sub-blocks can be found in Figs. 22-24 as well as Figs. 26-27.

FLOW SOLUTIONS

The block mesh generated by INGRID is used as input for the MELINA flow solver. The multigrid, multi-block Euler integration algorithm MELINA may be classified as a 'Jameson-type' cell vertex scheme with central differences, artificial dissipation and explicit Runge-Kutta time-stepping. MELINA is able to run on any grid that fits to a certain standard of block mesh topology. It automatically creates all auxiliary mesh planes that are necessary for the exchange of flow data between different blocks. Thus the interface from mesh generation to the flow solver is given by a file that contains the mesh block coordinates as well as a set of indices that fully describe the topology of the mesh. Both are created by INGRID.

Conservation Equations for Inviscid Flow

Inviscid flow can be modelled by the continuum mechanic's conservation equations for mass, momentum and energy. In 3D these are five scalar integral equations which are augmented by the thermic and caloric gas equations for ideal gas as material laws to form a closed system.

If $\rho, \vec{v}, E$ and κ are the density, velocity vector, total energy per unit volume and ratio of specific heats of the fluid, respectively, the pressure p can be calculated from

$$p = (\kappa - 1)\rho \left[E - \frac{\|\vec{v}\|^2}{2} \right].$$

With $H = E + \frac{p}{\rho}$ as definition of the total enthalpy, for any control volume V the conservation of mass, momentum and energy result in the system of Euler equations mass (continuity):

$$\frac{\partial}{\partial t}\int_V \rho\, dV = -\oint_{\partial V} \rho(\vec{v}\cdot\vec{n})\, dO \qquad (16)$$

momentum:

$$\frac{\partial}{\partial t}\int_V \rho\vec{v}\, dV = -\oint_{\partial V} \rho\vec{v}(\vec{v}\cdot\vec{n}) + p\vec{n}\, dO \qquad (17)$$

energy:

$$\frac{\partial}{\partial t}\int_V \rho E\, dV = -\oint_{\partial V} \rho H(\vec{v}\cdot\vec{n})\, dO \qquad (18)$$

∂V is the surface of the control volume V and $\vec{n}$ is the outward unit normal vector on the boundary ∂V.

Equations (16)-(18) can be abbreviated by the vector valued equation

$$\frac{\partial}{\partial t}\int_V \vec{W}\, dV = -\oint_{\partial V} \bar{\bar{F}}(\vec{W})\cdot\vec{n}\, dO \qquad (19)$$

where $\bar{\bar{F}}$ is the flux tensor and $\vec{W} = (\rho, \rho\vec{v}, \rho E)$ is the vector of dependent variables.

Cell Vertex Discretization

In order to discretize equation (19), a boundary-fitted finite volume mesh is used where the vertices of the finite volumes are taken as locations for the unknown variables $\vec{W}$. The method of lines decouples time and space directions which leads to a coupled system of five ordinary differential equations in time for any grid cell V.

Equation (19) is attached to finite volumes rather than grid points. For the cell vertex approach a summation of the equations is then applied which takes into account every eight cells connected to one common vertex. Thus with the discretization of the flux integrals one gets

$$\frac{\partial}{\partial t}\int_{V_S} \vec{W}\, dV = -\sum_{l=1}^{8}\oint_{\partial V_l} \bar{\bar{F}}(\vec{W})\cdot\vec{n}\, dO. \qquad (20)$$

Equation (20) is not stable in the sense that the stationary solution admits oscillatory components which are not physically correct. Therefore it is necessary to eliminate these components from the solution. In the context of central difference schemes as we have here this is usually done by adding an appropriate filter term to the system of equations. This artificial filter or dissipation operator D in equation (21) consists of two parts: The so-called background dissipation aims at the elimination of high frequency disturbances. It is necessary to ensure convergence for the whole range of flow speeds to be treated. In Jameson's approach it consists of undevided 4th differences of the solution vector multiplied by an approximation of the largest eigenvalue of the Jacobian matrix $\frac{\partial \bar{F}}{\partial \vec{W}}$ and a user-defined constant. The second dissipation term is an undevided second difference. It is designed to avoid oscillations of the solution near shocks and is switched on only in those regions by the use of a special sensor function.

The complete semi-discrete equations then read

$$\frac{\partial}{\partial t} \int_{V_S} \vec{W}\, dV = -\left[\sum_{l=1}^{8} \oint_{\partial V_l} \bar{\bar{F}}(\vec{W}) \cdot \vec{n}\, dO + D\vec{W}\right]. \tag{21}$$

The time discretization depends on the treatment of the volume integral in equations (20) or (21). In our case a hybrid, explicit 5-stage Runge-Kutta time stepping scheme is chosen where the integral is defined by the relation

$$\frac{\partial}{\partial t} \int_{V_S} \vec{W}\, dV \approx |V_S| \frac{\partial \vec{W}}{\partial t}\Big|_{centre}.$$

Any iterate $\vec{W}^{n+1}$ is calculated from the previous one $\vec{W}^n$ by

$$\begin{aligned} \vec{U}^0 &= \vec{W}^n \\ \vec{U}^k &= \left(\vec{U}^0 + \frac{\alpha_k \Delta t}{\|V\|} \vec{R}^{k-1}\right), k = 1,2,3,4,5 \\ \vec{W}^{n+1} &= \vec{U}^5 \end{aligned} \tag{22}$$

where

$$\vec{R}^k = -\left[Q(\vec{U}^k) + \sum_{m=0}^{k} \gamma_{k,m} D(\vec{U}^m)\right] \tag{23}$$

represents the right hand side of equation (21). The coefficients $\gamma_{k,m}$ are chosen such that we obtain a hybrid (5,2) scheme that is to evaluate the filter terms only in the first two of the five Runge-Kutta steps. The parameters α_k are set to

$$\alpha_1 = \frac{1}{4},\ \alpha_2 = \frac{1}{6},\ \alpha_3 = \frac{3}{8},\ \alpha_4 = \frac{1}{2},\ \alpha_5 = 1.$$

For convergence acceleration additional techniques like local time stepping, implicit residual averaging with variable coefficients, enthalpy damping (if applicable) and multigrid acceleration are used.

A general outline of the multigrid procedure for the special case of a 3-level W-cycle with 1-stage local sub-block refinement is presented in Fig. 21. There are various possibilities for the implementation of a multigrid algorithm in the context of a multi-block environment which can lead to extrem discrepancies in the convergence behaviour or even to divergence if not properly done, Rossow [10], [11]. The reasons for these convergence problems are mainly due to the time lags that may occur between the grid blocks during a multigrid cycle.

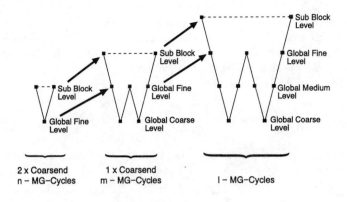

Fig. 21 Full multigrid procedure.

Boundary Conditions

Within the multi-block flow solver MELINA, different boundary conditions have to be treated. In principal, one can destinct between physical and purely algorithmic boundary conditions. Any of those is characterized by a certain index which is contained in the topological information that has been defined by INGRID.

Different types of <u>physical</u> boundary conditions arise because the computational domain may be bounded by solid walls, the farfield, fan face and exhaust planes of the engine and the symmetry plane if one exists.

On a <u>solid body</u> the velocity component normal to the wall is zero. Hence, the flux integral parts along the faces of the control volume that are aligned with the body reduce to the pressure integrals of the momentum equations.

At the artificial <u>farfield</u> the boundary conditions have to bring the information of the undisturbed flow into the computational domain and have to let pass disturbances emenating from the model into the free stream without reflection. The theory of characteristics, which can be applied to the hyperbolic locally linearized system of Euler equations, leads to a set of conditions for the characteristic variables. A transformation of these variables to the dependant variables $\vec{W}$ used in the rest of the flow field results in proper equations for the values of $\vec{W}$ at the far field.

The boundary condition of <u>symmetry</u> is enforced with the use of guard cells which are mirror images of the cells abutting the symmetry plane.

The boundary condition at the <u>fan face</u> is defined to yield the mass flow, specified by the engine data. The temperature, density, static pressure and flow direction that satisify the massflow requirement are used to determine the value of $\vec{W}$ at the fan face using the same characterisic procedure as in the case of the farfield condition.

The boundary condition at the <u>exit of the engine</u> is formulated such that the jump in stagnation pressure and stagnation temperature given in the engine data is realized.

Using isentropic relations and 1D energy conservation, the density and static pressure can be computed from stagnation density, stagnation temperature and speed at the exhaust plane.

The second group of boundary condition types is related to the algorithmic boundaries between grid blocks. In the present code seven different boundary condition types of this category are implemented which depend on the cell size relation between abutting blocks. They are one to one, fine to coarse and coarse to fine in both, only the first or only the second cyclic direction.

These inner cut conditions manage the exchange of grid coordinates and dependent variables at the nodes from one block to the other depending on the type of blocks that meet at a common segment part of a block face. Practically this is done by introducing guard cells for each block face which leads to an overlapping of blocks. In the simple case of a one to one correspondance of points on both abutting block faces, the grid coordinates and data at the nodes are simply copied to the guard points from the corresponding points of the neighboring block. In order to optimize the distribution of nodes, the options of enrichment and coarsening from one block to an other have been implemented. Enrichment means the splitting of each finite volume into two in each desired coordinate direction and interpolation of the coarse node data, whereas in the case of coarsening every other grid node is dropped in the specified direction.

RESULTS FOR LOCAL SUB-BLOCK REFINEMENT

The combination of local sub-block refinement and multigrid methods has already been suggested by Brandt in the seventies and eighties [3]. The target here is on the one hand to demonstrate the capability of the technique of local sub-block refinement to improve the spatial resolution at relatively low cost. On the other hand, the enormous convergence acceleration to steady state through the application of the multigrid method is demonstrated. No degradation of performance has been observed when local sub-blocks were present.

The new version of MELINA was applied to several configurations from which two will be presented here: a quasi 2D NACA0012-wing as a preliminary generic test case and a wing/body combination of a modern transport aircraft as a first production type application.

The mesh for the NACA0012-wing test case was generated by konformal mapping. A 2D O-mesh around the NACA0012-airfoil was copied and shifted in the spanwise direction to form a 3D mono-block mesh around an unswept NACA0012-wing. In spanwise direction, two-dimensionality of the flowfield is enforced by symmetry conditions at the wing "tips".

In order to study the influence of mesh block splitting on the performance of the overall algorithm, the 3D NACA0012-mesh can be split into multi-blocks at every desired position. It is also possible to add locally refined sub-blocks to test the influence of sub-block refinement. Various block topologies without sub-block refinement have been computed with MELINA and compared with respect to convergence rate and robustness of the computation; we observed only very limited differences between multi-block and mono-block meshes concerning these criteria.

A typical configuration of the NACA0012-wing test case is presented in Fig. 22. The pressure distribution is projected onto the surface of the wing and one wing tip symmetry plane for the test conditions of $M_\infty = 0.8$ and 1.25^0 angle of attack. The mesh is split in an upper and lower mesh block along the horizontal symmetry plane of the wing. Also two locally refined sub-blocks, indicated by their boundary surfaces, are put into place in the region of the wing leading edge. Like the global mesh blocks they are devided by the horizontal symmetry plane and are connected to each other via inner cut boundary conditions on their common block faces.

Fig. 22 NACA0012-wing, sub-block orientation and pressure distribution.

Let us first focus on the problem of spatial accuracy and investigate the solution quality that can be obtained on a coarse mesh without sub-block refinement. Fig. 23 shows a lines of constant total pressure loss for the coarse mesh with 80 cells around the cross-section, 16 cells normal to the wing and 4 cells in spanwise direction. The total pressure gains and losses at the leading edge are no physical phenomena and result exclusively from spatial discretization errors. In this case the total pressure losses, that are convected downstream, add up to a peak value of 7.3 percent.

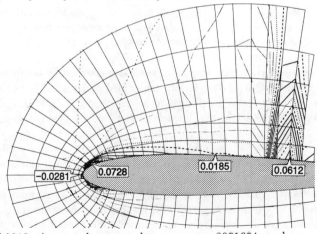

Fig. 23 NACA0012-wing, total pressure loss on coarse 80*16*4 mesh.

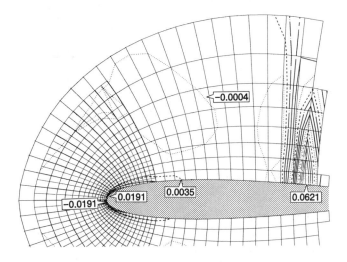

Fig. 24 NACA0012-wing, total pressure loss on 80*16*4 mesh with local sub-blocks.

If we compare the latter result with the total pressure losses on the same mesh but with local sub-block refinement at the leading edge (Fig. 24), we can observe a general reduction of the total pressure losses by a factor of approximately 4. This is the improvement that one would expect for a second order accurate scheme (Doubling of the grid density leads to a reduction of discretization errors by a factor of 4). A comparison with a global fine calculation with $160 * 32 * 8$ finite volumes, leading to the same spatial resolution at the leading edge as in the sub-blocks, confirms that it is possible to obtain locally the same accuracy with sub-block refinement as with a global fine mesh.

To investigate the convergence behaviour of the new version of MELINA with the techniques of multigrid acceleration and local sub-block refinement beeing applied, we compare the convergence history for three NACA0012 test cases in Fig. 25. The results correspond to running the code in a 4-level multigrid mode without sub-blocks and in 3-level multigrid mode with sub-block refinement as well as on a single mesh. The convergence behaviour is plotted as the reduction of the residual $\vec{R}_h^5$ (eqn. (23)) on the finest mesh and the development of the lift coefficient as a function of CPU time. In all three cases, the mesh density of the finest grid-level corresponds to the density of a 160*32*8 mesh. The full multigrid strategy was applied by starting the iteration on the medium iteration level.

Concerning the convergence of lift, we come up with a reduction of at least 70 percent in CPU time with the multigrid method relative to the single mesh application. A noticable improvement of the 3-level multigrid mode with sub-block refinement relative to the 4-level multigrid mode has not been observed. Also it seems to be somewhat fortitious that the lift converged to exactly the same values on the global fine meshes (4-level multigrid and single mesh) compared to the coarse mesh with local sub-block refinement, since instead of the good resolution at the leading edge, with local refinement there are still some discrepancies in the solutions especially in the shock region.

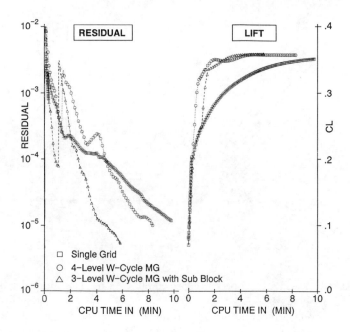

Fig. 25 NACA0012-wing, convergence history.

The mesh for the production type case of the wing/body configuration of a modern transport aircraft configuration was generated with INGRID and has an H-type structure in streamwise direction and is of O-type normal to that direction. All computations on the wing/body combination were performed at cruise conditions with $M_\infty = 0.76$ and 2.2^0 angle of attack.

In order to optimize the distribution of finite volumes, the option of coarsening and enrichment from one mesh block to the other is used extensively. Because wing/body computations are usually performed as baseline computations for engine installation prediction at DA, a high mesh density is required in block no. 1 from which the engine block may be cut out and reconfigurated. The lower external block is coarsened relative to block no. 1 in all three coordinate directions. The upper side blocks are treated likewise except that both are coarsened in j-direction relative to the lower side blocks. The strategy of coarsening and enrichment should not be mixed up with the technique of local sub-block refinement, but means that the individual blocks of the fine mesh level have different grid densities in the three coordinate directions.

As before we adress convergence acceleration using multigrid technique and the improvement of spatial resolution indepently from each other. The mesh on the fine multigrid level consists of approximately 700000 cells in 4 blocks which corresponds to 48 cells in spanwise direction, and 104 cells around the wing (Fig. 6).

Because of the coarsening it was only possible to use 3 multigrid levels. Again we applied the full multigrid approach starting with 40 W-cycles on the medium iteration level and 100 W-cycles on the fine iteration level. The lift converged after 70 cycles on the fine iteration level and the residual convergence rate was determined to be 0.89. With respect to CPU time we achieved a reduction from 120 min on a VP200 vector computer on the single mesh for this configuration to 43 min, which corresponds to a reduction of 65 percent.

To evaluate the potential of sub-block refinement to improve the spatial resolution of the discretization for the wing/body combination, we performed various computations on coarse and fine meshes as well with different sub-blocks. In the fine mesh we discretized the wing in spanwise direction with only 32 cells and with 96 cells around the wing in order to reduce the computational cost for the investigation. The coarse mesh is obtained from the fine mesh by skipping every other cell in each of the coordinate directions.

Critical regions with respect to spatial accuracy are regions of strong gradients in the solution that typically are found at large surface curvatures like at the wing leading edge or wing/body intersection. This is confirmed in Fig. 26 where the total pressure loss and the isomach lines for the coarse mesh are plotted for a wing section plane at midspan. The total pressure losses, generated at the leading edge, are a consequence of insufficient spatial resolution. They add up to roughly 14 percent and have a strong influence on the distribution of the Mach number. We observe a strong reduction of the Mach number in the regions of total pressure losses near the wall, an effect that is sometimes called 'Euler boundary layer'.

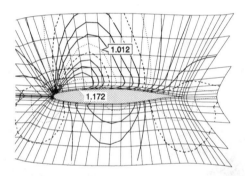

Fig. 26 Total pressure loss and isomach lines on coarse mesh at midspan station.

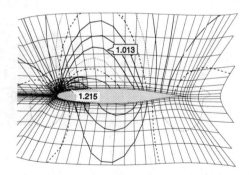

Fig. 27 Isomach lines on coarse mesh with sub-block refinement at the leading edge.

In order to get rid of these disturbances, we patched sub-blocks onto the coarse mesh in the regions of interest. Three different sub-block configurations have been tested. The first one can be viewed as 'minimal solution' and is made up of two small sub-blocks in the leading edge region of the wing, one in the upper coarse mesh block and the other sub-block in the lower coarse block. Although the problem of the 'Euler boundary layer' has been solved with this approach, the global features of the flow field, especially the supersonic region, are still not predicted accurately (Fig. 27).

This result leads to the extension of the sub-blocks to boxes, completely containing the wing. Two characteristic extended sub-block orientations are depicted in Figs. 28 and 29.

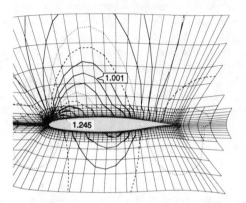

Fig. 28 Isomach lines on coarse mesh with 'small' sub-block extension.

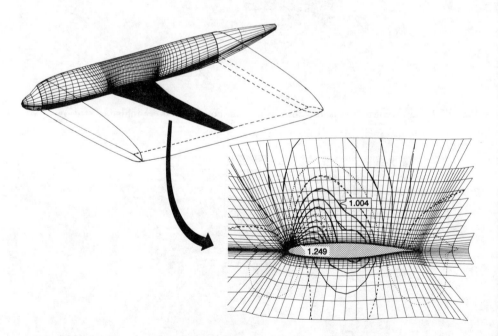

Fig. 29 Isomach lines on coarse mesh with 'large' sub-block extension.

The results of the Euler simulations on these two meshes reveal that in general the method of sub-block refinement works nicely. But in order to capture the important flow features on very coarse meshes, like the one we have been using here, the sub-blocks should be extended that far that flow regions like leading and trailing edge and supersonic region are contained within the sub-blocks.

CONCLUSION AND FUTURE ASPECTS

The INGRID system has been successively enhanced in order to cope with realistically complex airplane configurations. A component definition and handling system has been developed based on which surface meshes can be generated. In further steps, 3D multi-block meshes can be generated using relatively simple explicit algebraic techniques. The system has been augmented by an interactive structured block mesh embedding which has been validated in connection with the cell vertex multi-block multigrid Euler flow solver MELINA.

The experience with the above system in industrial application and the validation results for the successive improvements described above have proven that INGRID is very useful for CFD simulations for complex configurations. It is easily extendable with respect to more complete aircraft and CFD mesh quality requirements. The latter shall be explored within the context of structured solution adaptive mesh refinement in the near future. This means close algorithmic coupling between at least some kernel functions of the mesh generator and the flow solver. Strategic considerations are necessary for an efficient connection between flow evaluation by appropriate sensors and the generation of additional refined local mesh blocks. Fortunately, the multigrid philosophy opens an easy way for the interaction of grids with different levels of refinement. Thus the overall strategy will be based on that principle.

ACKNOWLEDGEMENTS

We like to thank H. Sobieczky from the Institute of Theoretical Aerodynamics at DLR Goettingen for the for his ideas concerning the ongoing improvement of the definition and handling of airplane components within the INGRID system. Additionally, we thank the people at the Institute of Design Aerodynamics at DLR Braunschweig for their valuable help with the development of the MELINA code. This work has been partly conducted as BRITE/EURAM area 5, CEC funded, applied research. We are grateful for this support.

References

[1] Becker. K.: "The Interactive Grid Generation System INGRID - Version 3.0", DA-report, to appear.

[2] Becker, K.: "Interactive Algebraic Mesh Generation for Twin Jet Transport Aircraft", Proc. 3rd Int. Conf. Num. Grid Gen., Barcelona, June 1991.

[3] Brandt, A.: "Multigrid Techniques: 1984 Guide with Applications to Fluid Dynamics", GMD-Studien No. 85, Gesellschaft fuer Mathematik und Datenverarbeitung mbH, Bonn, 1984.

[4] Carcaillet, R.; Kennon, S.R.; Dulikravich, G.S.: "Optimization of Three-Dimensional Computational Grids", AIAA-Journal of Aircraft, Vol. 23, No. 5, May 1986.

[5] "Initial Graphics Exchange Specification (IGES)", Version 3.0, Publication No. NB-SIR 86-3359, National Bureau of Standards, Department of Commerce, Washington, D.C., 1986.

[6] Jameson, A.; Schmidt, W.; Turkel, E.: "Numerical Solutions of the Euler Equations by Finite Volume Methods Using Runge-Kutta Time Stepping Schemes", AIAA Paper 81-1259, 1981.

[7] Radespiel, R.; Rossow, C.: "Konvergenzbeschleunigung bei der Berechnung der reibungsfreien Umströmung eines Tragflügels", DGLR-Bericht 88-05, 1989, 301-312.

[8] Rill, S.; Becker. K.: "Simulation of Transonic Flow over a Twin Jet Transport Aircraft", AIAA Paper 91-0025, 1991.

[9] Rill, S.; Becker. K.: "MELINA - A Multi-Block, Multi-Grid 3D Euler Code with Local Sub-Block Technique for Local Mesh Refinement", ICAS Paper 92-4.3.R, ICAS Conf., Beijing, Sept. 1992.

[10] Rossow, C.: "Berechnung von Strömungsfeldern durch Lösung der Euler-Gleichungen mit einer erweiterten Finite-Volumen Diskretisierungsmethode", DLR-FB 89-38, 1989.

[11] Rossow, C.: "Efficient Computation of Inviscid Flow Fields Around Complex Configurations Using a Multi-Block Multigrid Method", 5th Copper Mountain Conference on Multigrid Methods, Copper Mountain, Colorado, USA, March 31 - April 5, 1991.

[12] Sobieczky, H.: "Geometry Definition and Grid Generation for Wing-Fuselage Configurations and Turbomachinery Components", DFVLR-IB 221-86/A20, 1987.

[13] Sobieczky, H.: "Geometry Generation for Transonic Design", in: Advances in Computational Transonics, Series on Recent Advances in Numerical Methods in Fluids, Vol. 4, Ed. W. G. Habashi, Pineridge Press, Swansea, 1985.

DEVELOPMENT OF 3D MULTI-BLOCK MESH GENERATION TOOLS

J.Oppelstrup and O.Runborg,
C2M2, KTH, S 100 44 STOCKHOLM

P.Mineau and P.Weinerfelt,
CERFACS, 42 Av. G. Coriolis, F 31057 TOULOUSE

R.Lehtimäki,
Laboratory of Aerodynamics, HUT, SF 02150 ESPOO

B.Arlinger,
Saab Aircraft Division,
S 581 88 LINKÖPING

SUMMARY

The paper describes the development and investigation of tools for some of the critical steps in the creation of a multi-block structured grid. We consider elliptic-interactive generation of meshes on solid surfaces, algebraic-interactive mesh generation for individual blocks, and smoothing procedures for multi-block meshes. Interactivity is used wherever suitable to improve the user interface, and object-oriented programming is proposed as an excellent vehicle for creation of software for complex geometrical objects such as grids. The tools are applied to test configurations, and a pilot multi-block mesh generator is developed which can serve as a skeleton for further tool development.

INTRODUCTION

The Euromesh project has fostered cooperation between aerospace industry, universities, and research institutes. The project described here is typical in the sense that groups from all three classes of partners have contributed actively. Tangible results of such development work is the exchange of software or data - in this case geometry data, computational grids and solutions to flow problems - between the partners, and this has been achieved in this sub-project to the extent that geometry definitions and grids created by other packages have been imported and further developed by the tools.

Surface grid generation is well suited to interactivity and elliptic schemes are known to produce smooth meshes. Taking the generation system of [Tak] as a starting point, we have extended the technique by multi-grid acceleration, automatic control function computation, and an interactive final step in which the user adjusts the grid by the direct manipulation of control points, [Opp].

Algebraic generation of single-block volume grids is fast enough to allow interactive use, and a technique which uses interior control surfaces and a non-iterative smoothing procedure for grid line control was developed [Leh].

Finally, a multi-block mesh generation code [Sca], [Run], and a multi-block elliptic mesh smoothing procedure, [Min], were developed and merged. The mesh generator is a test vehicle for mesh generation tools and was designed to investigate a consistent object-oriented approach to representation and manipulation of the geometric objects common in mesh generation.

INTERACTIVE ELLIPTIC SURFACE MESH GENERATION

Overview of Technique
A program module is developed for elliptic surface mesh generation on a surface which is topologically rectangular, i.e. is described by a map from $[0,1]^2$ to $\mathbf{R}^3$

(u,v) in $[0,1]^2 \rightarrow \mathbf{r}(u,v) = (x(u,v), y(u,v), z(u,v))$.

The grid on the surface is found as the image of a regular grid on the unit square by solution of an elliptic system [Tho], [Tak]:

$$|\mathbf{r}_t|^2 (u_{ss} + Pu_s) + |\mathbf{r}_s|^2 (u_{tt} + Qu_t) - 2(\mathbf{r}_t \cdot \mathbf{r}_s) u_{st} + J^2 F = 0$$
$$|\mathbf{r}_t|^2 (v_{ss} + Pv_s) + |\mathbf{r}_s|^2 (v_{tt} + Qv_t) - 2(\mathbf{r}_t \cdot \mathbf{r}_s) v_{st} + J^2 G = 0$$

where J is the Jacobian, $J = u_s v_t - u_t v_s$ and F and G are somewhat complicated functions of derivatives of (x,y,z) w r t u and v. P and Q are the source functions used to control the grid. P influences the grid distribution in the u-direction, and Q in the v-direction.
The solution gives a composite mapping $(s,t) \rightarrow (u,v) \rightarrow (x,y,z)$. The technique formally requires x, y and z to be C^2 functions of (u,v). Thus, an (IL+1) x (JL+1) grid is obtained as

$(\mathbf{r}(u(s_{ij}, t_{ij}), v(s_{ij}, t_{ij})))$, $i = 0,\ldots,IL), j = 0,\ldots,JL$.

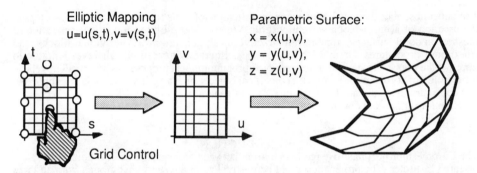

Figure 1: Schematic Representation of Elliptic Surface Mesh Generation

The whole procedure is as follows

1. The surface representation is generated. In our code, this is a bi-cubic spline surface constructed from interpolation to a rectangular set of data points, arranged as cross sections. The cross sections need not be parallel, or plane, or equispaced, but should not deviate too much from this ideal, lest the parametrization of the surface become a problem.

2. Fix the gridpoints along the edges of the surface.

3. Interpolation of the grid control functions P and Q from their values at the boundary by unidirectional or linear transfinite interpolation. The generation equations are solved and the resulting surface grid is displayed. This grid is now the image of an equi-spaced grid in the (s,t)-plane.

4. The pre-image in (s,t)-space of the grid is represented by B-spline functions the coefficients of which, the Control Net points, are shown and manipulated on the screen. The new (s,t) grid is computed from the Control Net points, the (u,v)-grid from interpolation of the new (s,t)-grid in the elliptic (u,v) grid, and finally the **r** grid from (u,v). The new grids in (u,v)-space and in (x,y,z)-space are displayed.

5. The Control Net points are adjusted manually and the (x,y,z)-grid recomputed (step 4) until a satisfactory grid is obtained.

Numerical Solution

The numerical solution is carried out by point relaxation, with optional multi-grid acceleration. $P = Q = 0$ is expected to give a unique solution to the continuous problem, see below. For the discretized version used to produce the grids, there are indications that the Euler-Lagrange equations used for relaxation may sometimes be expected to have multiple solutions. This manifests itself in lack of convergence in the iterative solution procedure. This behavior has recently been analyzed in [Ste] in the even simpler setting of producing a grid on a curve, equi-spaced in arclength. The conclusions are that coarse discretisations may be expected to have convergence trouble if the coarseness of the grid is such as not to resolve the geometry properly, and that many more points are often necessary than expected from visual inspection of grids on the curve.

Increasing the magnitude of P and Q eventually creates a situation where even the continuous problem possesses no solution, in the sense that the Euler-Lagrange PDEs of the associated minimization problem have no solution well-behaved enough to qualify for a grid. The question of robustness of the generation algorithms has to be addressed. The surface generation system can be compared with the plane generation equations with non-zero control functions. It has been found that the point iteration (with under- or over-relaxation) does not always converge in this case. The following analysis indicates that this is indeed to be expected, but that *if a solution does exist, then it will guarantee a well-formed grid, i.e. the mapping will be a bijection.*

One can show, that when the elliptic system used comes from transforming the generation equations

$$\Delta_x s = P(x,y), \Delta_x t = Q(x,y),$$

where the subscript (x or s) on the Laplace operator Δ indicates the independent variables, on the physical domain onto the computational space, with bounded P and Q, and this system has a solution, then the grids will always be bijections.

Thus, what can go wrong is that in the case of non-zero source functions there may be no solution. The 1D case can be handled explicitly and shows how non-existence comes about.
Let x be physical space, $0 \le x \le L$, s computational space, $0 \le s \le 1$. Now, the Poisson system is

$$d^2s/dx^2 = P(x), s(0) = 0, s(L) = 1$$

which is formally transformed into ($x' = dx/ds$, etc.)

$$- x''/(x')^3 = P(x(s)), x(0) = 0, x(1) = L .$$

If this problem has a solution, it is clear that x' cannot be zero, unless x" is also. Indeed it is required that x" be of order $(x - x^*)^3$ if x' is zero at $x = x^*$. This means that interior extrema are impossible, and x must be monotonic.
Now there remains the question of the equivalence of the original and transformed equations. Take P constant. It follows that

$$s = P/2\, x^2 + (- PL/2 + 1/L)\, x$$

is the unique solution, but that becomes non-monotonic, hence without well-defined inverse, when |P| is big enough. Such a solution cannot come out of the transformed equations. It follows that for $|P| > 2/L^2$ there is no solution to the transformed equations. *Thus, the control functions must be suitably limited to guarantee a solution.*

Computation of control functions P and Q in the numerical experiments.

Taking $P = Q = 0$ corresponds to a minimization of a functional which is related to the Beltrami operator, which is much like the Laplacian taking the curvature of the surface into account. The resulting grids are very smooth and, by the maximum principle, do not fold over.

Non-zero P and Q are used to cluster the grid to interesting regions. There are options to use either P and Q which vary linearly - usually one sets them to zero in this case - with the computational coordinates, or to compute P and Q from boundary point distributions. The latter is the more practical, since intuitive appreciation of how a change in P or Q effects the grid requires extensive experience.

Given grid points on the bounding curves of the surface patch, one first computes (part of) a grid by (bilinear) transfinite interpolation. When the algebraic grid is substituted into the discretized PDE, P and Q can be evaluated along the patch edges. After possible smoothing, these boundary data are used to compute P and Q over the whole patch by unidirectional or transfinite interpolation. The smoothing performed in the code is a simple limiting,

$$P_{smooth} := \max(P_{min}, \min (P_{max}, P_{original})),$$
similarly for Q

since it was usually enough to avoid the very large values calculated at singular points.

Interactive grid control

A tool which allows movement of B-spline grid control points and display of the grid in the computational rectangle has been developed. The interpolation to the elliptic mapping by bicubic Hermite basis functions is very simple to implement since the grid constitutes an equidistant table.

Figure 2 shows a 4 by 4 net of B-spline control points interpolated to a 12 x 12 grid. The control net is (almost) square, except for point (3,3) which has been moved down and right. The B-spline basis functions used ensure that the outermost points in the net are the only ones which influence the boundaries. Thus, fixing those guarantees that the grid matches the boundary curves.

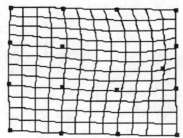

Figure 2: Relation between Control Net Points and mesh

The use of polynomial basis functions with compact support means that changes to control points have effect only locally and are computationally efficient. By using more control points the user can exercise precise control. However, since the basis functions are polynomials of low degree (three), the exponential type stretchings necessary for the resolution of boundary layers cannot be made with few control points. The variation diminishing property of B-splines imply that no special care need be taken to avoid inadvertently creating cusped or folded nets. The drawback is that large excursions of the control points are necessary for strong effects on the grid. In the figures, the control net in (s,t)-space is shown superimposed on the resulting grid in (u,v)-space to the left, and the physical grid on the right.

Test Cases

The HERMES Workshop Double Ellipsoid was used to see the effect of an inflection in the surface on a simple geometry and to test the effect of the multi-grid - acceleration. The double ellipsoid test case was run with a 16 x 16 grid, fig. 3, $P = Q = 0$ and a max. residual reduction of three orders of magnitude indicating convergence.

The multi-grid acceleration gives a substantial improvement in convergence rate, about a factor of three, see fig. 4. The control point influence is shown in fig. 5. The initial grid is the algebraic grid of fig 3. The interior control points have to be moved very strongly to produce clustering to a point.

The HERMES Workshop ALEX Delta Wing is more difficult, having discontinuous curvature, and a sharp edged apex. It is made up of two planes and a circular cylinder. The apex is a polar-type singularity. The boundary point distributions were clustered to the leading edge, and control functions P and Q interpolated from these distributions and limited. The limiting was active only at the apex. An example of a grid generated is given in fig 6. The control points were then further adjusted to produce a grid with about equal angular distribution on the leading edge cylinder. The result is shown in fig. 7.

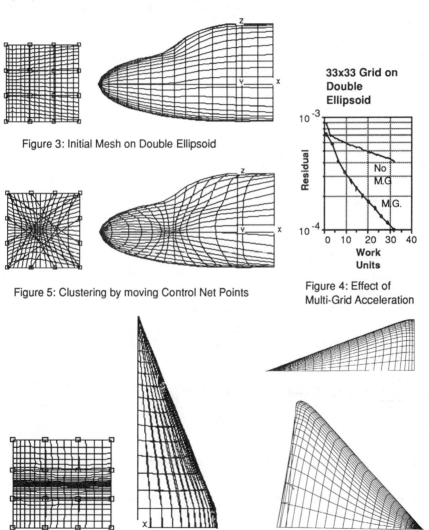

Figure 3: Initial Mesh on Double Ellipsoid

Figure 5: Clustering by moving Control Net Points

Figure 4: Effect of Multi-Grid Acceleration

Figure 6: Initial clustered grid on ALEX.   Figure 7: Final clustered grid on ALEX, two views.

THE SINGLE-BLOCK ALGEBRAIC MESH GENERATOR

The block is seen as the image of a mapping of the unit cube,
$$(x,y,z) = \mathbf{r}(\xi,\eta,\zeta), \ 0\leq\xi\leq1, \ 0\leq\eta\leq1, \ 0\leq\zeta\leq1.$$

The grid on the block is supposed to be indexed by i, j and k, $0 \leq i \leq IL$, $0 \leq j \leq JL$, $0 \leq k \leq KL$. The gridpoints will be
$$(\mathbf{r}_{ijk}) = ((x,y,z)_{ijk}), \ x_{ijk} = x(\xi(i),\eta(j),\zeta(k)), \text{ etc.}$$

An index surface is a subset of the grid defined by one of i, j, or k being constant. The generation of the volume grid in the block proceeds from transfinite interpolation to a set of prescribed index surfaces. One can specify not only the exterior surfaces but also interior surfaces. This feature is then used also to enforce orthogonality of grids to faces, when so desired.

The standard case is for the six surfaces i = 0 and IL, j = 0 and JL, and k = 0 and KL to be specified. In many cases some of the surfaces can be generated by interpolation to edges of already defined surfaces, and in this case it is enough to specify the surfaces which cannot be so derived. A case in point is the creation of a grid around a wing from only the wing surface and the farfield surface by unidirectional interpolation. Conditions, such as periodicity, that may be desirable in creating blocks that wrap around and become degenerate in some fashion have to be specified in terms of prescribed index surfaces and grid orthogonality.

Grid control is exercised by specification of the stretching functions,
$$\xi = \xi(i), \ \eta = \eta(j), \ \zeta = \zeta(k),$$
which are mappings from grid index values to parameter space, by controlling orthogonality to prescribed index surfaces, and by index surface smoothing.

Interpolation Functions

The computation of the volume grid by transfinite interpolation may be written as a sequence of three unidirectional interpolations, where we assume that the index surfaces
$$\mathbf{r}(\xi_i,\eta,\zeta), \ i = 1,...,\text{Isurf},$$
$$\mathbf{r}(\xi,\eta_j,\zeta), \ j = 1,...,\text{Jsurf, and}$$
$$\mathbf{r}(\xi,\eta,\zeta_k), \ k = 1,...,\text{Ksurf}$$
are given, with surface stations ξ_i, η_j, and ζ_k prescribed. Then

$$\mathbf{F}_1 = \sum_{i=1}^{\text{Isurf}} \phi_i(\xi) \, \mathbf{r}(\xi_i,\eta,\zeta), \qquad \mathbf{F}_2 = \sum_{j=1}^{\text{Jsurf}} \psi_j(\eta)[\mathbf{r}(\xi,\eta_j,\zeta) - \mathbf{F}_1(\xi,\eta_j,\zeta)],$$

$$\mathbf{F}_3 = \sum_{k=1}^{\text{Ksurf}} \theta_k(\zeta)[\mathbf{r}(\xi,\eta,\zeta_k) - \mathbf{F}_1(\xi,\eta,\zeta_k) - \mathbf{F}_2(\xi,\eta,\zeta_k)],$$

$$\mathbf{r}(\xi,\eta,\zeta) = \mathbf{F}_1(\xi,\eta,\zeta) + \mathbf{F}_2(\xi,\eta,\zeta) + \mathbf{F}_3(\xi,\eta,\zeta)$$

where the blending functions ϕ, ψ and θ are the basis functions for Lagrange interpolation, e.g.

$$\phi_i(\xi) = \prod_{j \neq i} \frac{\xi - \xi_j}{\xi_i - \xi_j}, \text{ which satisfy the interpolation conditions}$$

$$\phi_i(\xi_j) = 1, \text{ if } i = j, \ 0 \text{ if } i \neq j.$$

Grid Clustering

The stretching functions must be monotone, increasing and smooth. The hyperbolic tangent stretching

$$\xi = A + B \tanh(Ci + D), \text{ etc.},$$

has been chosen here because it allows specification of cell sizes at both ends or at one, and of the number of grid points. It has been found useful even for wall normal stretchings to accommodate boundary layers, etc. The definition and computation follows [Tho] closely. The final mapping from [0,IL] (say) to [0,1] is composed of a sequence of tanh-stretchings, each covering a specified set of grid points. An example is given in fig. 8, where two tanh-stretchings are combined.

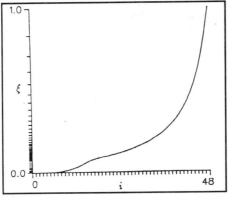

Figure 8: Stretching by combining two tanh-functions.

Note that the cell sizes and the clustering points are specified in the computational spaces (i,j,k) and (ξ,η,ζ). The physical (x,y,z)-space cell sizes and locations of clustering points are affected by the interpolation procedure, and when relevant by the grid smoothing procedures. The definition of the final stretchings is therefore an iterative procedure.

Orthogonality to an Index Surface

Orthogonality of the grid to a prescribed index surface is achieved by interpolation to an auxiliary index surface computed from the prescribed one by offsetting it along approximate normals. The offset distance is computed by transfinite interpolation to the known distance at the surface edges, where the grids are fixed.

Choosing the true normal over the whole surface creates jump discontinuities of grid line slopes at the face edges, since the faces need not be orthogonal to each other. This jump is smoothed by choosing the offset direction **t** to be a blending of the normal direction **n** and a grid direction **g** computed by bi-linear interpolation from grid line directions in the four neighbour faces, say for a surface i = const.:

$$\mathbf{t}_{jk} = w(j/JL, k/KL)\, \mathbf{n}_{jk} + (1 - w(j/JL, k/KL))\, \mathbf{g}(j,k).$$

The weight function $w(\eta,\zeta)$ is a composition of power functions of η and ζ which vanish at the edges and become one for $\eta = \zeta = 0.5$. The width of the transition regions is controlled by choice of the powers. The normals are computed from cross products of central differences of grid coordinates.

Grid Index Surface Smoothing; Grid Quality

The interpolation propagates possible boundary slope discontinuities into the volume grid. To alleviate this problem, an algebraic smoother can be applied which moves the gridpoints in an index surface, say $i = i0$, towards "ideal" positions. The ideal positions are computed by Hermite interpolation to positions and slopes taken from neighbour index surfaces $i = i0 + 1$ and $i = i0 - 1$. The amount of movement allowed is decreased close to boundaries by applying weight functions similar to the ones used in normal smoothing so that the boundaries are not affected.

The volume grid quality can be assessed by the computation of the Eriksson [Erik] cell skewness value ρ. This is an indicator for individual cells, and does not take into account possible grid problems relating to neighbour cell variations like too violent stretchings, nor does it penalize high aspect ratios. But it provides a simple check that the grid does not fold over. For a cell, the three vectors **a**, **b**, and **c** connecting the centroids of opposing faces are computed. The skewness value is defined to be

$$\rho = \frac{\mathbf{c} \cdot \mathbf{a} \times \mathbf{b}}{|\mathbf{a}| |\mathbf{b}| |\mathbf{c}|}.$$

Now, $-1 \leq \rho \leq 1$, with $\rho = 1$ for an orthogonal cell of the right handedness, and $\rho < 0$ for the opposite handedness. In practical computations with cell centered finite volume codes, $\rho > 0.01$ usually seems not to create problems.

Application to the DLR F5 wing

The wing surface was generated by the E88F5 code published by Sobieczky. The wing tip and trailing edge are left open by E88F5 and were closed arbitrarily but reasonably by addition of an extra point. The farfield surface was generated by transfinite interpolation. The grid topology is O-O, with two parabolic singular lines emanating from the wing tip. The i-index runs chordwise, the j-index spanwise with $j = 0$ at the root and the k-index radially, $k = 0$ at the wing surface.

The first run generates a 48 x 16 x 16 grid by specifying all six of the exterior faces of the block. A single tanh-stretching in the k-direction is performed with a cell thickness (in computational space) of 0.003 at wing surface, and the grid is required to be orthogonal to the wing. The result is depicted in fig. 9.

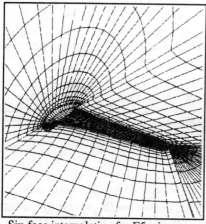

Six-face interpolation for F5 wing.
Figure 9.

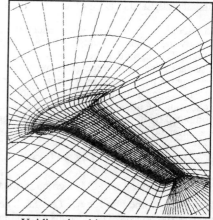

Unidirectional interpolation for F5 wing
Figure 10.

Next a grid is generated from only the wing surface and the farfield surface. The clustering is tightened to a cell size of 0.0003, fig. 10. The slope discontinuities at the singular lines are subsequently smoothed by applying the algebraic smoother to j = 14, 15, 13 and 14 again. The result of course depends on the order of the operations, and one could use the smoother to work on a whole slab of the grid by repeated application. The number of iterations must be limited, however, since there will be a net drift of the grid towards convex boundaries and away from concave ones, much like is found with simple grid smoothers working by averaging in the physical space. The results of the smoothing are displayed in figures 11 a, b, showing the i = 0 and 24, k = 0 wing surface, and j = 16 grid planes. Note that the smoother does not work across the wrap-around boundaries, hence the j = 16 grid surface is unchanged. The change has moved the points only on the index surfaces j = 13, 14 and 15.

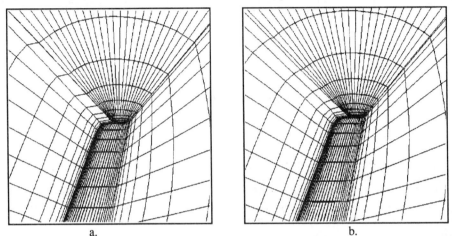

a.   b.

Figure 11 a, b : Detail of F5 grid before (a) and after (b) smoothing.

THE MULTI-BLOCK VOLUME GRID SMOOTHER/GENERATOR

Since the block grids are generated from face data, they match up in position at the block boundaries, but not necessarily in grid line slope or cell sizes. For this reason, it is usually beneficial to perform a smoothing operation which works across block boundaries and is not restricted to modify only interior points. Indeed, some of the block boundaries, such as those on solid walls, will be treated as fixed, whereas the block boundaries in the field will be allowed to move to improve smoothness of the grid. The grid could also be allowed to move in the solid surfaces. Such experiments were carried out only for the simple case of planar boundaries, where it is sufficient to restrict the control points to a plane.

The methods used are variants of the functional minimization schemes. If run for sufficiently many iterations, they converge to a grid which is independent of the initial algebraic grid, but when the initial block shapes are favorable, only a few iterations are needed to smooth out the most serious kinks across the block boundaries.

The starting point is the definition of functionals which are measures of smoothness, and other desirable properties of grids, such as given in [Jacq]. The blocks are seen as mappings from the unit cube to the physical space, and the functionals may be computed by integrating transformed integrands over the blocks in computational space and summing the results. In this work, a few of the more commonly used functionals were tried.

The functional minimization corresponds to optimizing the shapes of bounding faces and the "interiors" of the grid blocks, with the obvious restrictions of some block faces being fixed. Two distinct approaches were investigated for the representation of block shapes, the analytical solid modelling, or control point, approach and the more traditional gridpoint approach.

Analytic Solid Modelling for block shape optimization

The control point representation has substantially fewer degrees of freedom by virtue of using basis functions which are polynomials of degree at least three. An experimental code which allowed easy experimenting with different functionals was written. It employed conjugate gradient methods which do not require actual computation of the Hessian matrix. The analytic determination of optimal stepsize, which requires the Hessian, is replaced by a line search. For some of the functionals, the line search is actually a minimization of a quadratic function and so could be performed analytically, [Jac]. It was conjectured that the economy of unknowns would more than offset the complicated functional and gradient computations and thus would be efficient enough to allow interactive response on graphical workstations. The conclusion after several serious attempts at speeding up the algorithms was that interactive use was not practical. Figures 12 a) - d) illustrate the use of three different functionals on a grid on a square with stretched grids along the edges.

Laplace Functional:

$$\phi = \iiint dx\,dy\,dz\, (\,\|\nabla_x \xi\|^2 + \|\nabla_x \eta\|^2 + \|\nabla_x \zeta\|^2),$$

Orthogonality Functional, where $\mathbf{r} = (x,y,z)$:

$$\phi = \iiint d\xi\,d\eta\,d\zeta\,((\mathbf{r}_\xi \cdot \mathbf{r}_\eta + \mathbf{r}_\eta \cdot \mathbf{r}_\zeta + \mathbf{r}_\zeta \cdot \mathbf{r}_\xi\,)).$$

The Laplace functional used in 12 a) promotes equidistribution of cell area; 12 b) is the result of optimizing orthogonality. The Euler equations for this variational problem are the Cauchy-Riemann equations and we obtain an orthogonal grid. Figure 12 c) shows the effect of simply reversing ξ and x in the Laplace functional while the last grid 12 d) is the result of treating the control points (6 x 6) as if they were grid points and solving the elliptic generation system (see below).

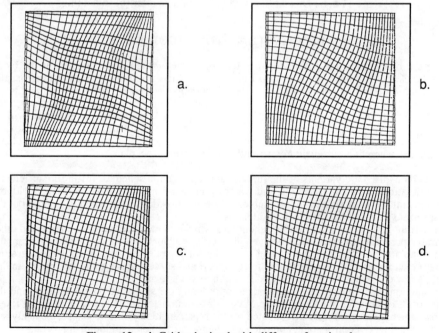

Figure 12 a-d: Grids obtained with different functionals.

The Gridpoint Representation

The gridpoint-level representation is simpler to program and work with. For instance, the control point approach needs a control point representation of the blocks, and thus cannot use grids or surfaces defined by only gridpoints. As discussed above it turned out that it was also actually faster than using functional minimization techniques on the control point representation. For these reasons, we concentrated on the gridpoint representation using functionals which correspond to the elliptic Poisson generation systems of [Tho], expressed with the physical coordinates as independent variables:

$$\Delta_x \xi = P(\xi,\eta,\zeta)$$
$$\Delta_x \eta = Q(\xi,\eta,\zeta) \qquad \text{(P)}$$
$$\Delta_x \zeta = R(\xi,\eta,\zeta)$$

on a region consisting of all the blocks, and with boundary conditions corresponding to fixing the gridpoints on some faces.

This system (P) is solved by reversing the roles of dependent and independent variables so that the geometry becomes a collection of cubes endowed with equidistant grids. The resulting equations are quasi-linear, the coefficients are the same in the x, y and z-equations, and they can be solved by point iteration. For $P = Q = R = 0$ they reduce to the Euler equations for the minimization of the "smoothness" functional and the numerical solutions show remarkable robustness.

Grid Source Functions

Taking $P = Q = R = 0$ is generally unacceptable because the smoothing will destroy the grid distribution desired. Thus, we need to compute source functions such that the grid distribution is not too distorted by the iterative smoothing. We will assume that the algebraic grids already generated have the correct distribution. The procedure will be to evaluate initial source functions corresponding exactly to the algebraic grid and subsequently modify them. Large or even infinite values of the source functions are obtained very locally where the grid is clustered, and the source functions must be modified near strong stretchings. The modification is a matter of smoothing and limiting. Smoothing by averaging spreads the (desired) local peaks and is limited to regions where the original source functions do not vary greatly. For this reason, an adaptive limiting procedure was developed. It will be described by application to the one-dimensional case, i.e.

$$\frac{d^2\xi}{dx^2} = P(x), \; \xi(0) = 0, \; \xi(L) = 1;$$

which is reversed by interchanging the roles of x and ξ to

$$\frac{d^2 x}{d\xi^2} = -\left(\frac{dx}{d\xi}\right)^3 P(x), \; x(0) = 0, \; x(1) = L \; .$$

A positive value of P makes $\xi(x)$ concave, thus pushes gridpoints towards $x = L$. As was shown above, if P is constant, $|PL^2| < 2$ is necessary for a solution to the generation equations to exist. P thus must scale inversely with the the square of the physical dimensions of the region. It is well known that the successive cell sizes $h_{i+1} = x_{i+1} - x_i$ may not vary more rapidly than, say

$$1 - \delta < h_{i+1}/h_i < 1/(1 - \delta), \; \delta \approx 0.2 \; .$$

The difference approximation to the reversed equation becomes

$$P_i = -8 \frac{h_{i+1} - h_i}{(h_{i+1} + h_i)^3} = -8 \frac{Q-1}{(Q+1)^3} (h_i)^{-2}.$$

with $Q = (h_{i+1}/h_i)$, and the final limits on P are
$$-.175 < Ph^2 < .275.$$
For the multidimensional case, h is interpreted to be the distance along the i (viz. j or k) -line to the next gridpoint. The limiting of P, Q and R by the above heuristic was generally successful but occasionally deformed the block boundaries from their initial shapes. (The solid boundaries, of course, are not affected by the deformation)

Examples: MUBBLE

A small interactive 3D mesh generator **MUBBLE**, [Run], is being developed to test the object oriented language C++ , [Sca], [Str], as a vehicle for handling geometrical objects. The block smoother was interfaced to MUBBLE and a simple test geometry resembling a rough model of an aircraft with wings, body, and tail assembly was created. A number of block topologies were generated, and we show here the grids generated in one where the block system is essentially symmetric around the horizontal mean plane, and one block covers the wings and the lower part of the body between the wings, with a companion block joining from the upper side. The grid thus is of H-type chordwise. Figures 13 a - b show the surface grid on the model and a shaded rendering of the model surface and the under-wing block;
The grid is clustered - for illustrative purposes - on the forepart of the body towards the junction to the under-and-over wing blocks and a substantial cell size jump as well as a grid line slope discontinuity appears. The smoothing procedure described above was applied and Figures 14 a - b show the grid in the symmetry plane before and after application of the smoothing. The features of the model are smooth enough that no problems appear with the heuristic source function limitation.

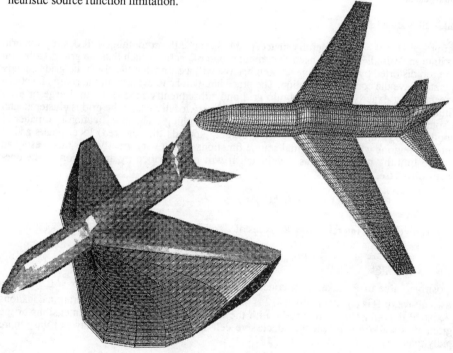

Figure 13 a,b: Surface grid and part of block system on simple aircraft configuration.

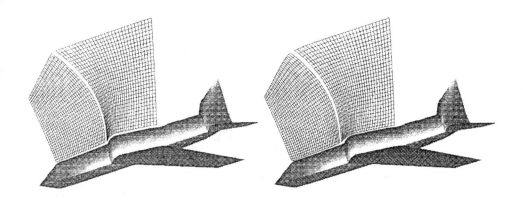

Figure 14 a,b: Smoothing of block boundary in symmetry plane: a) before, b) after.

REFERENCES

[Eri.] ERIKSSON,L.E.: Generation of Boundary-Conforming Grids around Wing-Body Configurations using Transfinite Interpolation, AIAA Journal, Vol 20, no. 10, pp1313-1320,1982

[Jac] JACQUOTTE,O.P.: Recent Progress on Mesh Optimization, p 581 ff, Numerical Grid Generation in Computational Fluid Dynamics and Related Fields, Ed:s, A.S.Arcilla et al., North-Holland, 1991.

[Leh] LEHTIMÄKI,R.: Single-Block Algebraic Grid Generation by Transfinite Interpolation, BRITE-EURAM Mesh Generation Subtask 1.3.2, Final Report, HUT, 1991.

[Min] MINEAU,P.: Smoothing of Grid Discontinuities Across Block Boundaries, BRITE-EURAM Mesh Generation Subtask 1.3.3, Final Report, Cerfacs, 1991.

[Opp] OPPELSTRUP,J., WEINERFELT,P.: Interactive Elliptic Surface Mesh Generation, BRITE-EURAM Mesh Generation Subtask 1.3.1, Final Report, KTH, 1991.

[Run] RUNBORG,O.: A C++ Skeleton Program for Multi-Block Mesh Generation, TRITA-NA-E-92xxx, KTH, 100 44 STOCKHOLM, Sweden (Diploma Thesis).

[Sca] SCATENI,R., OPPELSTRUP,J.: An Object Oriented Framework for Interactive Multiblock Topology Generation, TR/RF/91/54, Cerfacs, TOULOUSE.

[Ste] STEINBERG,S., ROACHE,P.: Anomalies in Grid Generation on Curves, J.Comp.Phys (91), pp255-277, 1990.

[Str] STROUSTRUP,B.: The C++ Programming Language, Addison-Wesley, 1986.

[Tak] TAKAGI,T., MIKI,K., CHEN,B.C.J, and SHA,W.T.: Numerical Generation of Boundary Fitted Curvilinear Coordinate Systems for Arbitrary Curved Surfaces, J.Comp.Phys (58), pp67-79, 1985.

[Tho] THOMPSON,J.F., WARSI,Z.U.A., and MASTIN,C.W.:Numerical Grid Generation, Foundations and Applications, North-Holland,1985

MULTI-BLOCK MESH OPTIMIZATION

T. Fol, V. Tréguer-Katossky
AEROSPATIALE, Aircraft Division
316 Route de Bayonne, Toulouse Cedex 03, France.

SUMMARY

The optimization package OPTIM3D has been extensively tested within an industrial environment. Conclusions concerning single-block and multi-blocks optimization are given. The tuning of the parameters is synthetically explained. Operationnal procedures for improving the efficiency of the program are proposed.

INTRODUCTION

The evaluation work reported here deals with the 3D multiblock mesh optimization code OPTIM3D. The optimization is based on a variationnal method due to O.P. Jacquotte and J. Cabello [1] and should be coupled with an algebraic mesh generator. OPTIM3D which first existed in a single-block version, has been extended to multi-blocks [2], as part of Task 1.2.: "Mesh Optimization" of Brite/Euram "Euromesh" project.

Single-block optimization have been evaluated first, with local and global procedures. It appeared clearly that a global improvement of the mesh quality, as indicated by a decrease of the cost function, may deteriorate the mesh locally. It was felt that more information was needed for understanding the mesh quality.

Basic metrics have been defined for this purpose [3]. They quantify separately properties of the mesh, such as orthogonality, skewness or regularity that can be combined into flexible measures of mesh quality.

Multi-blocks optimization requires the specification of inter-blocks boundaries and the prescription of degrees of freedom. Since this is a time consuming operation it is recommended to use a dedicated piece of software. The code developped at Aerospatiale for this purpose [4] will be commented. Examples will show the results of local and global optimization around complex configurations.

1. PRINCIPLE OF OPTIMIZATION

The optimal mesh is solution of a minimization problem, solved iteratively with Polak-Ribiere's Conjuguate Gradient Algorithm. The funtionnal to be minimized is analogue to the total mechanical deformation between the cells of the input mesh and reference cells. The optimizer is parametrized, so as to put weights in the cost function, or to choose the convenient reference cells for smoothness, or regularity.

The input data consist in the MESH file (containing the initial mesh in a VOIR3D format) and the DATA file (containing at least the parameters, and other multi-blocks information, if needed). The output data consist in the RESUL file (containing the optimal mesh in a VOIR3D format) and the INFO file (diagnostics and error reports).

2. SINGLE BLOCK OPTIMIZATION

The multi-blocks optimizer operates the same way, whatever the number of blocks. Thus, single-block meshes are sufficient for understanding the tuning of the parameters.

2.1. Examples

Example 1 shows 2 successive tunings of the optimizer. A global/local strategy is also applied.

The initial HH mesh, around a wing-fuselage configuration, shows an invalid grid line, accidentally produced by the algebraic mesh generator. The deffect is located within the flow field, far from the body.

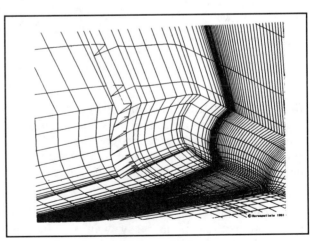

Figure 1(a)

First step:

Search for orthogonality, throughout the entire mesh:

after 150 iterations, with the cost function $\sigma = I1 + I2 - 6J$, and with orthogonal reference cells, the grid line of the converged mesh looks much better.but the grid spacing is still locally very irregular.

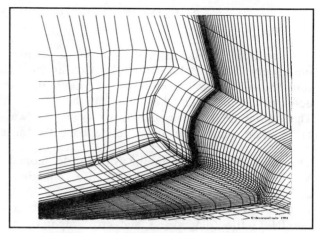

Figure 1(b)

Second Step:

Search for regularity, locally, inside a box of 3*5*7 cells, all nodes outside the box being given 0 degree of freedom:

after 150 more iterations, with the same cost function and with reference cells all identical and orthogonal, the grid spacing is improved.

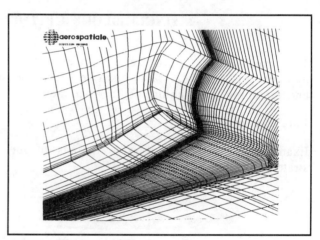

Figure 1(c)

The cost of the intervention is 550 sec CPU on a CRAY Y-MP, but the final grid is still not satisfactory, because the grid spacing has been locally smoothed in a region where clustering was intended. From this point, it is impossible to restore the grid spacing, since it would require adaptive facilities rather than optimization. Obviously, if remeshing had been possible, it would have been more appropriate in this case.

The following example is more relevant to mesh optimization: it shows that optimization is an easy way for improving algebric meshes. The optimizer is tuned for orthogonality, with control of volumes. The cost of the intervention is 886 sec on CRAY Y-MP.

Cross-section of the initial algebraic mesh around a wing-fuselage configuration:

the mesh lacks orthogonality in the region of the leading edge, because of its H-H topology.

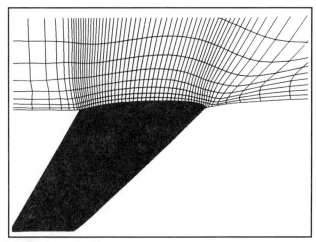

Figure 2(a)

Search for orthogonality, throughout the entire block of 54760 mesh points:

after 300 iterations, with the cost function $\sigma = I1 + I2 - 6J + (J-1)^2$, and with orthogonal reference cells, the mesh is more orthogonal, in the region of the leading edge.

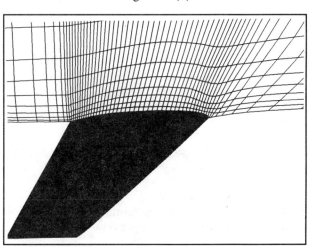

Figure 2(b)

2.2. Tuning of the parameters

Parameters ct1 to ct4 in the DATA file allow to tune the cost function. By default, use I1+I2-6J. If negative volume cells must be corrected, better use I1+I2-6J+(J-1)^2, although it slows down the convergence (see Figure 7).

Since the convergence of residues is not smooth enough, the influence of parameter eps is best frozen, by taking eps ≤ 10^{-9}. Optimization is thus stopped after itmax iterations.

Parameter kref determines the reference cells used. By default use kref=3, that is search for orthogonality only, because this choice allow to

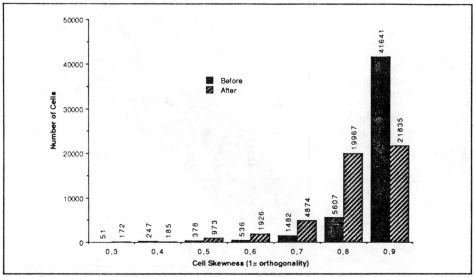

Figure 3: Effect of optimization on the measure of skewness

preserve the initial mesh spacing. If smoothing is also wanted, better use `kref=1`. However, this choice is not always satisfactory near the boundaries, due to the absence of moving nodes facilities on the surfaces. Any other choice of `kref` is forbidden and would result in using the unit cube as the reference cell!

By default use `alf=0`, that is uniform weighting of the cells. Choosing `alf=-1` increases the weight of small cells. Better not proceed large cells at the same time.

By default use `nint=4` as number of Gauss points in the integration rule, since `nint=8` brings no improvement, though it costs twice more.

3. BASIC METRICS FOR MESH QUALITY

The value of the overall cost function is insufficient for understanding in details the quality of a mesh. For example, the previous optimization made the average skewness of the cells decrease: Figure 3 shows 20000 more cells in the range from 0.5 to 0.8, but at the expense of 20000 cells in the range from 0.9 to 1, and 120 more cells under 0.3.

A set of 9 basic metrics for mesh quality has been developed [3]. Every metric proceeds one cell or one vertex at a time. Use them for coloured graphical display on selected mesh surfaces or for statistical analysis. Coloured display is best to correlate mesh quality with CFD results.

The 9 metrics allow for example, to detect non convex cells, to measure deviation of the mesh cells from orthogonal cells or from parallelograms, to check the stretching functions, aspect ratio, and volumes.

4. MULTI-BLOCK OPTIMIZATION

All multi-blocks optimizations, in this section (except Figure.5) deal with the same wing+pylon+engine configuration shown on Figure 4. The mesh is made of 21 blocks, with C blocks around every component, with either 100000 or 200000 mesh points. Other examples are reported in [4].

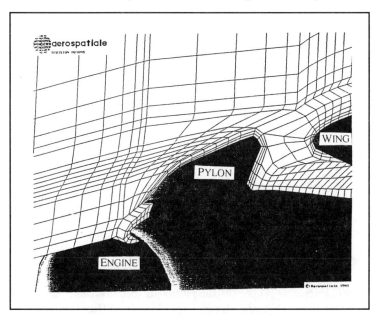

Figure 4: close-up of the optimized mesh (100000 mesh points)

4.1. Degrees of Freedom

Beyond the specification of the optimization parameters, Optim3D DATA file may contain multi-block information that prescribes 0 to 3 degrees of freedom (DoF) to the mesh points. The default is 3 DoF WITHIN EACH BLOCK that contains the point. Thus, by default, the domain boundaries are NOT respected and the neighbouring blocks are NOT connected.

For each number of DoF, ranging from 0 to 3, index regions may or not be specified. Each specified index region is assigned the current number

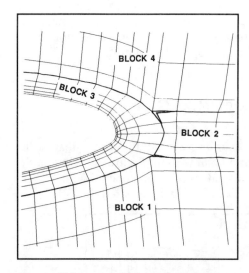 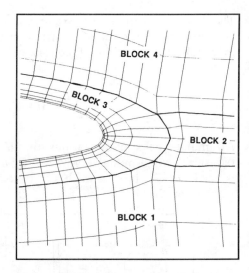

Figure 5(a): cross section of 3D mesh around a wing; holes are embedded inbetween blocks 1-2-3 and 2-3-4

Figure 5(b): with two more 1D inter-blocks boundaries declared, the blocks remain connected

of degrees. An index region is described, within a block, by start and end values of the variable indices. The specification of 0 DoF overwrites any forthcoming specification. Optimization boundaries (0, 1 or 2 DoF) and inter-blocks boundaries need one index region specification per block which contains them. ALL inter-blocks boundaries (2D, 1D and 0D) must be explicitly declared in order to prevent holes inbetween the blocks, as shown on Figure 5.

If the blocking structure is described elsewhere, it is advisable to generate default DATA files automatically. The program written at Aérospatiale prescribes 0 DoF to the domain boundaries, and 3 DoF to inter-blocks boundaries (for the latter, specify 2 index regions per block face, 1 index regions per block edge and per block which contains it, and 1 index region per block vertex and per block which contains it). If moving nodes facilities are available on surfaces and curves, better prescribe 2 DoF on the domain boundaries, and 1 DoF to intersection lines between components of the body.

4.2. Operating procedure

Better take `itmax>50`, except for very small meshes. Several shorter runs are more effective than one longer (Figure 6). A first inspection of the mesh is performed after a run with `itmax=200`, in order to decide to refine the tuning of the optimizer, or to use local strategies, or both. The behavior of the convergence of residues after 200 iterations (Figure 7) will indicate the best value of `itmax` between successive restarts. The

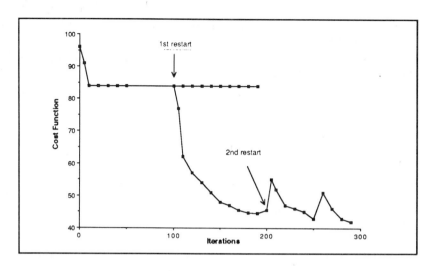

Figure 6: Convergence behavior (200000 mesh points)

admissible range is from 100 to 300.

4.3. Performance

Above 25000 mesh nodes and 150 iterations, the assymptotic behavior of the CPU time consumed (in seconds on a CRAY Y-MP) is:

$$CPU = (-21.09 \cdot 10^{-4} + 56.44 \cdot 10^{-6} \times \text{NbIterations}) \times \text{NbNodes}$$

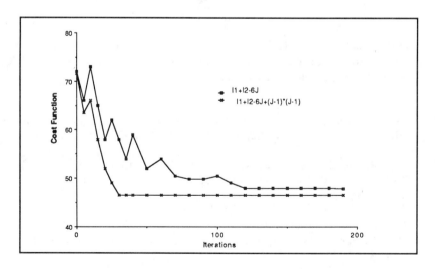

Figure 7: Convergence behaviors (100000 mesh points)

The best size for meshes to be optimized ranges from 20000 to 50000 nodes. On the one hand, there must be enough degrees of freedom in order to achieve orthogonality, and on the other hand, for large meshes, the convergence rate decreases with the size of the mesh.

CONCLUSION

For all the complex configurations tested, it has been possible to improve the mesh quality with the multi-block optimizer.

It is impossible to operate the multi-block optimizer in a "black box" mode for industrial configurations. A high level of expertise is required.

The optimization of the mesh costs nearly as much as the flow solution

Local strategies should be better supported: independant tuning of the parameters in different blocks, treatment of nodes with 0 degree of freedom.

The choice of reference cells lacks flexibility, for respecting prescribed mesh spacing during optimization.

There is a strong need for surface optimization.

REFERENCES

[1] O.P. Jacquotte, J. Cabello, "A mechanical model for a new mesh generation method in computationnal fluid dynamics", Comp. Methods Appl. Mech. Eng., 77, 6-212 (1989).

[2] O.P. Jacquotte, C. Gaillet, "Construction de maillages multiblocs tridimensionnels par une méthode variationnelle", ONERA technical report N° 7/3519 AY 424 A, 1991 (in English).

[3] T. Fol, V. Tréguer-Katossky, "Brite/Euram Euromesh Sub-Task 1.2: Basic Metrics for Mesh Quality", Aerospatiale, Technical Report N°443-558/91.

[4] T. Fol, V. Tréguer-Katossky, "Brite/Euram Euromesh Sub-Task 1.2: Multi-block Mesh Optimization", Aerospatiale, Technical Report N°443-559/91.

SMOOTHING OF GRID DISCONTINUITIES ACROSS BLOCK BOUNDARIES

Pascal Mineau

C.E.R.F.A.C.S.

European Centre for Research and Advanced Training in Scientific Computing
42 avenue Gustave Coriolis, F-31057 Toulouse Cedex, France

SUMMARY

This article summarizes the work completed at CERFACS in the sub-task 1.3.3 concerning the problem of slope discontinuities occurring at block interfaces. The starting point was a multi-block three-dimensional mesh generator developed by H. Ewetz, at KTH, Stockholm, Sweden. It was based on a solid modeling approach and a variational formulation to optimize the shapes of the blocks (references [1] and [2]). This code has been completely rewritten in order to increase its speed and to add new functionals. Graphic features have also been added to permit an interactive visualization of the generated mesh. Unfortunately, the results obtained with the new code were not always of the best quality, and moreover, computing time was very long. A more classical approach based on elliptic smoothing was then used with more success and will be also presented hereafter.

1 INTRODUCTION

A great number of problems arising in physics and engineering can be modeled by a set of partial differential equations. Unfortunately, most often, these equations can not be solved in an analytical way for realistic problems, particularly because of the complexity of the boundary conditions. A numerical approach, in addition to the fast increase in the power of super computers, have proven to be very powerful. For the past few decades, researchers and industrialists have been making a deep investment in the development and validation of numerical codes. We have now reached a point where the calculation of the flow around a complex configuration (for example, a complete aircraft), has become a realistic problem. The quality of a numerical solution depends both on the technique used for solving the partial derivative equations and the mesh. The requirements for the mesh vary according to the numerical method, but, in all cases, the mesh should fit the geometrical boundaries with great accuracy . The best method, with regard to this last point, is to build non-structured meshes, unfortunately leading to complex data structures which are difficult to handle, and which are responsible for a loss of efficiency, especially on vector computers. Most often, structured meshes are preferred (associated with finite volume methods). Nevertheless, all shapes can not be fitted by a single mesh. In order to deal with the more complex shapes, we have to divide the whole domain into several pieces (blocks), which can be fitted by a single structured mesh. The blocks are connected in an unstructured way. As there are usually a small number of blocks, the connections are not difficult to handle, and it is not a strong limitation for computational efficiency. It is easy to insure the continuity of mesh lines across the blocks' boundaries, but the continuity of their slopes is more difficult to build. The present work addresses the problem of smoothing discontinuities of the mesh line slopes. This is a crucial step in mesh generation : to have accurate results using the finite volume method requires the mesh to be smooth. We should note that slope discontinuities arise not only at block interfaces, but also inside blocks, particularly because one block can be obtained by the merging of several blocks (in order to reduce their total number). The two following sections will present two different approaches used to solve this problem.

2 OPTIMIZATION OF SMOOTHNESS BY MOVING CONTROL POINTS

In this section, an analytic solid modeling approach to the mesh generation and mesh smoothing is presented.

2.1 Solid Modeling Definition of the Blocks

Let us define the blocks (in physical space) as the mapping of a cube in the computational space :

$$\begin{cases} x(\xi,\eta,\zeta) = \sum_{u=v=w=0}^{3} x_{u+int(\xi),v+int(\eta),w+int(\zeta)}.B_u(\xi - int(\xi)).B_v(\eta - int(\eta)).B_w(\zeta - int(\zeta)) \\ y(\xi,\eta,\zeta) = \sum_{u=v=w=0}^{3} y_{u+int(\xi),v+int(\eta),w+int(\zeta)}.B_u(\xi - int(\xi)).B_v(\eta - int(\eta)).B_w(\zeta - int(\zeta)) \\ z(\xi,\eta,\zeta) = \sum_{u=v=w=0}^{3} z_{u+int(\xi),v+int(\eta),w+int(\zeta)}.B_u(\xi - int(\xi)).B_v(\eta - int(\eta)).B_w(\zeta - int(\zeta)) \\ \text{with } (\xi,\eta,\zeta) \in ([1, NI-2] \otimes [1, NJ-2] \otimes [1, NK-2]). \end{cases} \tag{1}$$

The points $P_{i,j,k}$, whose coordinates are $(x_{i,j,k}, y_{i,j,k}, z_{i,j,k})$, are called control points. It is possible to control the shape of the solid by moving these control points. NI (resp. NJ, NK) is the number of control points in the direction ξ $(resp.\ \eta,\ \zeta)$. Increasing the number of control points enables us to have a finer control of the behavior of the mapping. The B_i functions are the B-Splines basis functions :

$$\begin{cases} B_0(t) = (1-t)^3/6 \\ B_1(t) = (3t^3 - 6t^2 + 4)/6 \\ B_2(t) = (-3t^3 + 3t^2 + 3t + 1)/6 \\ B_3(t) = t^3/6 \\ t \in [0,1]. \end{cases} \tag{2}$$

This is an extension to the classical B-Spline representation of surfaces (further information can be found in reference [6]). We extend this definition to the volume $([0, NI-1] \otimes [0, NJ-1] \otimes [0, NK-1])$ of the computational space, and we use a modified set of basis functions on the edges in order to force the corners of the blocks to coincide with the extreme control points.

To connect the faces of two blocks, the control points of the given faces must be made equal. Then, the continuity of the mesh lines is automatically satisfied. The blocks must be connected by complete faces. For complicated configurations, this may lead to a large number of blocks. This representation will be used *only* during the generation of the mesh. After the completion of this phase, some of the blocks can be merged.

2.2 An Optimization Method for Minimizing Grid Discontinuities Across Block Boundaries

The mapping introduced in the previous subsection describes a solid's interior as well as its boundaries. Let us suppose that we use N control points on each edge (N^3 control points in total). The information on the interior of the solid is provided by $3(N-2)^3$ parameters, which would not exist if only the six faces were defined. The philosophy of the method is to exploit these extra degrees of freedom to optimize the shape of the blocks (i.e. : to insure C^1 continuity at block interfaces). It is interesting to note that until this point we don't work at the mesh level. The mesh will be built in a simple and direct fashion as an ultimate phase, after optimizing the blocks' shapes (see figure 1). This points out two strong advantages of such a method :

- We will have to optimize a functional in a space with a small number of dimensions (The number of control points will be less than the number of mesh points).

- It will be possible to change the mesh (e.g. to change the density of the mesh) without repeating the optimization phase.

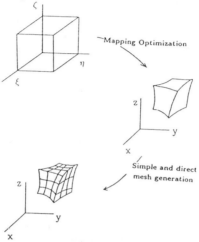

Figure 1: Mesh generation.

Functionals (which measure a property to be minimized) must be found depending on the coordinates of the control points. The three functionals that are used are:

$$\Phi_1 = \int\int\int d\xi d\eta d\zeta (\|\vec{\nabla_\xi} x\|^2 + \|\vec{\nabla_\xi} y\|^2 + \|\vec{\nabla_\xi} z\|^2) \tag{3}$$

$$\Phi_2 = \int\int\int dx dy dz (\|\vec{\nabla_x}\xi\|^2 + \|\vec{\nabla_x}\eta\|^2 + \|\vec{\nabla_x}\zeta\|^2) \tag{4}$$

$$\Phi_3 = \int\int\int d\xi d\eta d\zeta ((x_\xi x_\eta + y_\xi y_\eta + z_\xi z_\eta)^2 + (x_\xi x_\zeta + y_\xi y_\zeta + z_\xi z_\zeta)^2 + (x_\eta x_\zeta + y_\eta y_\zeta + z_\eta z_\zeta)^2). \tag{5}$$

Φ_2, is derived from a weak formulation of the Laplace equation. The Laplace equation, as other elliptic equations, is well known for its smoothing properties, and is often used in mesh generation. So, it is natural to use Φ_2 as a measure for the smoothness of the mesh. Φ_1 is simply derived from Φ_2 by changing the (x, y, z) and the (ξ, η, ζ) variables. Φ_3 measures the orthogonality of the mesh, and is null if the mesh lines are orthogonal everywhere.

In the computations, integrals in the definitions of the functionals are replaced with discrete summations. In general, it was unnecessary to use more integration points than control points in the summation. This result is not very surprising, because the basis functions used in the mapping are of low order.

Conjugate gradient methods (references [3] and [4]) are well suited for finding the solution of our optimization problem. Their main advantage is that only the gradient is needed, not the Hessian matrix. The Fletcher-Reeves or the Polak-Ribiere method is used. The minimization along the search direction can be performed exactly for the functional Φ_1 because it is quadratic. For Φ_2 and Φ_3, we use the simple Golden Section Search method (reference [5]).

2.3 Resolution of the Laplace equation at the control point level

The code built using the previous method is very slow, despite the acceleration attempts. It is impossible to use in an interactive environment. Furthermore, the results were not always of the best quality, especially in the case of strong initial perturbations.

Keeping with the same idea of a solid modeling definition of the blocks, in order to reduce the number of degrees of freedom, we will now smooth the mapping by directly solving the Laplace system at the control points level (i.e., considering the structured set of control point as if it were a mesh). The derivatives will be calculated by simple finite differences using the position of the control points rather than using the exact definition of the mapping. The system to be solved is :

$$\begin{cases} \triangle_x \xi = 0 \\ \triangle_x \eta = 0 \\ \triangle_x \zeta = 0 \, . \end{cases} \quad (6)$$

Well known, fast, and stable methods can be used for the resolution of this system.

2.4 Results

In this subsection we will present some results obtained with the two preceding methods. We will only treat very simple one- or two-blocks configurations. In using simple test cases (instead of realistic ones) the behavior of the method is better understood. In all the examples, the mesh shape is invariant in the z-direction and, most often, only a *slice* of the mesh is displayed.

2.4.1 A two-block configuration

A simple two-block configuration is used as a first test. Each block is defined by $6 \times 6 \times 4$ control points. Inner control points are free. The control points which define the external faces are allowed to move in a plane, in order not to deform the faces (504 degrees of freedom). The mesh is smoothed by minimizing ϕ_2 (cf : figures 2 and 3). A drift of the mesh can be seen, as well as very skewed cells (near the lower edge of the common face).

2.4.2 A one-block configuration with initial perturbation

In this example, the initial mesh has been strongly perturbated, then ϕ_2 is minimized. The algorithm converges through a non-smooth state. Such problems of local minimum are often encountered when we perturb the initial mesh. The local minimum may result from a non-accurate calculation of the functional. Nevertheless, this indicates a limitation of the method (cf : figure 4).

2.4.3 Comparison between the methods

The initial block is defined by $6 \times 6 \times 4$ control points. Figure 5 compares the results obtained using the methods previously presented.

2.4.4 Stability of the Laplace method

This last example shows the stability of the Laplace method. A smooth mesh is obtained starting from a strongly perturbated two-block mesh. The great number of control points insures a good control of the mesh. This is why there is not a strong drift near the convex and the concave boundaries (cf : figure 6).

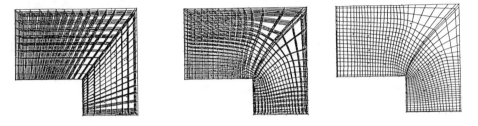

Figure 2: Initial state and final state after 500 iterations. .

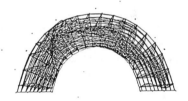

Figure 3: Final state after 121 iterations.

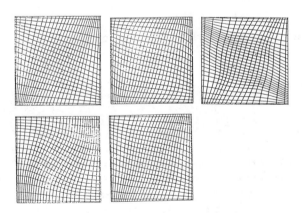

Figure 4: From left to right : Initial state, final state obtained minimizing ϕ_1, ϕ_2, ϕ_3, and final state obtained using the Laplace equation at the control points level.

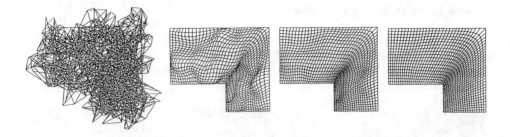

Figure 5: Initial mesh, and after 10, 20, and 40 iterations.

3 ELLIPTIC SMOOTHING

In this section, we will present a classical elliptic smoothing approach. (A good reference for elliptic mesh generation is reference [7]). The starting point is the generalized Laplace system including grid control functions P, Q and R. (This transforms the system into a Poisson type).

$$\begin{cases} \triangle_x \xi = P(\xi, \eta, \zeta) \\ \triangle_x \eta = Q(\xi, \eta, \zeta) \\ \triangle_x \zeta = R(\xi, \eta, \zeta) \,. \end{cases} \quad (7)$$

The grid control functions can provide the flexibility required to control mesh point spacing, and to prevent the mesh from drifting near convex or concave boundaries. The general difficulty is to compute satisfactory grid control functions. Here we compute P, Q and R from the initial mesh, and we introduce them in the Poisson system after a smoothing. The Poisson system is then solved by a Gauss-Seidel type iterative method. This procedure insures that the mesh point density is nearly preserved by the smoothing.

The key point of our method lies in the smoothing of the grid control functions. The following scheme is used for several iterations (typically ten):

$$\begin{cases} P_{i,j,k}^{(n+1)} = \omega P_{i,j,k}^{(n)} + \frac{1-\omega}{4} \left(P_{i,j+1,k}^{(n)} + P_{i,j-1,k}^{(n)} + P_{i,j,k+1}^{(n)} + P_{i,j,k-1}^{(n)} \right) \\ Q_{i,j,k}^{(n+1)} = \omega Q_{i,j,k}^{(n)} + \frac{1-\omega}{4} \left(Q_{i+1,j,k}^{(n)} + Q_{i-1,j,k}^{(n)} + Q_{i,j,k+1}^{(n)} + Q_{i,j,k-1}^{(n)} \right) \\ R_{i,j,k}^{(n+1)} = \omega R_{i,j,k}^{(n)} + \frac{1-\omega}{4} \left(R_{i+1,j,k}^{(n)} + R_{i-1,j,k}^{(n)} + R_{i,j+1,k}^{(n)} + R_{i,j-1,k}^{(n)} \right) \,. \end{cases} \quad (8)$$

Of course, the values $P_{i+1,j,k}^n$ and $P_{i-1,j,k}^n$ are not used in this procedure, because smoothing P in the ξ direction would lead to a smoothing of the mesh density. (the same for Q and R). Huge or infinite values of the grid control function can arise in a grid (e.g., discontinuities of the mesh point spacing, or parabolic stretching). During the smoothing of the function, these huge values must not be transferred to other mesh points. The first solution to this problem is to modify the smoothing algorithm : the average uses only points where the value of the function is not very different from the local value. This gives generally good results, except when the mesh contains parabolic singularities. In this case, bad results were obtained (in particular, mesh folding). Another solution consists in truncating the huge values of the grid control functions before smoothing them. A good truncation value must be found. If it is too large, our problem remains unsolved. On the other hand, if it is too small, the P and Q and R will be small everywhere and the final mesh will be nearly the Laplace solution. (The mesh density will also be smoothed). Some tests clearly show that the truncating value must depend on the mesh. This can be easily understood using a one-dimensional approach. From the 1-D Poisson equation, we see that P scales like x^{-2}. Let us suppose that P_{trunc} is a correct truncating value for a given mesh. If we consider the mesh obtained by multiplying all the dimensions by 10, a correct truncating value is $\frac{P_{trunc}}{100}$. The truncating value depends on the size of the mesh (the size of the mesh step).

For accuracy reasons, finite volume methods require that the mesh step $h_i = x_{i+1} - x_i$ varies slowly. $\alpha = \frac{h_{i+1}}{h_i}$ should be very close to 1. Typically, we can choose $0.8 < \alpha < \frac{1}{0.8}$. This can be used to derive a truncating value for P. The Poisson equation can be reversed :

$$\frac{d^2x}{d\xi^2} = -(\frac{dx}{d\xi})^3 P(x) \tag{9}$$

which can be discretized using centered finite differences :

$$P_i = -8\frac{x_{i+1} - 2x_i + x_{i-1}}{(x_{i+1} - x_{i-1})^3} = -8\frac{h_i - h_{i-1}}{(h_i + h_{i-1})^3} . \tag{10}$$

The inequality on α yields :

$$-\frac{0.274}{(h_i)^2} < P_i < \frac{0.175}{(h_i)^2} . \tag{11}$$

This gives a maximum and a minimum value for P_i. For the three-dimensional case, this is more complicated because P also depends on y and z. Nevertheless, it possible to use the previous expression by replacing h_i by $\sqrt{(x_{i+1} - x_i)^2 + (y_{i+1} - y_i)^2 + (z_{i+1} - z_i)^2}$. Using this method for our test cases permitted the smoothing of the mesh lines across the boundaries and avoided the problem of mesh folding. Unfortunately, truncating the grid control functions on the blocks' boundaries often leads to a rather large deformation of these boundaries.

3.1 Results

3.1.1 A three block configuration

This example contains two difficulties : the singular point in the center of the mesh, and a parabolic stretching which gives infinite values of the control grid functions on the common boundaries of the blocks. (cf : figure 9)

3.1.2 Stability test

This example shows the stability of the Poisson method. The grid control functions are calculated from an initial two-block mesh, with a non-homogeneous point density. These functions are smoothed and the mesh is perturbed (except on external boundaries). Then the Poisson equation is solved. Several dozen iterations are needed to obtain a smooth mesh. (cf : figure 10)

4 CONCLUSION

Several smoothing methods for 3-D meshes were investigated. It seemed that our first approach, based upon control points, was a good way to generate smooth meshes. Unfortunately,

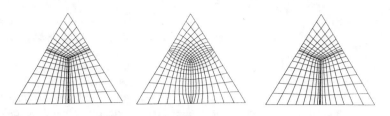

Figure 6: Initial mesh, mesh obtained by solving the Laplace equation (i.e. : setting the control grid functions at zero), mesh obtained by solving the Poisson equation (control grid functions are truncated depending on the local mesh step ; then, they are smoothed). In the last figure, the truncation of the high values of the grid control functions is responsible for the deformation of the common boundaries. This phenomenon disappears if the stretching is not parabolic (so that the grid control functions are not infinite on the common boundaries).

Figure 7: Initial mesh, perturbed mesh, and final mesh after 100 iterations (Jacobi method).

the minimization technique requires very large computing time (mainly for the computation of the gradient). Therefore, we are forced to use a small number of control points and integration points. This often leads to bad quality results. New functionals requiring only a few calculations should be found. (For example, functionals computed using only the position of the control points, not the very fine description of the mapping). Solving the Laplace equation at the control points level is fast because the derivatives are calculated by simple finite differences on the position of the control points. Better results could be obtained by introducing P, Q, and R, and replacing the Laplace equation by the Poisson equation, as is done at the mesh point level. (For example, this could avoid the skewed cells appearing near the lower edge of the common face in example **2.4.1**).

Finally, the Poisson equation was used at the mesh point level for smoothing existing meshes while preserving the mesh point density. The P, Q, and R grid control functions, computed from the initial mesh, can have very high values (especially in the cases of a parabolic stretching, a discontinuity in the mesh point density, or the existence of very curved mesh lines). The mesh is very sensitive to the modification of the grid control functions near these high values. Therefore, to smooth them is very difficult ; a satisfactory solution has not been fount yet. A more detailed study should be initiated in order to have a better understanding of the effect of the grid control functions in 3-D.

REFERENCES

[1] H. Ewetz and J. Oppelstrup. *Multiblock Mesh Generation for 3D CFD Applications*. Numerical Analysis and Computing Science, KTH S 100 44 STOCKHOLM, Sweden. (September 1987).

[2] J. Oppelstrup and J. Hörnfeldt. *Multiblock Mesh Generation and a General 3D Steady Flow Code*. Proceedings of the First Scandinavian Symposium on Viscous Fluid in Hydraulic Machinery, Trondheim, 1987.

[3] W.H. Press, B.P. Flannery, S.A. Teukolsky, and W.T. Vetterling. *Numerical Recipes*. Cambridge University Press. p.301, 307.

[4] Gill, Muray, and Wright. *Practical Optimization*. Academic Press, New York, 1981.

[5] W.H. Press, B.P. Flannery, S.A. Teukolsky, and W.T. Vetterling. *Numerical Recipes*. Cambridge University Press. p.277, 282.

[6] J.D. Foley, A. van Dam, S.K. Feiner, and J.F. Hughes. *Computer Graphics*. Second Edition. The System Programming Series - Addison Wesley. p.471, 532.

[7] J.F. Thompson, Z.U. Warsi, and C.W. Mastin. *Numerical Grid Generation*. North Holland, New York, 1985.

V. GRID OPTIMIZATION AND ADAPTION METHODS

GRID ADAPTION IN COMPUTATIONAL AERODYNAMICS

R. Hagmeijer and K.M.J. de Cock
National Aerospace Laboratory NLR
P.O. Box 90502, 1006 BM Amsterdam, The Netherlands

SUMMARY

A recently developed algorithm to adapt computational grids [1] is applied to aerodynamic problems. The algorithm is briefly described and the main features are discussed. Applications to both inviscid and viscous flows around two-dimensional airfoils show that the algorithm is robust and almost fully automatic, and that shocks, expansion zones, boundary layers and shear layers are well resolved by the adapted grid.

INTRODUCTION

To eliminate the need for a priori qualitative estimates of solutions to physical problems, considerable effort has been directed towards the development of solution-adaptive grid generation methods, see e.g. the surveys of Eiseman [2] and Thompson [3]. Dwyer [4] applied the one-dimensional equidistribution principle along one family of grid lines in two-dimensional problems. Nakahashi and Deiwert [5] applied the principle to both families of grid lines in a tension spring analogy, and added torsion springs in order to gain control over the orthogonality of grid cells. Anderson [6] and Anderson and Steinbrenner [7] noticed the similarity between the equations of adaptive equidistribution and elliptic grid generation in one space dimension, and used this similarity to formulate a scheme for adaption in two space dimensions in the format of Poisson equations used for conventional grid generation. They showed that the adaption scheme can be interpreted as equidistribution of a weight function along grid lines, that measures orthogonality, curvature and solution derivatives.

Brackbill and Saltzman [8] presented a variational approach in which conflicting properties such as smoothness, orthogonality and adaption are explicitly taken account for in a compound functional consisting of three integrals. The integral that measures smoothness was earlier used by Winslow [9], who modified the integrand by introduction of a single weight function that measures the gradient of the flow solution, which resulted in isotropic diffusion equations. Recently Hagmeijer [1] developed an alternative formulation by introducing two separate weight functions in the smoothness integral that measure derivatives of the flow solution in separate directions. The resulting equations can be interpreted as anisotropic diffusion equations. In order to retain the assumed positive characteristics of the initial grid w.r.t. geometry resolution, the idea of Lee and Loellbach [10] was utilized, i.e. the grid adaption equations were formulated in the so-called parametric domain to determine the mapping from the computational to the parametric domain. The equations were finally modified to obtain desirable properties w.r.t. orthogonality and boundary layer resolution.

The present paper briefly describes the adaption algorithm presented in [1], and several applications to problems in computational aerodynamics are presented.

ADAPTIVE GRID GENERATION IN THE PARAMETRIC DOMAIN

In [1] it is assumed that the initial grid to be adapted has desirable properties w.r.t. geometry resolution and orthogonality. In order to retain these properties in the adapted grid the concept of grid adaption in the parametric domain of Lee and Loellbach [10] was utilized, illustrated in figure 1.

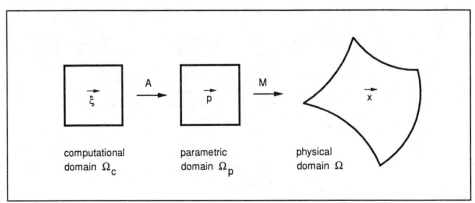

Figure 1. Grid adaption by using an adaptive mapping A that maps the computational domain onto the parametric domain.

Let the initial grid in the physical domain Ω be the image of a uniform cartesian grid in the unit square $[0,1]^2 \subset \mathbb{R}^2$ under a mapping M. The cartesian coordinates in the unit square are denoted by $\vec{p} = (p,q)^T$ and serve as parameters to describe the physical domain. Hence the unit square is called parametric domain denoted by Ω_p in the present paper. The idea of Lee and Loellbach [10] is to adapt the cartesian grid in Ω_p and to apply the mapping M to generate the adapted grid in the physical domain Ω.

The problem of grid adaption in Ω_p can conveniently be formulated in terms of finding a suitable mapping A that maps a cartesian grid in the computational domain Ω_c to the adapted grid in Ω_p, see figure 1. Finally the adapted grid in the physical domain Ω is the image of the cartesian grid in Ω_c produced by the compound mapping M ∘ A. Hence the problem is to construct the mapping A determined by the parametric coordinates p and q as functions of the computational coordinates ξ and η.

BASIC ADAPTION MAPPING

The integral that was used by Brackbill and Saltzman [8] as a measure of smoothness can be formulated in the parametric domain:

$$\iint_\Omega (\|\vec{\nabla}_p \xi\|^2 + \|\vec{\nabla}_p \eta\|^2)\, dpdq, \qquad (1)$$

where ξ and η are the computational coordinates, p and q are the parametric coordinates, and ∇_p is the nabla operator in the parametric domain. In order to minimize this integral, the Laplace equations have to be satisfied:

$$\nabla_p^2 \xi = 0, \quad \nabla_p^2 \eta = 0. \tag{2}$$

It can be noted that by regrouping the quadratic terms, the integrand of (1) can be rewritten as:

$$\|\vec{\nabla}_p \xi\|^2 + \|\vec{\nabla}_p \eta\|^2 = \|\vec{\xi}_p\|^2 + \|\vec{\xi}_q\|^2, \tag{3}$$

where $\vec{\xi} = (\xi, \eta)^T$, and the subscripts p and q denote differentiation w.r.t. p and q respectively. The two terms in the right hand side of equation (3) can be interpreted as squared norms of vector tangents to curves of constant p or q in the computational domain, illustrated in figure 2.

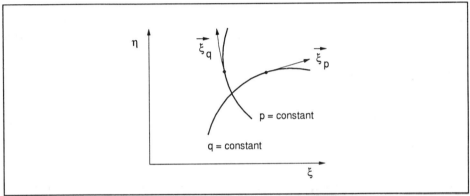

Figure 2. Vector tangents to curves of constant p and q in the computational domain.

In order to obtain adaptive features, the smoothness integral (1) is altered by dividing both squared norms of the vector tangents by two separate weight functions resulting in the following functional to be minimized:

$$K[\xi(p,q), \eta(p,q)] = \int\int_{\Omega_p} (\frac{\|\vec{\xi}_p\|^2}{w_1(p,q)} + \frac{\|\vec{\xi}_q\|^2}{w_2(p,q)}) dpdq, \tag{4}$$

where the weight functions w_1 and w_2 are assumed to satisfy

$$w_1 \geq c, \quad w_2 \geq c, \tag{5}$$

with the constant c conveniently taken as c = 1. The functional (4) strongly suggests to define the two weight functions w_1 and w_2 as appropriate measures of the flow solution derivatives in p and q direction respectively. To illustrate this, suppose that at some point of the physical domain flow solution derivatives in p direction are relatively large providing a large value of w_1, then if the functional (4) assumes its minimal value this leads to a relative large value of $\|\vec{\xi}_p\|$. As a result *the grid spacing in p direction is relatively small*. It should be emphasized that at this stage grid spacing is not interpreted as arc-length increment along curves of constant ξ or η in the parametric domain, but as the reciprocal $\|\vec{\zeta}_p\|^{-1}$ along curves of constant p or q in the computational domain.

It can be shown [1] by interchanging the dependent and independent variables in the functional K given by equation (4), that when K reaches its minimum value, the functions $p(\xi,\eta)$ and $q(\xi,\eta)$ satisfy generalized Cauchy-Riemann equations, e.g. see Oskam and Huizing [11]. If these Cauchy-Riemann equations are satisfied, the curvilinear coordinate system $p(\xi,\eta),q(\xi,\eta)$ will not necessarily be orthogonal. This implies that minimization of the functional K defined by equation (4) does not minimize orthogonality. For special cases

(i) $w_1 = w_2 = 1$, see Brackbill and Saltzman [8], and
(ii) $w_1 = w_2$, see Winslow [9],

non-orthogonality is minimized by minimizing K. The advantage of the present formulation is that not only the magnitude of the flow gradient but also its orientation is taken into account. The disadvantage of not minimizing non-orthogonality is believed to be acceptable because the functional K is formulated on the square parametric domain Ω_p and the occurrence of excessive skew cells is less probable.

The Euler equations that have to be satisfied to minimize the functional K given by equation (4) are:

$$\frac{\partial}{\partial p}(\frac{1}{w_1}\xi_p) + \frac{\partial}{\partial q}(\frac{1}{w_2}\xi_q) = 0,$$
$$\frac{\partial}{\partial p}(\frac{1}{w_1}\eta_p) + \frac{\partial}{\partial q}(\frac{1}{w_2}\eta_q) = 0. \qquad (6)$$

These partial differential equations are linear and decoupled and can be interpreted as anisotropic diffusion equations with diffusion coefficients w_1^{-1} and w_2^{-1} in p and q direction respectively. The isotropic diffusion equations of Winslow [9] can be recovered by taking $w_1 = w_2$. Equations (6) determine a basis for specification of the inverse adaption mapping A^{-1} in figure 1.

MODIFICATION OF THE INVERSE ADAPTION MAPPING

As is explained in [1] action has to be taken to prevent that the adapted grid in the physical domain incorporates excessive skewness when in the initial grid cells occur with extreme large or small aspect ratios, i.e. in boundary layers. The diffusion equations (6) are modified into

$$\lambda_1\frac{\partial}{\partial p}(\frac{1}{w_1}\xi_p) + \lambda_2\frac{\partial}{\partial q}(\frac{1}{w_2}\xi_q) = 0,$$
$$\lambda_1\frac{\partial}{\partial p}(\frac{1}{w_1}\eta_p) + \lambda_2\frac{\partial}{\partial q}(\frac{1}{w_2}\eta_q) = 0, \qquad (7)$$

where λ_1 and λ_2 are functions of p and q called modification functions, defined as

$$\lambda_1(p,q) = w_1^2 \|\vec{x}_q\|^2, \quad \lambda_2(p,q) = w_2^2 \|\vec{x}_p\|^2, \tag{8}$$

where $\vec{x}_p$ and $\vec{x}_q$ are the arclength derivatives along curves of the initial grid. Application of these modified anisotropic diffusion equations (7) with Neumann boundary conditions imposed along boundaries that represent solid surfaces in the physical domain, incorporates the following important properties in boundary layers if the initial grid has highly stretched cells in flow direction [1]:

(i) preservation of orthogonality,
(ii) essentially one-dimensional equidistribution adaption in normal direction to the surface, and
(iii) adaption in flow direction controlled by the outer flow.

These properties originate from the fact that $\lambda_1 \ll \lambda_2$ in the boundary layer due to the highly stretched initial grid so that the first terms in equations (7) can be neglected w.r.t. to the second terms. Hence in the boundary layer one-dimensional adaption for both ξ and η occurs resulting in properties (i) and (ii). Because outside the boundary layer $\lambda_1 \sim \lambda_2$ and $w_1 \gg w_2$ the first terms in equations (7) are dominant resulting in property (iii).

Finally the boundary value problems that upon solution for the functions $\xi(p,q)$ and $\eta(p,q)$ determine the inverse mapping A^{-1}, can be conveniently formulated by means of a linear differential operator L_p defined as

$$L_p = \Lambda \vec{\nabla}_p \cdot W^{-1} \vec{\nabla}_p, \tag{9}$$

where Λ and W are diagonal matrices

$$\Lambda = \begin{bmatrix} \lambda_1 & 0 \\ 0 & \lambda_2 \end{bmatrix}, \quad W = \begin{bmatrix} w_1 & 0 \\ 0 & w_2 \end{bmatrix}. \tag{10}$$

Both modified equations (7) incorporate the linear operator L_p and the boundary value problems for ξ and η are finally formulated as

$$L_p[\xi] = 0, \quad (p,q)^T \in \Omega_p,$$
$$\xi(0,q) = 0, \quad \xi(1,q) = 1, \quad \xi_q(p,0) = \xi_q(p,1) = 0, \tag{11}$$

and

$$L_p[\eta] = 0, \quad (p,q)^T \in \Omega_p,$$
$$\eta_p(0,q) = \eta_p(1,q) = 0, \quad \eta(p,0) = 0, \quad \eta(p,1) = 1. \tag{12}$$

Upon specification of the weight functions w_1 and w_2 in (11) which is described in the next section, this concludes the formulation of the adaption equations in the parametric domain.

WEIGHT FUNCTIONS

To complete specification of the boundary value problems (11),(12) the weight functions have to be defined. Two questions are of importance:

(i) in what way can the weight functions be chosen such that the numerical flow solution is improved upon re-calculation on the adapted grid, and

(ii) in what way is specification of the weight functions influenced by the fact that the adaption problem is formulated in the parametric domain instead of the physical domain.

Both questions have been analyzed to some extent in [1] for one-dimensional problems with the flow solution characterized by a single scalar function.

First it can be derived that when the scalar flow solution, say Q, is a monotone increasing function of coordinate x in the physical domain, one-dimensional equidistribution of a weight function $w = Q_x$ results in a vanishing second derivative in the computational domain: $Q_{\xi\xi} = 0$. This is a desirable property if it is assumed that numerical truncation errors, due to discretization of the flow equations by second order accurate finite differences, are proportional to $Q_{\xi\xi}$.

Secondly a completely similar equidistribution scheme can be formulated in the parametric domain, which is equivalent with formulation in the physical domain when w is defined as $w = Q_p$.

In general problems Q will not be a monotone increasing or decreasing function of x, and grid spacing becomes infinite at points where $Q_x = 0$, therefore the weight function is altered into $(1 + Q_p^2)^{\frac{1}{2}}$. Backward transformation of the equidistribution scheme in the parametric domain to formulation in the physical domain shows that the equivalent weight function is $(p_x^2 + Q_x^2)^{\frac{1}{2}}$. Hence these equivalent formulations are not similar. The term p_x^2 in the weight function in the physical domain expresses that when a uniform flow solution is imposed, Q = constant, then the stretching of the initial grid is retained by the mapping M of figure 1 which is determined by the function x(p).

The above described analysis of one-dimensional problems where the flow solution is a single scalar function can be used as a guide for extension to multiple-dimensional problems where the flow solution is represented by a vector function, say $\vec{Q}$. In a natural way, the weight functions w_1 and w_2 that determine the linear operator L_p given by equations (9),(10) are chosen as

$$w_1(p,q) = \sqrt{1 + \|\vec{Q}_p\|^2}, \quad w_2(p,q) = \sqrt{1 + \|\vec{Q}_q\|^2}, \qquad (13)$$

where $\vec{Q} \in \mathbb{R}^n$, $n \geq 1$, represents the flow solution and $\|\cdot\|$ denotes the L_2 norm.

C-TOPOLOGY GRIDS

At this stage it is convenient to describe a set of additional mappings that are necessary to adapt C-topology grids around two-dimensional airfoils, see figure 3.

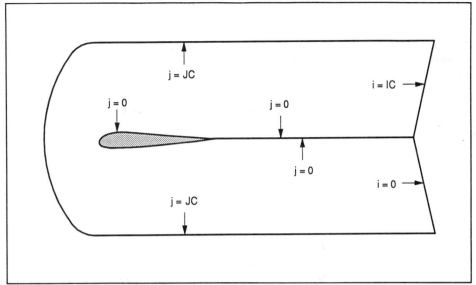

Figure 3.   C-topology around two-dimensional airfoil.

The problem is that grid lines of constant $i = i_0$ must be connected to grid lines with $i = IC - i_0$ (IC is the number of cells in i-direction) if the end points are lying on the wake line (the part of the boundary $j = 0$ which does not belong to the airfoil). It is not possible to impose a Dirichlet condition at the wake line instead of a Neumann condition because the wake is a viscous layer where the grid must be orthogonal.

In order to fulfil the requirement of grid continuity at the wake line two additional correction mappings C^* and C^{**} are applied, see figure 4.

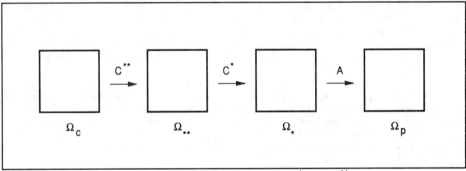

Figure 4.   Two additional correction mappings C^* and C^{**}.

The first mapping C^* provides that two grid lines with $i = n_w$ and $i = IC - n_w$, where n_w is the desired number of cells along the wake line, have common end-points at the trailing edge of the airfoil. The second mapping C^{**} provides common end points along the rest of the wake line. Both correction mappings are described in detail in [1].

NUMERICAL SOLUTION METHOD

The boundary value problems (11),(12) for the functions $\xi(p,q),\eta(p,q)$ that determine the inverse mapping A^{-1} in figure 1, are solved numerically on a uniform computational grid in the parametric domain depicted in figure 5.

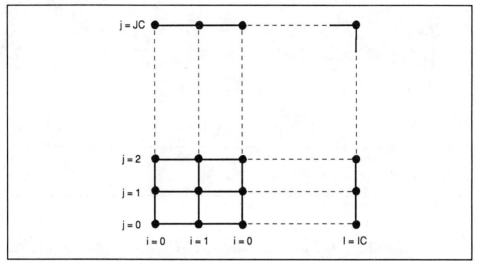

Figure 5. Schematic impression of the initial uniform grid in the parametric domain Ω_p.

The initial computational grid in Ω_p consists of IC cells in p-direction and JC cells in q-direction. The linear operator L_p is approximated by a linear difference operator by replacing derivatives by central differences. The Neumann boundary conditions are approximated by use of one-sided differences. As a result the continuous boundary value problems (11),(12) are replaced by second order accurate approximations each consisting of a system of linear algebraic equations. Both linear systems are solved by point Red-Black Successive-Over-Relaxation. A Correction-Storage multi-grid technique [11] with fixed V-cycles is used to increase the rate of convergence.

The solutions of the above described linear systems provide values for ξ and η in all points of the computational grid in the parametric domain Ω_p and represent a discrete approximation of the inverse mapping A^{-1}. To invert this mapping to the mapping A in figure 1 that maps a uniform grid in the computational domain to the adapted grid in the parametric domain, the following set of non-linear algebraic equations has to be solved for the set of points $\vec{p}_{ij} = (p_{ij}, q_{ij})^T$

$$\vec{\xi}(\vec{p}_{ij}) = (\frac{i}{IC}, \frac{j}{JC})^T, \quad i \in [0, IC], \ j \in [0, JC], \tag{14}$$

where $\vec{\xi}(\vec{p}_{ij}) = (\xi(p,q), \eta(p,q))^T$ is a piece wise bilinear interpolation of the values of ξ and η on the uniform grid in Ω_p. Solution of equations (14) gives values of p and q on a uniform grid in the computational domain Ω_c. In order to solve the non-linear system (14), the following iteration procedure is applied:

$$\vec{p}_{ij}^{n+1} = \vec{p}_{ij}^n - \Delta t [\vec{\xi}(\vec{p}_{ij}^n) - (\frac{i}{IC}, \frac{j}{JC})^T], \quad n=1,2,\ldots,\qquad(15)$$

which can be interpreted as an explicit time-stepping scheme with time-step Δt ($\Delta t = 0.1$ in this paper). The stationary solution of (15) satisfies the non-linear system (14).

Finally the two correction mappings of figure 5 are applied, and the obtained adapted grid in the parametric domain is mapped to the physical domain Ω, see figure 5. At all stages piece wise linear or bilinear interpolation is used to approximate the mappings C^*, C^{**}, A and M. This concludes the description of the numerical solution method.

NUMERICAL EXAMPLES

The above described grid adaption algorithm is applied to two aerodynamic problems.

Problem I : Inviscid flow around NACA0012 airfoil.
Flow equations: Euler
Flow identification: $M_\infty = 0.85$, $\alpha = 1°$
Grid identification: C-topology, IC = 128, JC = 32

Views of both the initial grid and the adapted grid are shown in figures 6[a] and 6[b]. Two adaptions have been performed with the first adapted grid taken as the initial grid for the second adaption. A cell centered central-difference scheme has been used to calculate the flow solution [13]. Clearly the strong shock at the upper side as well as the weaker shock at the lower side of the airfoil have induced strong adaption of the grid without causing serious depletion of grid cells at the leading edge. The distributions of the pressure coefficient along the surface of the airfoil are shown in figures 7[a] and 7[b]. Both shocks are crisp on the adapted grid whereas the upstream pressure distribution remaines almost unaltered. Because the initial grid at the trailing edge is rather coarse the local stagnation point remaines undetected. This has a significant influence on the shock positions causing an over-prediction of the lift coefficient: on the initial grid $C_l = 0.3396$, and on the twice adapted grid $C_l = 0.4314$.

Problem II: Viscous flow around NACA0012 airfoil.
Flow equations: Reynolds-averaged Navier-Stokes,
Flow identification: $M_\infty = 2.0$, $Re_\infty = 10^4$, $\alpha = 0°$, $T_\infty = 300K$
Grid identification: C-topology, IC = 256, JC = 64

Views of both the initial grid and the adapted grid are shown in figures 8[a] and 8[b]. Two adaptions have been performed with the first adapted grid taken as the initial grid for the second adaption. The bow shock induced strong adaption of the grid and very skew cells are introduced. Inspection of figures 9[a] and 9[b] however shows that the flow solution, i.e. the Mach number distribution, has been significantly improved w.r.t. shock resolution. It is fair to acknowledge that this is due to the use of a cell-vertex based Navier-Stokes solver [14]. In figures 10[a] and 10[b] close-up views of the leading edge section are presented. Besides the strong clustering of cells at the shock position also the boundary layer section is significantly improved. The initial grid contains some deliberately generated but yet undesirable stretchings at the edge of the boundary layer, which are eliminated in the adapted grid. Figures 11[a] and 11[b] show details of the trailing edge section. Grid point distribution in the wake is significantly changed in the adapted grid and a weak oblique shock region has developped. A closer inspection of the boundary layer resolution is given in figures 12[a] and 12[b], which show the velocity profile at $x/L = 0.1$ in normal direction. Clearly the adapted grid incorporates a strong clustering around the bow shock and the excessive fine grid at the edge of the boundary layer is depleted, although boundary layer resolution is preserved. Figures 13[a] and 13[b] show the wake velocity profiles at $x/L = 2$. The excessive clustering of points at the edge of the wake is eliminated upon adaption and resolution inside the wake is significantly improved.

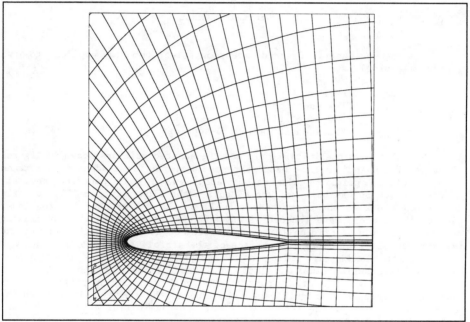

Figure 6[a]   Initial grid for problem I

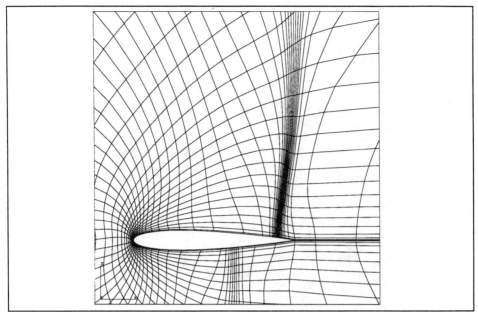

Figure 6^b Adapted grid for problem I after two adaptions.

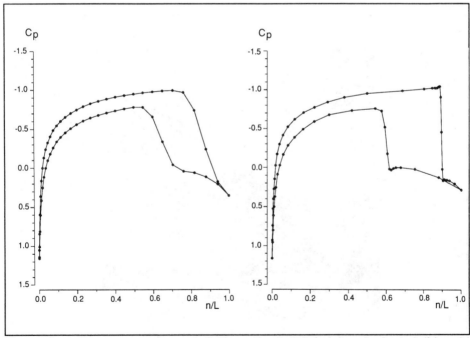

Figure 7 C_p contours along airfoil on the initial (a) and adapted (b) grid of problem I.

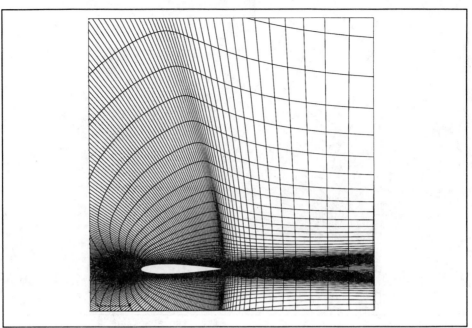

Figure 8ᵃ   Initial grid for problem II.

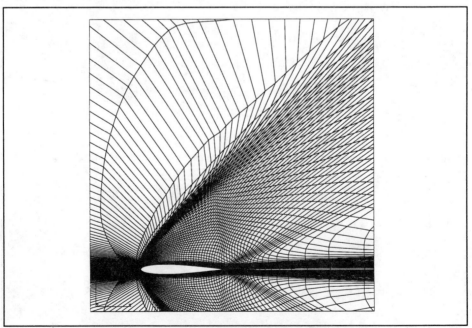

Figure 8ᵇ   Adapted grid for problem II after two adaptions.

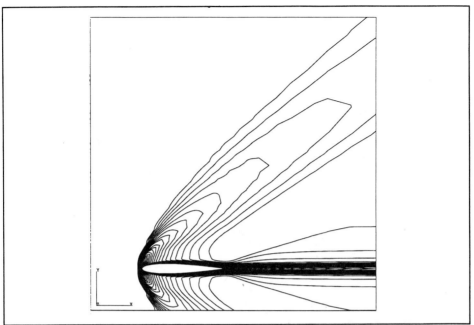

Figure 9ª Mach number distribution on initial grid of problem II.

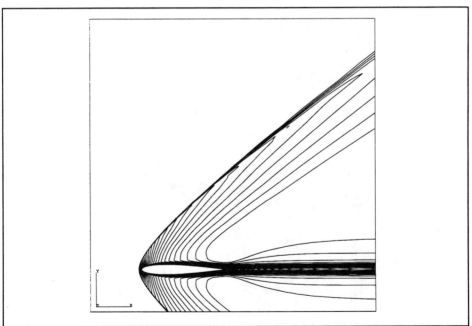

Figure 9ᵇ Mach number distribution on adapted grid of problem II.

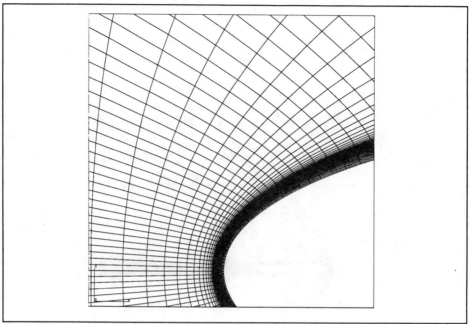
Figure 10^a Leading-edge region of initial grid of problem II.

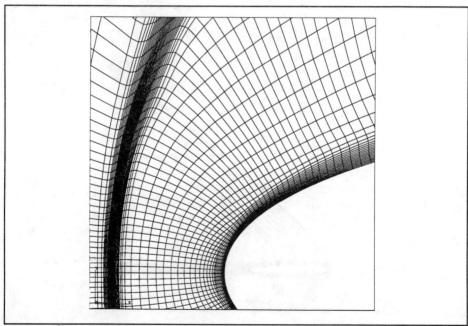
Figure 10^b Leading-edge region of adapted grid of problem II.

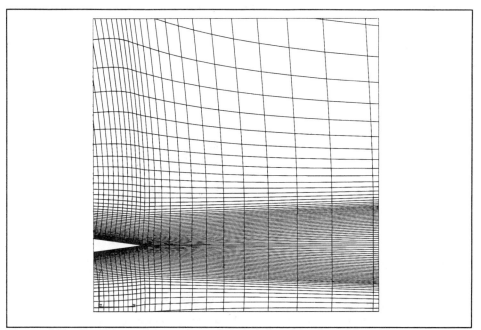

Figure 11[a] Trailing-edge region of initial grid of problem II.

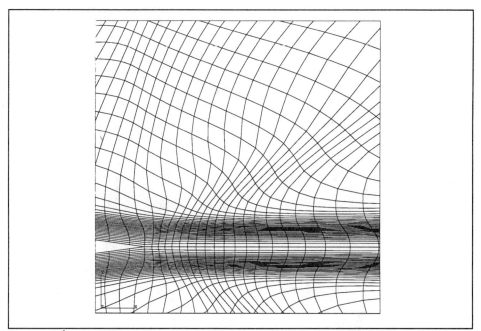

Figure 11[b] Trailing-edge region of adapted grid of problem II.

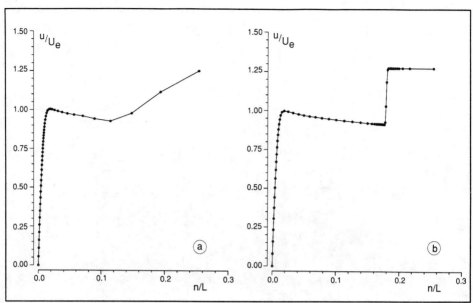

Figure 12 Boundary-layer/ shock-layer velocity profile in normal direction at x/L = 0.1 on the initial (a) and adapted (b) grids of problem II.

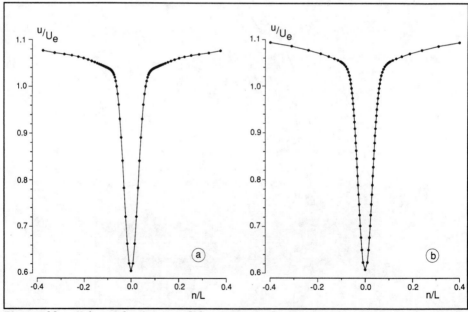

Figure 13 Wake velocity profile in normal direction at x/L = 2 on the initial (a) and adapted (b) grids of problem II.

CONCLUSIONS

A recently developed new grid adaption algorithm has been briefly described. Applications to two aerodynamic problems show that the algorithm is robust. Strong adaptions are obtained in only two adaption cycles, while the adapted grids are regular in the sense that overlapping grid cells do not occur. Also the algorithm is almost fully automatic, only one parameter for grid correction at the trailing edge has to be specified [1]. Although skew cells may be generated, the recalculated flow solutions are significantly improved w.r.t. both shock resolution as well as boundary-layer and wake resolution. Strong clustering of points around the shock position does not result in excessive depletion of points in the boundary layer. Unnecessary stretchings that are present in the initial grid are eliminated upon grid adaption.

ACKNOWLEDGEMENT

The authors wish to thank Dr. B. Oskam for significant contribution to the part on orthogonality analysis and for many useful comments.

REFERENCES

[1] Hagmeijer, R., Grid Adaption Based on Modified Anisotropic Diffusion Equations Formulated in the Parametric Domain, submitted to Journal of Computational Physics, 1992.

[2] Eiseman, P.R. Adaptive Grid Generation, Computer Methods in Applied Mechanics and Engineering, Vol.64, pp 321-376.

[3] Thompson, J.F., A Survey of Dynamically-Adaptive Grids in the Numerical Solution of Partial Differential Equations, AIAA-84-1606, Snowmass, Colorado, USA, 1984.

[4] Dwyer, H.A.,Grid Adaption for Problems in Fluid Dynamics, AIAA Journal, Vol.22, No.12, December 1984.

[5] Nakahashi, K., and Deiwert, G.S., Three Dimensional Adaptive Grid Method, AIAA Journal, Vol.24, No.6, pp 948-954, June 1986.

[6] Anderson, D.A., Equidistribution Schemes, Poisson Generators and Adaptive Grids, Applied Mathematics and Computation, Vol.24, 211-227, 1987

[7] Anderson, D.A., and Steinbrenner, J., Generating Adaptive Grids with a Conventional Scheme, AIAA-86-0427, Reno, Nevada, January 1986.

[8] Brackbill, J.U., and Saltzman, J.S., Adaptive Zoning for Singular Problems in Two Dimensions, Journal of Computational Physics 46, 342-368 (1982).

[9] Winslow, A., Adaptive Mesh-Zoning by the Equipotential Method, UCID-19062, Lawrence Livermore National Laboratories, University of California, 1981.

[10] Lee, K.D., and Loellbach, J.M., A Mapping Technique for Solution Adaptive Grid Control, AIAA-89-2178, Proceedings of AIAA 7th Applied Aerodynamics Conference, pp 129-139, Seattle, WA, July 1989.

[11] Oskam, B., and Huizing, G.H., Flexible Grid Generation for Complex Geometries in Two Space Dimensions Based on Variational Principles, AGARD CP412, Applications of Computational Fluid Dynamics in Aeronautics, Aix-en-Provence, 7-10 april, 1986.

[12] Brandt. A., Multi-Level Adaptive Solutions to Boundary value Problems, Mathematics of Computation, Volume 31, number 138, 1977, pp 333-390.

[13] J.I. van den Berg, and J.W. Boerstoel, Theoretical and Numerical Investigation of Characteristic Boundary Conditions for Cell-Centered Euler Flow Calculations, NLR TR 88124 L.

[14] F.J. Brandsma, and J.G.M. Kuerten, The ISNaS Compressible Navier-Stokes Solver; First Results for Single Airfoils, Proceedings of the 12th International Conference on Numerical Methods in Fluid Dynamics, Springer-Verlag, 1990.

Embedding Within Structured Multi-Block Computational Fluid Dynamics Simulation

S.N. Sheard and M.C. Fraisse
BAe Regional Aircraft Ltd.
Comet Way, Hatfield, United Kingdom.

Summary

Within a structured multiblock environment for CFD simulations, embedding capabilities have been developed by uniformly embedding whole blocks of grid by powers of 2 in each direction independantly, leaving discontinuities in grid density at block boundaries. The problem has two aspects which have been treated independantly. Firstly, the placement of new grid nodes in enriched blocks of grid. Secondly the modification of the cell centered finite volume flow simulation to cater for discontinuities in grid density at block boundaries.

New grid nodes are placed within enriched blocks so as to retain the curvature and stretch of the initial coarser grid. Variants of the Thompson (reverse Laplacian) system of elliptic partial differential equations are used with control functions defined on the initial coarse grid rescaled and applied to the update of new grid node positions of the resulting fine grid.

The flow simulation has been extended to cater for embedded block boundaries by modifying the way in which block boundary (halo) data is set. Halo data for fine blocks is set by iterpolation within neighbouring coarse blocks. That for coarse blocks is set by matching the total flux out of the cell faces bounding the fine block that match a cell face within the coarse block.

1 The Grid Generator

Mesh embedding involves the addition of new grid nodes in locations compatible with the old grid nodes, which are not moved. This is done by variations on the theme of the Thompson algorithm. There are two aspects to this. Firstly the algorithms. Within the block core, the standard 3D Thompson algorithm is used. On block faces only a two dimensional template is available, and a variant of the 2D Thompson algorithm is used. Similarly, on block edges only a one dimensional template is available, and a variant of the 1D Thompson is used. In these variants, the control functions are extended from scalars to 3-vectors and used to position the new nodes in 3-space, thereby defining the shape of the line or surface, in addition to the stretching and curvature within it. The second aspect of the embedding algorithms is the definition of the control functions. Esentially these are defined on the original coarse grid and used on the new fine grid nodes, but they must be rescaled to apply to the new grid density. These algorithms are defined in the following sections.

To summarise, the algorithms used for mesh embedding are as follows.

- Coarsening by node subtraction
- Free field embedding by Thompson point field update. Nodes from the initial grid remain unchanged.
  - block core nodes from 3D Thompson. Local calculation of control functions.
  - block face nodes from 3D extension of 2D Thompson. Vector control functions calculated on initial grid and interpolated.
  - block edge nodes from 3D extension of 1D Thomson. Vector control function calculated on initial grid and interpolated.

– block corner nodes from initial grid
- Geometrical suface embedding by geometrical constraint of motion within free field update algorithm

1.1 Thompson Algorithms

In the following section, we begin by describing the basic 3D Thompson algorithm to define notation, and then go on to describe the face and edge variants used for embedding and singular lines. The definition of control functions, which is a seperate issue, is treated separately in the next sub-section.

1.1.1 Core algorithm

The Thompson grid generation equations are

$$f(\mathbf{r}) = 0 \tag{1}$$

$$f(\mathbf{r}) = \eta_{ij} \frac{\partial \mathbf{r}}{\partial \xi_i \partial \xi_j} + \phi_i \frac{\partial \mathbf{r}}{\partial \xi_i} \tag{2}$$

$$\phi_i = 2\eta_{ii} P_i \tag{3}$$

where P_i are the control functions and the metric is

$$\eta_{ij} = \frac{\partial \xi_i}{\partial x_\alpha} \frac{\partial \xi_j}{\partial x_\alpha} \tag{4}$$

with implied summation on the repeated index α, so that its inverse

$$\eta_{ij}^{-1} = \frac{\partial x_\alpha}{\partial \xi_i} \frac{\partial x_\alpha}{\partial \xi_j} \tag{5}$$

has a discrete approximation which can be expressed in terms of differences between grid nodes positions. The point relaxation method drives the discrete approximation to function f to zero at first order. The Taylor expansion with respect to motion of one node $\mathbf{x}$ from $\mathbf{x}^n$ to $\mathbf{x}^{n+1} = \mathbf{x}^n + \Delta \mathbf{x}^n$ is

$$f\left(\mathbf{x}^{n+1}\right) = f(\mathbf{x}^n) + f'(\mathbf{x}^n) \Delta \mathbf{x}^n + O(\Delta \mathbf{x}^n)^2. \tag{6}$$

Requiring $f(\mathbf{x}^{n+1}) = 0$ specifies point motion

$$\Delta \mathbf{x}^n = -\frac{f(\mathbf{x}^n)}{f'(\mathbf{x}^n)}. \tag{7}$$

The Thompson equations are discretized by the following replacements: first derivatives are replaced by central differences

$$\Delta_{\xi_i} \mathbf{r} = \mathbf{r}(\xi + \Delta_i \xi) - \mathbf{r}(\xi - \Delta_i \xi) \tag{8}$$

$$\rightarrow 2 \frac{\partial \mathbf{r}}{\partial \xi_i} \tag{9}$$

so that second differences have the continuum limit

$$\Delta_{\xi_i \xi_j} \mathbf{r} \rightarrow n \frac{\partial^2 \mathbf{r}}{\partial \xi_i \partial \xi_j} \tag{10}$$

$$n = 1 \text{ for diagonal terms} \tag{11}$$

$$n = 4 \text{ for off-diagonal terms}. \tag{12}$$

The inverse metric discritizes as

$$gsb_{ij} = (\Delta_{\xi_i}\mathbf{r})(\Delta_{\xi_j}\mathbf{r}) \qquad (13)$$

$$\longrightarrow 4\eta_{ij}^{-1} \qquad (14)$$

this discrete form can be inverted by first taking the cofactor matrix so that

$$g = Cof(gsb) \qquad (15)$$

$$\longrightarrow \frac{1}{4}\det\left(4\eta^{-1}\right)\eta \,. \qquad (16)$$

As the discrete form of function f we take

$$\text{residr} = \frac{1}{n} g_{ij}\Delta_{\xi_i\xi_j}\mathbf{r} + g_{ii}P_i\Delta_{\xi_i}\mathbf{r} \qquad (17)$$

$$\longrightarrow \frac{1}{4}\det\left(4\eta^{-1}\right)\left\{\eta_{ij}\frac{\partial^2 \mathbf{r}}{\partial \xi_i \partial \xi_j} + 2\eta_{ii}P_i\frac{\partial \mathbf{r}}{\partial \xi_i}\right\}. \qquad (18)$$

The derivative of this with respect to the central node position is

$$-2\Sigma_{i=1}^{3} g_{ii} \qquad (19)$$

so that the discrete form of the point update is

$$\Delta \mathbf{r} = \frac{\text{residr}}{2(g_{11} + g_{22} + g_{33})} \,. \qquad (20)$$

1.1.2 Edge algorithm

The simplest form of the control functions defined at a boundary assumes orthogonality of all three coordinate directions, and neglects curvature terms.

$$P = -\frac{\Delta_\xi \mathbf{r} \cdot \Delta_{\xi\xi}\mathbf{r}}{\Delta_\xi \mathbf{r} \cdot \Delta_\xi \mathbf{r}} \,. \qquad (21)$$

This can be regarded as the dot product of the normalized first and second differences

$$P = -\hat{\mathbf{r}}_\xi \cdot \hat{\mathbf{r}}_{\xi\xi} \qquad (22)$$

where the size of the first difference is used to normalize both vectors so that

$$\hat{\mathbf{r}}_\xi = \frac{\Delta_\xi \mathbf{r}}{|\Delta_\xi \mathbf{r}|} \qquad (23)$$

$$\hat{\mathbf{r}}_{\xi\xi} = \frac{\Delta_{\xi\xi}\mathbf{r}}{|\Delta_\xi \mathbf{r}|} \,. \qquad (24)$$

When used to calculate the residual as defined above, equation (17), this sets a target value of the component of the second difference in the ξ direction. In one dimension this gives

$$\text{residr} = \frac{1}{n} g \mathbf{r}_{\xi\xi} - g\left(\hat{\mathbf{r}}_\xi \cdot \hat{\mathbf{r}}_{\xi\xi}\right)_t \mathbf{r}_\xi \,. \qquad (25)$$

Where the subscript t indicates a target value. Thus, in one dimension with this simple control function definition, the Thompson equations simply drive the deviation from the target second difference to zero. In three dimensions, the second difference is a vector. Along an edge the tangental component specifies the point distribution along the line, and the two normal components specify the line curvature. In order to extend this 1D Thompson algorithm into

3D, we simply use the target normalized second derivative as a vector control function, rather than using one component of it as a scalar control function.

$$\hat{\mathbf{P}} = -\hat{\mathbf{r}}_{\xi\xi}|_{target}. \tag{26}$$

This gives us the residual

$$\text{residr} = \Delta_{\xi\xi}\mathbf{r} + \hat{\mathbf{P}}|\Delta_\xi \mathbf{r}| \tag{27}$$

so that the point update is simply

$$\Delta \mathbf{r} = \frac{\text{residr}}{2}. \tag{28}$$

This reduces to the 1D Thompson equations in quasi-1D cases. It has been found to be robust. It is fast in the sense that it converges in two or three line relaxation iterations in a fixed background mesh. It is expensive in that both the control function and residual calculations involve square roots. The reason the edge update does not converge fully in one line relaxation iteration is that this square root makes the update of x y and z components interdependant. The first line relaxation step will move all x components simultaneously. Line relaxation on y, and z components are then performed separately.

1.1.3 Face algorithm

The two dimensional variant is derived in the same way as the one dimensional variant described above. The basic control terms

$$P_i = -\frac{\Delta_{\xi_i}\mathbf{r} \cdot \Delta_{\xi_i\xi_i}\mathbf{r}}{\Delta_{\xi_i}\mathbf{r} \cdot \Delta_{\xi_i}\mathbf{r}} \tag{29}$$

are extended to the vector control terms

$$\hat{\mathbf{P}}_i = -\hat{\mathbf{r}}_{\xi_i\xi_i}|_{target} \tag{30}$$

giving the residual

$$residr = \frac{1}{n}g_{ij}\Delta_{\xi_i\xi_j}\mathbf{r} + g_{ii}\hat{\mathbf{P}}_i|\Delta_{\xi_i}\mathbf{r}| \tag{31}$$

so that the discrete form of the point update is

$$\Delta \mathbf{r} = \frac{\text{residr}}{2(g_{11} + g_{22})}. \tag{32}$$

1.2 Control functions

The proposed edge and face algorithms have now been fully specified except for the location of the target at which the control function is calculated. The proposed embedding algorithm would calculate the control function on the coarse grid and interpolate to the new grid nodes on the fine grid.

Whilst we are using the Thompson algorithm for grid embedding in order to retain compatibility with the core algorithm, we feel that the local formulation of control functions has an adequately similar resultant grid. Further, using a local formulation whereby the layout of the initial coarse grid determines both the stretching and curvature of the resultant embedded fine grid, the method may produce adequate embedded grids even when the initial grid generation did not involve the Thompson algorithm.

1.2.1 Scaling

Control functions defined on the coarse grid cannot be directly applied to the fine grid. They must be rescaled. A simple dimensional analsys shows that the control functions are proportional to the cell width in parametric space. A refinement by a factor of 2 would thus result in a rescaling of the control functions by a factor of $\frac{1}{2}$. However, with a stretched grid with no curvature, the simple control term formulation 21 can be expressed in terms of the stretch factor s as

$$P = -\frac{s-1}{s+1} \qquad (33)$$

with an embedding factor of n the stretch factor should be rescaled to $\sqrt[n]{s}$ so that the control term is rescaled as

$$P \rightarrow \frac{\sqrt[n]{P-1} - \sqrt[n]{P+1}}{\sqrt[n]{P-1} + \sqrt[n]{P+1}} . \qquad (34)$$

For small stretch factors, this is equivalent the scaling by a factor of $\frac{1}{2}$.

1.2.2 Core Embedding Control Functions

Embedding within the core of a block is done by the full 3D core Thompson algorithm. In this case, the simplified form of the control functions equation (21) is not sufficient. The control functions are defined by requiring that the residual, equation (17) is zero on the coarse grid. Taking the dot product of equation (17) with the three first difference vectors $\Delta_{\xi_k}\mathbf{r}$ gives

$$P_i g_{ii}(\Delta_{\xi_i}\mathbf{r} \cdot \Delta_{\xi_k}\mathbf{r}) = -\frac{1}{n} g_{ij}\Delta_{\xi_i\xi_j}\mathbf{r} \cdot \Delta_{\xi_k}\mathbf{r} . \qquad (35)$$

This can be expressed as a matrix equation to be solved for the three control function values, as follows.

$$P_i m_{ik} = v_k \qquad (36)$$

$$m_{ik} = g_{ii} g s b_{ik} \qquad (37)$$

$$v_k = -\frac{1}{n} g_{ij}\Delta_{\xi_i\xi_j}\mathbf{r} \cdot \Delta_{\xi_k}\mathbf{r} . \qquad (38)$$

The resultant control functions take account of the curvature and stretch terms in the three grid directions between them.

1.3 Surface Geometry Fix

The algorithms described above operate in physical three-space. The resulting displacement will in general move boundary nodes off geometrical surfaces. In order to constrain points to geometrical surfaces, we need a mapping from physical space displacements to parametric space displacements. The parametric space displacement can then been applied to the parametric coordinates, and the transformation to physical space applied to find the new physical node coordinates. The resulting displacement of physical coordinates should be normal to the surface. Two methods of mappling physical space displacements to parametric space displacements have been tried, and are described below.

1.3.1 Tangental Projection

$$\Delta s = \Delta \mathbf{r} \cdot \frac{\partial \mathbf{r}}{\partial s} \bigg/ \frac{\partial \mathbf{r}}{\partial s} \cdot \frac{\partial \mathbf{r}}{\partial s} \tag{39}$$

$$\Delta t = \Delta \mathbf{r} \cdot \frac{\partial \mathbf{r}}{\partial t} \bigg/ \frac{\partial \mathbf{r}}{\partial t} \cdot \frac{\partial \mathbf{r}}{\partial t} . \tag{40}$$

By tangental projection, displacement $\Delta \mathbf{r}$ is projected onto the surface, converted to a parametric grid displacement $(\Delta s, \Delta t)$ and applied to update the parametric grid.

$$\begin{pmatrix} s \\ t \end{pmatrix} \rightarrow \begin{pmatrix} s \\ t \end{pmatrix} + \begin{pmatrix} \Delta s \\ \Delta t \end{pmatrix} . \tag{41}$$

This was found to overshoot, becoming unstable when the normal displacement exceeds the radius of curvature of the surface.

1.3.2 2D Newton Raphson

This was found to be a more stable alternative. We wish to drive the 2-vector function

$$\mathbf{f}(s,t) = \begin{pmatrix} \Delta \mathbf{r} \cdot \partial \mathbf{r}/\partial s \\ \Delta \mathbf{r} \cdot \partial \mathbf{r}/\partial t \end{pmatrix} \tag{42}$$

to zero by displacing s and t. Linearizing around old parametric coordinates $s_i = (s,t)$ gives

$$\mathbf{f}(s_i + \Delta s_i) = \mathbf{f}(s_i) + \Delta s_j \frac{\partial \mathbf{f}}{\partial s_j} + \mathcal{O}(\Delta s^2) . \tag{43}$$

This two dimensional linear system is solved for $(\Delta s, \Delta t)$, and the displacement applied to the parametric grid coordinates. If $\mathbf{n}$ is the new node position given by the face algorithm, off the surface, and $\mathbf{r}(s,t)$ is the old node position, in the surface, we wish to move $\mathbf{r}(s,t)$ over the surface to a point 'underneath' $\mathbf{n}$. The equations for the function $\mathbf{f}$ and its derivative matrix are as follows.

$$\Delta \mathbf{r}(s,t) = \mathbf{n} - \mathbf{r}(s,t) \tag{44}$$

$$\begin{pmatrix} f_s \\ f_t \end{pmatrix} = \begin{pmatrix} (\mathbf{n} - \mathbf{r}) \cdot \frac{\partial \mathbf{r}}{\partial s} \\ (\mathbf{n} - \mathbf{r}) \cdot \frac{\partial \mathbf{r}}{\partial t} \end{pmatrix} \tag{45}$$

$$\begin{pmatrix} \frac{\partial f_s}{\partial s} & \frac{\partial f_t}{\partial s} \\ \frac{\partial f_s}{\partial t} & \frac{\partial f_t}{\partial t} \end{pmatrix} = \begin{pmatrix} (\mathbf{n}-\mathbf{r}) \cdot \frac{\partial^2 \mathbf{r}}{\partial s^2} - \frac{\partial \mathbf{r}}{\partial s} \cdot \frac{\partial \mathbf{r}}{\partial s} & (\mathbf{n}-\mathbf{r}) \cdot \frac{\partial^2 \mathbf{r}}{\partial s \partial t} - \frac{\partial \mathbf{r}}{\partial s} \cdot \frac{\partial \mathbf{r}}{\partial t} \\ (\mathbf{n}-\mathbf{r}) \cdot \frac{\partial^2 \mathbf{r}}{\partial s \partial t} - \frac{\partial \mathbf{r}}{\partial s} \cdot \frac{\partial \mathbf{r}}{\partial t} & (\mathbf{n}-\mathbf{r}) \cdot \frac{\partial^2 \mathbf{r}}{\partial t^2} - \frac{\partial \mathbf{r}}{\partial t} \cdot \frac{\partial \mathbf{r}}{\partial t} \end{pmatrix} . \tag{46}$$

Three iterations of the 2D Newton Raphson procedure have been found sufficient for a normal projection within the context of grid embedding.

1.4 Results

The grid embedding was tried out on the DLR F5 isolated wing configuration, and gave acceptable results. Figure 1 shows a zoom on the wing tip, with the wing surface, and the boundary surfaces of an embedded block drawn. The figure illustrates how the stretch and curvature of the grid lines in the embedded block are continuous and smoothly varying

2 The Flow Code

The required alteration of the flow code involves the treatment of block boundaries where the level of embedding in directions along the block face is different between blocks. The approach taken is to replace the routine setting the 'halo' data for each block from the flow variables of the neighbouring block. Figure 2 shows a 2D section through a coordinate plane in the region of a block boundary where the level of embedding increases by a power of 2.

The halo data for the coarse block comprises flow field data at locations indicated by letters H1 and H2. That for the fine block comprises flow field data at locations indicated by letters h1 and h2. The primary requirement for the scheme is to set these halo data is the conservation of flux across the interblock boundaries. That is the sum of the fluxes f_1 and f_2 across the fine block faces should equal the flux F_T across the coarse block face. The five flux components are a non-linear function of the five flow variables at neighbouring faces.

2.1 Fine Block Halos

Halo data for the fine block is set by interpolation of the values in the coarse blocks, taking these to be located at the cell centers. The interpolation to the first halo values also involves the values in the first fine block cells. At block edges and corners, some extrapolation is done to avoid the use of flow variables from neighbouring blocks, with the associated data structure management overhead.

2.2 Coarse Block Halos

Having calculated the fine block halos, we now have a flux $f_1 + f_2$ defined across the coarse block face.

$$F_T = f_1 + f_2 . \qquad (47)$$

This can be considered to be a non-linear function of the five flow variables at the first halo position H1. This can be inverted by a 5D Newton Raphson procedure analagous to the 2D Newton Raphson procedure used for geometrical constraint in the grid generator, described above, section 1.3.2. However, the Newton Raphson scheme has a narrow radius of stability. The initial node location in the grid generation procedure is suficiently close for stability, but the initial flow across block boundaries can be outside the basin of stability for the 5D Newton Raphson procedure. Alternative schemes were tried; a point relaxation sheme was found to have numerical problems associated with division by a flux velocity; a pseudo time step approach was found to be more numerically stable, but achieved a lower level of convergence. The scheme implemented in the end involved the 5D Newton Raphson, with a bypass whereby a simple interpolation was used in its place when the Newton Raphson scheme would not converge. The interpolation scheme is not conservative, but should provide the initial stability required to allow the Newton Raphson scheme to take over.

The second level of halo data are set by a similar inversion scheme, ensuring that the flux calculated across the cell faces between halos H1 and H2 is the same whether calculated using halo data at H1 and H2, or using the cell center flow variables between them.

2.3 Results

Problems remain with the robustness of the scheme. Whilst promising results have been obtained with some test cases, convergence and solution quality problems remain with others. The embedded flow code has been tried on two quasi-2D stub-wing geometries, using the RAE5225, and the NACA0012 sections, and on the full 3D DLR F5 isolated wing configuration.

Geometry	Flow	Result
RAE 5225	$M = 0.7\ \alpha = 0.1$	Embedding around the nose. Converged to sensible solution. Level of convergence of maximum enthalpy not as good as the unembedded cases.
RAE 5225	$M = 0.7\ \alpha = 1.3$	Upper surface embedding. Failed to converge
NACA 0012	$M = 0.75\ \alpha = 2.0$	Upper surface embedding. Comverges to an acceptable solution. Slight problem apparent near surface with solution smoothness.
F5	$M = 0.82\ \alpha = 0.0$	Upper surface embedding. Converges to an unphysical solution

A pressure distributon for the NACA 0012 case is shown in figure 3. More validation work is required on quasi-2D configurations to resolve the apparent stability and solution quality problems before further work is done on fully 3D configurations.

3 Conclusions

Variations on the Thompson algorithm have been successfully applied to enrich regions of a three dimensional grid, providing an embedded grid suitable for running structured multiblock computational fluid dynamics simulations on. However, problems remain with the treatment of discontinuities in grid density at block boundaries in the flow code, and these will probably require a change of approach.

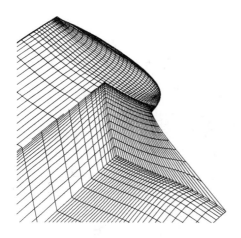

Figure 1: Zoom on wing tip

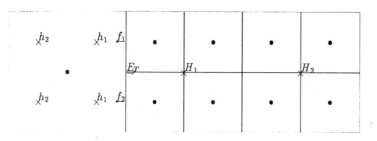

Figure 2: Halo Calculation on an Embedded Block Boundary

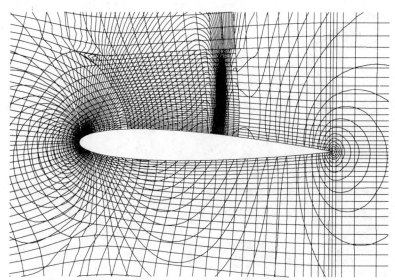
Figure 3: NACA0012 Pressure Distribution. Embedded grid.

ADAPTIVE MESH GENERATION WITHIN A 2D CFD
ENVIRONMENT USING OPTIMISATION TECHNIQUES

A.F.E.Horne
British Aerospace plc, Sowerby Research Centre, Filton, Bristol

SUMMARY

A method for generating optimal solution adaptive computational grids in terms of smoothness, local orthogonality and a volume weighted functional has been decribed. This method is 2 dimensional but it extends to 3 dimensions. The first operation in this grid adaption is to locate all regions which require grid refinement or coarsening. A procedure has been developed which detects where the grid needs adapting. It returns an indicator which has been termed the 'adaptive sensor'. This indicator is scaled and filtered and then used when adapting the grid. The second operation in this grid adaption is to generate the new grid which involves the development of a method to transfer information from the old grid to the new grid. This method is divided into two separate parts. The first part is a search to decide where the new grid point is positioned with respect to the old grid and the second part interpolates the information from the old onto the new grid. Both the interpolation scheme and the search method are discussed.

Using the adaptive grid generation method developed, varying degrees of coupling have been considered. Tests have been carried out on flow over a 10% bump in a channel, ranging from extremely loose coupling (producing adaptive grids externally to the flow code execution) to extremely close coupling (producing adaptive grids following each flow code cycle). The resulting solutions and execution times have been compared and a recommendation made as to the level of coupling to be used for the most efficient use of the grid adaption procedure.

INTRODUCTION

The field of Computational Fluid Dynamics is concerned with the solution of non-linear partial differential equations by numerical means. These equations describe the physics of the flow and because of their non-linear nature, the numerical computation for the flow around a given object must extend over the whole field of flow which is disturbed by the presence of the object. The numerical solution of these governing partial differential equations has two steps: grid generation and numerical integration. The accuracy of a numerical solution depends on both the solution technique and the grid.

It is essential that grid points should be positioned sufficiently close together to resolve the features of the flow to the desired level of accuracy and also that they are oriented relative to each other in such a manner that errors in the flow solution are minimised. On the other hand, computer speed and memory limit the number of grid points that can be used in the solution of a given problem. It is, therefore, necessary to distribute the grid points to maximise the overall accuracy but still to cover the entire region of interest. However, certain features of the solution, for example the location of shock waves, are in general unknown at the time the grid is generated and before the solution is obtained. In these regions the solution will be varying rapidly, but the grid points will not be particularly close together and so the solution errors will be large. There is therefore a requirement for grid generation techniques to be capable of moving points to adequately resolve the features of the flowfield. This has led to the area of adaptive grid generation in which the grid generator and flow solver interact, the subject of the present paper.

Both the grid generator and the flow solver are iterative in nature. Various degrees of coupling between the flow solver and the grid generator are investigated in order to decide where the optimum coupling lies. The different degrees of coupling considered ranged from very loose coupling (producing an adaptive grid externally to the flow code execution) to very close coupling (producing an adaptive grid following every cycle of the flow code).

GRID ADAPTION

Sensing the flow solution

The first operation in grid adaption is to locate all the regions which require grid refinement or coarsening. A method is discussed for locating these regions. This method returns an indicator which has been termed the 'adaptive sensor'. The value of the indicator varies according to the position of the grid point relative to the flow field. The method for calculating this adaptive sensor indicator is constructed so that it detects critical features of the flow field. Typically the solution error is greatest in regions where the flow field gradients are large and so the method chosen to identify important flow field features is a measure of the gradient of some dependent variable Q. Different methods are needed for Euler solvers and Navier-Stokes solvers and these are combined so that both Euler regions of the grid and Navier-Stokes regions which need refining can be located.

For Euler equations, some of the main flow features of the solution can be shock waves, stagnation points and vortices and any indicator should accurately identify these flow characteristics. Several different physical criteria have been tested to see which is the best at identifying the flow characteristics. The physical criteria which have been tested for the Euler equations are [1]:

$$|\, \mathbf{u}.\nabla Q\,|, \qquad |\,\frac{\nabla Q}{Q}\,|^2, \qquad \|\,\nabla Q\,\|$$

where $\mathbf{u}$ is the velocity and Q is Mach number, density or pressure.

For Navier-Stokes equations, the grid needs to be refined, not only in the vicinity of the shocks, but also in the boundary layer regions and so different criteria are used. The criteria tested for indicators of boundary layers and wakes are:

$$|\, \mathbf{u} \times \nabla Q\,|, \qquad |\,\nabla \times \mathbf{u}\,|$$

where $\mathbf{u}$ is the velocity and Q is Mach number, density or pressure.

The adaptive sensor should be able to detect the features of both Euler solvers and Navier-Stokes solvers. Since shear layers and shocks are very different flow phenomena, they are detected by different sensor criteria, therefore a combination of the Euler and Navier-Stokes criteria is necessary. The combined criteria used is $|\mathbf{u}.\nabla Q|$ $+ |\mathbf{u} \times \nabla Q|$. The values of the Euler sensor indicators and the Navier-Stokes sensor indicators may have completely different scales. In order that one criterion does not dominate the other, the indicators need to be scaled before being combined. Another reason for scaling the sensor values is that the range of variation of the calculated values can be too large (for example, from 10^{-8} to 10^3 for $|\mathbf{u} \times \nabla M|$) to be used directly in the grid adaption code.

A scaling procedure proposed by Brackbill and Saltzman [2] has been used. This is described as follows:
Let the adaptive sensor value at grid point (i,j) be W_{ij}
If σ_o is a user supplied parameter ($\sigma_o > 1$) then define:

$$\sigma = min(\sigma_o, \frac{MaxW}{MinW})$$

where $MaxW = max_{ij}W_{ij}$ and $MinW = min_{ij}W_{ij}$
The values of the adaptive sensors are scaled by resetting:

$$S_{ij} = \frac{(\sigma^2 - 1)W_{ij}}{\sigma MaxW} + \frac{1}{\sigma} \quad \forall \ (ij) \ .$$

This determines the range of variation for S.

As well as scaling the sensor value it may also need to be smoothed, since the flow field under analysis may contain unwanted transient features, for example, 'rough' numerical data may have been supplied by the solution or there may be very large localised variations of the adaptive sensor. To remove any 'wiggles' or abrupt changes, but to leave intact the basic results of the adaptive sensor field, a low pass filter is applied. Smoothing is achieved by direct action of a Laplace filter upon the sensor values at each grid point. This filter is applied, at most, a few times rather than being driven towards convergence. The Laplace filter which has been employed is given by the simple Gauss-Seidal relaxation of

$$S_{ij}^{n+1} = S_{ij}^n + \frac{v}{4}(S_{ij+1}^n + S_{ij-1}^n + S_{i+1j}^n + S_{i-1j}^n - 4S_{ij}^n)$$

where S is the sensor value and $0 < v < 1$ and $1 \leq n \leq 3$.

This new scaled and smoothed adaptive sensor value can now be accessed by grid adaption codes.

Grid modification code

A grid adapter has been developed based upon optimisation techniques. A grid generation method has been used which is based upon the optimisation of a global mesh quality functional [3]. This quality functional is defined as a composite weighted measure of departure from smoothness and orthogonality. It is composed of the sum of the local departures from smoothness and orthogonality of each master cell in the grid. In 2 dimensions, the master cell is made up of the four neighbouring elementary grid cells that have the central point $P(x_{ij}, y_{ij}) = P_{ij}$ in common, as shown in figure 1. A master cell is said to be smooth if it has minimal change in area from one of its

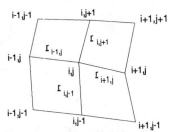

Figure 1: Two dimensional master cell

elementary grid cells to the next. A master cell is orthogonal if the coordinate grid lines passing through P_{ij} intersect at right angles. The new optimal objective function is given by :

$$F = \sum_i \sum_j \alpha ORT_{ij} + (1-\alpha)SM_{ij} + \beta VOC_{ij} \qquad (1)$$

where ORT_{ij} is the local orthogonality and is given by:

$$ORT_{ij} = (\mathbf{r}_{i+1j}.\mathbf{r}_{ij+1})^2 + (\mathbf{r}_{ij+1}.\mathbf{r}_{i-1j})^2 + (\mathbf{r}_{i-1j}.\mathbf{r}_{ij-1})^2 + (\mathbf{r}_{ij-1}.\mathbf{r}_{i+1j})^2.$$

SM_{ij} is the local smoothness which is given by:

$$SM_{ij} = \mid \mathbf{r}_{i+1j} \mid^2 + \mid \mathbf{r}_{i-1j} \mid^2 + \mid \mathbf{r}_{ij+1} \mid^2 + \mid \mathbf{r}_{ij-1} \mid^2$$

and α is a scalar weight parameter enabling a tradeoff between grid smoothness and local orthogonality.

VOC_{ij} is the volume control functional associated with cell ij and β is a scalar weight parameter which indicates how much adaption is required.

Using Brackbill and Saltzman [2] a volume control functional of the general form:

$$VOC_{ij} = A_{ij} \times W_{ij}$$

is sought, where A_{ij} is the area of the master cell centred at the grid point $P(x_{ij}, y_{ij})$ and W_{ij} is the value at P_{ij} of a suitable weight function. If the sum of VOC_{ij} over the master cells is minimised then the master cells will shrink where W is large and expand where W is small. A simpler formulation of VOC_{ij} that involves only the values of the weight function W at the grid points has been used. This simpler formulation of VOC_{ij} over a master cell is defined as follows:

$$VOC_{ij} = w_1 \mid \mathbf{r}_{i+1j} \mid^2 + w_2 \mid \mathbf{r}_{ij+1} \mid^2 + w_3 \mid \mathbf{r}_{i-1j} \mid^2 + w_4 \mid \mathbf{r}_{ij-1} \mid^2$$

where

$$w_1 = \frac{(W_{ij} + W_{i+1j})}{2}, \qquad w_2 = \frac{(W_{ij} + W_{ij+1})}{2},$$

$$w_3 = \frac{(W_{ij} + W_{i-1j})}{2}, \qquad w_4 = \frac{(W_{ij} + W_{ij-1})}{2}.$$

The averaging of the weight function values strongly couples adjacent grid points and therefore prevents the grid from responding to sudden local changes in the weight function on the scale of one grid cell only. This means that very large localised variations of the weight function should not dominate the adaption process. The values of the weight functions W_{ij} are the values of the scaled and filtered sensor indicators described above.

Transfer of flow sensor indicators to new grid

After one iteration of the adaptive optimiser, the grid will be perturbed from its original position and so the values of the adaptive sensor indicators will no longer correspond to grid points. There is therefore a need to transfer information from the old grid to the new grid. This is the process of interpolation of sensor values in the old grid to points in the new grid. The problem can be quantified as follows. Consider a data set of grid points with Cartesian coordinates x, y and grid indices i, j. Associated with each grid point is a sensor value S. Given a point R with cartesian coordinates x_1, y_1, not necessarily coincident with the cartesian coordinates of any of the grid points, a need exists to be able to calculate the value of S at R. The sensor value could be calulated using all the grid points in the data set, however it is felt that this would be very CPU intensive and hence very slow. In order to solve this problem, a small part of the data set near to the point R could be taken and this small number of points could be used to calculate the sensor value. The problem can therefore be divided into two separate parts. The first part is a search which will decide which points in the data set to use in the calculation and the second part will interpolate the sensor value.

Several search methods have been developed which calculate the sensor value at R. Each method used the same interpolation scheme for calculating the sensor value but used different search algorithms to decide which points in the data set to use in the calculation. The method, which was both quick and accurate and could be extended to three dimensions, was then used in the grid adaption code. The choice of points in the data set is determined by the interpolation scheme used for the sensor value and so this interpolation method is discussed first.

A linear model has been used to estimate the sensor value of the point R. For a linear model in two dimensions, three points from the data set are needed to approximate the sensor value at R. In order for the calculation to be accurate, the three points should surround R and so the best points to choose are three points which, when joined by straight lines, form a triangular region containing the point R. Let the three chosen points be $(x_1, y_1), (x_2, y_2), (x_3, y_3)$ with sensor values S_1, S_2, S_3 respectively. For a linear model there are three equations with three unknowns. These equations are of the form:

$$S = ax + by + c.$$

These can be solved for a, b and c and then the sensor value at R can be calculated using

$$S_R = ax_R + by_R + c.$$

This linear interpolation is accurate if R is inside the triangle or very close to it. It is also quick computationally. A bilinear interpolation method was also investigated but it was found that the extra work involved (especially when the method is extended to three dimensions) was too great to justify its use.

The interpolation routine uses a triangle of grid points to calculate the sensor values, but many search methods are based on quadrilateral cells. Thus, after a grid search and prior to interpolation the cell containing the new grid point is split into two triangles and each triangle is searched to see which one contains the point R.

The Newton-Raphson search method has been used in the grid adaption because tests showed that it was a fast and accurate method in both two and three dimensions. The x, y coordinates of the new point R (x_R, y_R) are known and the indices i, j for which the old grid point (x_{ij}, y_{ij}) is in the region of (x_R, y_R) need to be determined. Rather than using indices, continuous variables s, t are used where $s = i$, and $t = j$ at grid points. The old grid coordinates can be considered as functions of s and t and the following equations need to be solved for s and t;

$$x(s,t) = x_R, \qquad y(s,t) = y_R.$$

These can be solved using the iterative Newton-Raphson method

$$s_{n+1} = s_n - (\frac{\partial s}{\partial x}\delta x + \frac{\partial s}{\partial y}\delta y), \qquad t_{n+1} = t_n - (\frac{\partial t}{\partial x}\delta x + \frac{\partial t}{\partial y}\delta y)$$

where

$$\delta x = x(s_n, t_n) - x_R, \qquad \delta y = y(s_n, t_n) - y_R.$$

The Newton-Raphson algorithm is:

- Make a first guess R_o with s, t values of s_r, t_r

- Add $\frac{1}{2}$ to each of s, t values, so s_n, t_n are the grid value of a cell centre

- Calculate s_{n+1}, t_{n+1}

- Round new s_{n+1}, t_{n+1} to nearest odd half integer so that they are the grid value of a cell centre

- Continue until $\mid s_n - s_{n+1} \mid < 0.5$ and $\mid t_n - t_{n+1} \mid < 0.5$ i.e. the correct cell has been found

- Split the 2D cell into 2 triangles and find out which triangle R is in

- Interpolate sensor value for R.

For the majority of points in a grid, the Newton-Raphson search method will result in the correct cell being found. However if one of $\mid s_n - s_{n+1} \mid$ or $\mid t_n - t_{n+1} \mid$ is very close to 0.5 then, because these values are only approximations, it is possible that the next guess may move from one neighbouring cell to another and back again in an infinite loop. This is checked for and, if it occurs, both cells are searched to find the position of R and the sensor value interpolated. This method is very quick and usually finds the correct cell after only one or two iterations, however it will break down if the grid is too irregular because the approximations for the partial derivatives will not be accurate. In this case, it is better to use another search method, for example, the Simplex method [4].

Iteration of the adaption process

Once the values of the sensor indicators have been interpolated at the new grid points then the adaption process can continue. After each iteration of the grid adaption code, the sensor indicators are interpolated onto the new grid from the original sensor indicators calculated from the flow solution. This process ends when the adapted grid has converged.

COUPLED SOLUTION AND GRID ADAPTION

In order to improve the solution produced by the flow solver, it needs to be run on the adapted grid. When the grid adaption has finished, the flow solution on the original grid is interpolated onto the adapted grid. Using this new solution, the flow solver is restarted. Figure 2 gives a diagramatic description of the coupling procedure. Varying degrees of coupling are considered, ranging from the extremes of very loose coupling i.e. producing adaptive grids externally to the flow code execution, to very close coupling, i.e. producing adaptive grids following each flow code cycle, in order to decide what the best level of coupling is in terms of both efficiency and accuracy.

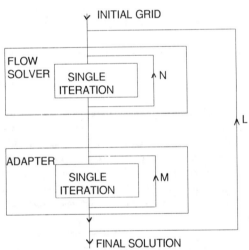

Figure 2: Overview of coupling

Quantification of Solution Accuracy

In order to be able to judge the improvements in the solutions on different adapted grids, a very accurate 'gold standard' solution is needed. The test case used to compare different degrees of coupling was a 17 ×65 mesh describing a 10% bump in a channel. For this test case a very fine grid was generated (65 ×257) and the flow solver run on this fine grid in order to produce the gold standard solution. The gold standard solution and the solution on an adapted grid are compared by defining the error over the whole adapted grid as :

$$error \approx \sum_i \sum_j (P_F - P)^2 \delta A$$

where P_F and P are the pressure at the point (x_{ij}, y_{ij}) of the gold standard solution and the solution on the adapted grid respectively and δA is the mean area of the grid cells of which (x_{ij}, y_{ij}) is a vertex. The error is calculated at various stages in the adaption process and a graph can be drawn of the error versus the number of iterations. The graphs for different adapted grids can be compared to see which degree of coupling gives the least error for the minimum number of cycles of the flow solver.

A better method of determining whether it is worth doing a lot of adaption is to consider the actual effort involved in running both the flow solver and the grid adapter. A graph of the error against the time taken to run both the grid adapter and the flow solver is plotted and the different levels of coupling compared.

Results from investigating optimum coupling

Various degrees of coupling were tested on a 17 × 65 mesh describing a 10% bump in a channel. The solution obtained was for an incoming Mach number of 0.675.

The flow solver used in this test-case was grid dependent to some extent. To prevent the grid from becoming too distorted for the flow solver, an extra weight parameter has been added to the $(1-\alpha)$ weight parameter for grid smoothness (Equation 1). The local smoothness at each grid point is now also multiplied by the adaptive sensor indicator at that grid point and so the weight parameter for smoothness SM_{ij} is now $(1-\alpha+S_{ij})$. This means that if the adaptive sensor indicator has a large value then the smoothness measure will have a greater effect than when the adaptive sensor has a small value and so a high quality grid will be generated in the regions of high activity.

Figure 3 shows the errors calculated against the actual effort involved in running both the flow solver and grid adapter for different degrees of coupling. Both extremes have been calculated i.e. no adaption (N=1500, M=0, L=0 in figure 2) and adaption after every cycle of the flow solver (N=1, M=50, L=1500), as well as several intermediate degrees. In timing tests, it was found that one cycle of the flow solver took seven times as long as one iteration of the grid adapter. Each time the grid was adapted, the grid adapter code ran for fifty iterations which is equivalent in time to seven cycles of the flow solver. Figure 3 was produced by increasing the number of cycles of the flow solver by seven each time the grid was adapted and plotting the error calculated against this new number of cycles. As can be seen, for the grid which was adapted after every cycle of the flow solver, it takes a lot longer for the solution error to become less than on a grid with no adaption. However for the other degrees of coupling, especially adapting every 45 cycles and every 100 cycles of the flow solver, the solution is better than the solution on the original grid with considerably less work.

Figure 4 shows the original grid, the final grid after adapting every 45 cycles and the fine gold standard grid and the Euler flow solutions in terms of Mach number contour lines on these three grids. The shock is much sharper on the gold standard grid and the adapted grid than on the original grid.

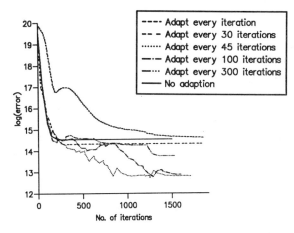

Figure 3: Graphs of error against total effort running flow solver and grid adapter

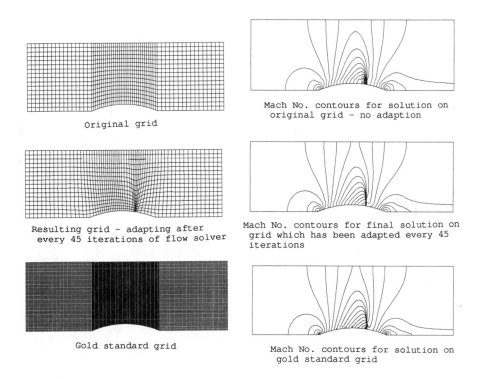

Figure 4: Comparison of grids and solutions for different degrees of coupling

CONCLUSIONS

A code has been developed which will identify regions which need to be adapted, adapt the grid, interpolate the flow solution on to the new grid points and return the solution data and new grid points to a flow solver so that it can be restarted. Tests have been carried out on the flow over a 10% bump in a channel, using varying degrees of couping ranging from the extremes of no adaption to adapting after every iteration of the flow solver and their solutions have been compared with a gold standard solution. The results show that an intermediate degree of coupling (in this particular case adapting after every 45 iterations of the flow solver) gives the best solution for the minimum amount of effort.

REFERENCES

[1] Bristeau,M.O. and Periaux,J. 'Finite element methods for the calulation of compressible viscous flows using self-adaptive mesh refinement', VKI Lecture Notes on CFD, March 1986.

[2] Brackbill,J.U. and Saltzman,J.S. 'Adaptive Zoning for Singular Problems in Two Dimensions',J.Comput. Physics 46(1982).

[3] Carcaillet,R. 'Generation and Optimisation of flow-adaptive computational grids', MSc Thesis University of Texas at Austin, 1986.

[4] Pierre,D.A, 'Optimization Theory with Applications', ISBN 0-486-65205-X, 1986.

TWO DIMENSIONAL MULTI-BLOCK GRID OPTIMISATION BY VARIATIONAL TECHNIQUES

M.R.Morris

British Aerospace Plc, Sowerby Research Centre, Filton, Bristol

SUMMARY

A method for generating optimal two dimensional structured grids based on a method developed by Kennon and Dulikravitch is described. The global quality of the grid is calculated as a linear weight of the sum of the local orthogonality and smoothness measures. A conjugate gradient method is used to optimise the quality function by varying the distribution of grid points. Special consideration is given at the boundaries to ensure that the grid points can move freely along the boundary to optimise the local quality measures. Results are presented for some basic test cases.

The basic algorithm is extended to a two dimensional Multi-Block structure. Connectivity across internal block boundaries is resolved to ensure that orthogonality and smoothness are retained between blocks. Results are given that illustrate the applicability of the method.

Both versions of the code can readily be extended to three dimensions with the minimum of extra effort.

INTRODUCTION

In order to solve the physics of the flow of a fluid around a geometry it is necessary to solve a set of non-linear partial differential equations (NLPDE). It is often not possible to solve the equations exactly, so some numerical approximation method is used. The differential equations are approximated by difference equations evaluated over small, discrete, intervals in both the x and y directions (in two dimensions). The physical domain is discretised apriori using a two dimensional structured mesh of points. Structured grids are used for their inherent connectivity, making it simpler for the numerical scheme to step through the mesh resolving the NLPDEs at each point by evaluating the difference equation using the current point and the surrounding neighbours.

The number of neighbouring points used is dependent on the particular numerical scheme. For basic two dimensional schemes, five point stencil (figure 1a) schemes are common. The NLPDE at (i,j) is resolved by solving the difference equation approximation using the surrounding discrete points $(i-1,j)$, $(i+1,j)$, $(i,j-1)$, $(i,j+1)$.

Approximation schemes, often by the very nature of their approximation, result in a solution that is slightly different from the analytic solution. These differences are called truncation errors. The magnitude of the truncation error is dependent on the scheme used and the level of discretisation.

The magnitude of the truncation error can be reduced through a number of methods, but there are three basic ones as defined by Anderson[1]:-

1. Grid points should be closely spaced in the physical domain where large flow changes are expected.

2. The rate of change of grid spacing should be as smooth as possible to provide continuous transformation derivatives.

3. Excessive grid skewness should be avoided, as this sometimes exaggerates truncation errors.

Point 1. relates to grid adaption which require some prior knowledge of where the flow features will occur. This topic is discussed in greater detail in other papers in this publication and in [2].

Points 2. and 3. relate to the physical placement of the grid points relative to the surrounding points. Grids with fewer points, but where these points are distributed in order to satisfy the two criteria, should provide solutions as accurate as finer grids but with a greatly reduced computational effort.

If the two factors, smoothness and orthogonality, were quantifiable measures then an overall quality function could be defined for the grid. An increase in the quality would result in an decrease in the magnitude of the truncation error.

Since the quality is defined as a function of the grid, then the value of the function will vary as each grid point is moved. Thus the quality function has (2*n) parameters (where n is the number of points and the 2 represents the x and y movement).

The problem of reducing the truncation error becomes one of maximising the quality function by varying the distribution of the grid points. The techniques to solve this problem are known as grid optimisation through variational techniques.

BASIC ALGORITHM

The method proposed in this paper is an optimisation method through grid variation. This procedure is based on a method devised by Kennon and Dulikravitch [3,4].

The basic aim is to optimise the quality function by moving the grid points within the physical domain.

The quality of the grid is defined as a global quality measure which, in turn, is defined as the sum of the quality measure at each grid point.

The quality measure of a grid can be calculated in a variety of ways, but for the optimisation purposes the following measures are used.

$$F = \alpha RHO + (1 - \alpha)SMO$$

where α is a linear weighting factor $(0 \leq \alpha \geq 1)$, RHO is a global orthogonality measure and SMO is a global smoothness measure.

For the measures chosen, maximisation of the grid quality is obtained by minimising function F.

There are a number of formulations that could be used to express the global orthogonality of the grid. The one chosen can be described as the sum of the local orthogonality measures.

$$RHO = \sum_i \sum_j RHO_{ij}$$

where the local orthogonality measure is the sum of the squares of the dot product

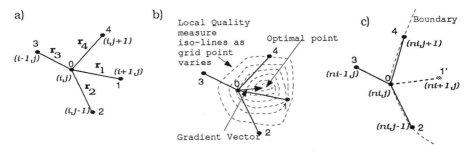

Figure 1: a) Five point stencil b) function as node point varies c) boundaries

of the pairs of vectors defined by the five point stencil shown in figure 1a.

$$RHO_{ij} = <r_1, r_2>^2 + <r_2, r_3>^2 + <r_3, r_4>^2 + <r_4, r_1>^2$$

In the physical sense each term in the formulation is the square of the angle between each pair of vectors. As the grid point is moved to the ideal location the measure reduces to zero (figure 1b). The resulting orthogonality measure is a fourth order function in x and y.

Similarly the global smoothness measure is the sum of the local smoothness measures.

$$SMO = \sum_i \sum_j SMO_{ij}$$

where the local smoothness measure is the sum of the lengths of the four vectors defined by the five point stencil. As the grid point moves to the ideal location the smoothness measure reduces (figure 1b). The resulting smoothness measure is a second order function in x and y.

$$SMO_{ij} = cl_1 <r_1, r_1> + cl_2 <r_2, r_2> + cl_3 <r_3, r_3> + cl_4 <r_4, r_4>.$$

The functions $cl_i(\forall i = 1, 4)$ is the clustering control function, it can be used to cluster the grid points to logical lines within the physical domains.

The above function, F, can be defined as a fourth order function with 2*ni*nj degrees of freedom. The aim of the variational method is to maximise the quality of a grid as efficiently as possible by minimising (optimising) the function F through the variation of the grid points. The most suitable optimisation process for systems with several degrees of freedom is the Conjugate Gradient (CG) method. There are a variety of CG algorithms [5] (e.g. Hestenes and Stiefel (HS), Davidon Fletcher Powell (DFP), Fletcher Reeves (FR))

The HS method was developed for the solution to a set of simultaneous linear equations with symmetric positive definite matrices which makes it unsuitable for systems as complex as the one defined above. The DFP method is probably the best CG optimisation process, but requires the storage of the Hessian matrix which can be costly to compute and store. Thus the advocated method for the system defined above is the FR method.

The advantages of using the FR method are the following:-

- It is a step-size method, therefore only gradient information of F is required.
- The convergence rate is increased by using mutually conjugate search directions, with information from the previous steps being used to determine the current search direction.
- There exists a defined convergence theorem that ensures that, for a quadratic function that is positive definite, a solution will be reached in df iterations where df is the degrees of freedom (2*ni*nj in the above system).

GRADIENT CONTRIBUTIONS

At each grid point the local quality measures RHO_{ij} and SMO_{ij} are differentiated with respect to the five points defining the stencil of figure 1a.
These differentiated values are then added to the relevant gradient value, to produce the gradient contribution for each grid point. This process ensures a distribution of information between neighbouring grid points. The differentiated values are summed to form the gradient values in the following manner.
The five differentiated values for point (i,j) or (x_0, y_0) are defined as :-

$$\frac{\partial SMO_{ij}}{\partial x_k}, \frac{\partial SMO_{ij}}{\partial y_k}, \frac{\partial RHO_{ij}}{\partial x_k} and \frac{\partial RHO_{ij}}{\partial y_k} \quad \forall k = 0, 1, 2, 3, 4 .$$

Since point 1 of (i,j) is the same as point 0 of (i+1,j) and point 2 is the same as point 0 of (i,j-1) etc. then the derivative value at point 1 forms part of the gradient contribution to point (i+1,j). Similarly the derivative value of the quality measure of point (i+1,j) differentiated with respects to point 3 (i.e $\frac{\partial SMO_{(i+1,j)}}{\partial x_3}, \frac{\partial SMO_{(i+1,j)}}{\partial y_3}$ etc.) is added to the gradient contribution of the grid point at (i,j). As in figure 2 The gradient contribution at point (i,j) is defined as the

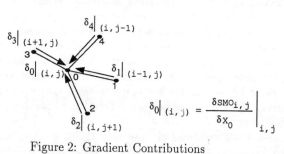

Figure 2: Gradient Contributions

following:-

$$g_x|_{(i,j)} = \frac{\partial SMO_{(i,j)}}{\partial x_0}|_{(i,j)} + \frac{\partial SMO_{(i+1,j)}}{\partial x_3}|_{(i+1,j)} + \frac{\partial SMO_{(i-1,j)}}{\partial x_1}|_{(i-1,j)} +$$
$$\frac{\partial SMO_{(i,j-1)}}{\partial x_4}|_{(i,j-1)} + \frac{\partial SMO_{(i,j+1)}}{\partial x_2}|_{(i,j+1)}$$
$$+the\ terms\ for\ RHO$$

where for example $\frac{\partial SMO}{\partial x}\big|_{(i,j)}$ represents the differentiation of SMO with respect to x and evaluated at point (i,j).

Similarly for g_y where x is replaced with y.

BOUNDARY CONDITIONS

There are two issues to consider along the boundary.

GRADIENT CONTRIBUTIONS ALONG THE BOUNDARY

The local quality measure and the gradient contribution require the surrounding points of a grid node for calculations. At boundaries the five point stencil is not complete (figure 1c), a piece of information is missing (two pieces at the corner nodes). Without this piece of information the gradient contribution tends to point out from the boundary, suggesting that the boundary point will move orthogonally out from the boundary.

Ideally, the boundary point should move along the boundary, implying that the missing grid point should be fixed such that the gradient contribution lies along the boundary.

It can be shown that the position of the grid point that satisfies this criterion is the result of a reflection of the point in the interior through the boundary (figure 1c).

MOVEMENT ALONG THE BOUNDARY

To allow points to move along the boundary it is necessary to first define the boundary as a mathematical function.

Initial definition was as straight line segments defined by the original grid distribution along the boundary. For simplistic (low curvature) geometries this definition is sufficient. For more complex geometries it is necessary to define the boundary as higher order polynomials. Ideally the geometry would be defined as an analytic function, the derivatives of which could be obtained by simple differentiation. In practice the definition of complex geometries by analytic functions is very difficult and often impossible.

The next best solution is to define the boundaries as cubic splines[6,7]. These are third order polynomials defining geometric patches. Only the point and first derivatives can be used to any level of accuracy which is sufficient for the needs of movement along boundaries. A higher order method would prove too computationally intensive to implement.

The gradient contributions of the quality function, mentioned previously, provide a vector along which the new grid node lies. This vector is tangential to the boundary at the current grid point (figure 3a). In most instances the grid point will move off

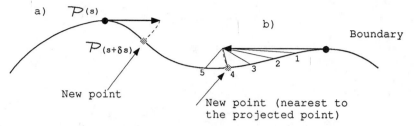

Figure 3: Boundary Movement and projecting points onto the boundary via:- a) Spline mapping b) Searching

the boundary, so it is necessary to project the point back onto the boundary. This can be performed in a variety of ways depending on the level of accuracy required.

The simplest method is to map the distance moved along the tangential vector into the spline space (figure 3a). $\mathcal{P}(s)$ is the current point, δs is distance along the vector and $\mathcal{P}(s + \delta s)$ is the new point on curve.

Another method is to step along the spline in small increments until the point on the curve nearest to the new point is found. This point on the curve is then set as the new point (figure 3b).

There are higher order methods but these generally involve lengthy computation.

MULTI-BLOCK CONTROL CODE

Discretising a realistic physical domain by a single structured grid can be complicated (or impossible). Even when the domain is discretised it is possible that regions may exist that have very skewed cells or un-smooth spacing. Depending on the physical nature of the geometry, it may be impossible to improve the grid.

A simpler solution is to discretise these types of domains by breaking the domain into smaller, easier to grid *chunks*. The only constraint imposed on these chunks or blocks is that the interfaces between blocks should be simple; and that the connectivity from block to block can be resolved. This technique is known as Multi-Blocked Structured Grids.

The basic grid optimisation code can be readily extended to a Multi-Block scheme. The global quality measure is calculated as the sum of the global quality measures of each block. It is necessary to ensure that internal block boundaries are resolved across blocks. This involves extending the five point stencil through blocks, and ensuring that coincident points have the same gradient values (figure 4).

Depending upon the constraints imposed by the block inter connectivity rules, singularity points can occur at the corners of blocks. A singularity point is a point that has an uneven number of curvilinear lines passing through it. The normal number of lines is two as in figure 5a. Figures 5b and 5c illustrate singularity points. It can be seen that the five point stencil is under or over specified at these points.

The displacement value is calculated for each singular point in each block, then the coincident singular points are averaged over the number of blocks and the values are updated for consistency.

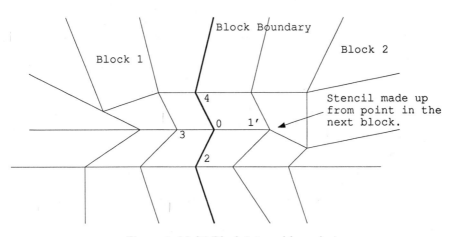

Figure 4: Multi-Block internal boundaries

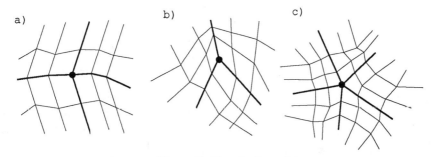

Figure 5: Singularity points

RESULTS

Results for the single block code are shown in figures 6, 7 and 8.

Figure 6 shows the robustness of the code. The initial grid is severely overlapped, the code unravels this within about twenty iterations. The grid is fully recovered after sixty iterations. The graph shows how the value of F decreases as the code progresses. It can be seen that after about forty iterations the grid is sufficiently optimised.

Figure 7 shows tests performed on a ten percent bump grid. The initial grid was developed by simple line interpolation between the boundaries (figure 7a). The next grid is that obtained after the optimisation process, the α linear weighting value is set to 0.5. Both smoothness and orthogonality have an equal effect on the quality function. The grid starts to fan out in order to improve the overall smoothness of the grid. Figure 7c is the optimised grid with $\alpha = 0.0$ (i.e. only smoothness imposed). By imposing just smoothness the fanning effect is larger as there is no orthogonality criteria at the upper boundaries. It appears that orthogonality has been imposed on the lower boundary around the bump but it is more likely that the necessity to produce a smooth grid in the interior causes a bulging around the bump.

In contrast Figure 7d is an optimised grid with $\alpha = 1.0$ which imposes just orthogonality on the grid. By looking at the grid it becomes quite apparent that no smoothness is imposed and the grid is truly orthogonal.

The convergence graph (figure 7e) illustrates one of the axioms of the Fletcher-Reeves optimisation process. Namely that the optimisation of quadratic functions (i.e. just smoothness) converges much quicker than for function of a higher order (i.e. orthogonality).

Figure 8 shows an initial and an optimised grid around a NACA0012 $\alpha = 0.5$ in the optimised function which improves orthogonality and smoothness.

Results for the Multi Block code are given for a group of four random blocks (figure 9), a Multi-Blocked car grid (figure 10) and a Multi-Blocked NACA0012 (figure 11). Notice how the Multi-Block NACA0012 is identical to the single block NACA0012 (figure 8) implying that the Multi-Block inter-block boundaries are functioning correctly.

CONCLUSIONS

The method of optimisation of grids through variational techniques is a simple algorithm to implement and quite good results can be obtained in a relatively short time. Problems occur when trying to impose geometric constraints on the boundary points. Movement of points along lines is very simple, higher order curves become more difficult which requires the introduction of various techniques to stop points from moving too far.

The Multi-Block version of the code illustrates how easy the concept of moving boundaries can be if no geometric constraints are imposed. The inter-block boundaries move through the physical domain with relative ease. It is when points are restricted that problems start to occur.

Extension to three dimensions or grid adaption is a straight forward procedure that involve the inclusion of an extra dimension or the addition of an extra quality measure respectively [2].

The algorithm is based on a fast optimisation method that is capable of improving poor quality grids in a known number of iterations (2*ni*nj). Once the problems associated with geometric constraints are solved the method will prove an invaluable tool for grid improvement.

REFERENCES

[1] Anderson, D.A., 'Computational Fluid Mechanics and Heat Transfer', ISBN 0-89116-471-5, 1984

[2] Horne, A.F.E., 'Sub-Task 4.3, Adaptive Mesh Generation within a 2D Computational Fluid Dynamic Environment using Optimisation Techniques', JS12016, March 1992.

[3] Carcaillet, R., 'Generation and Optimisation of flow-adaptive computational grids', Msc. Thesis University of Texas at Austin, 1986.

[4] Kennon, S.R. and Dulikravitch, G.S., 'Generation of Computational Grids using Optimisation', AIAA J., 24, 1069-1073, July 1986.

[5] Pierre, D.A., 'Optimisation Theory and Applications', ISBN 0-486-65205-X, 1986.

[6] de Boor, C. 'Bicubic spline interpolation', J. Math. Phys., 41, 212-218, 1962.

[7] Lancaster, P., 'Curve and Surface Fitting: An Introduction', ISBN 0-12-436061-0, 1986.

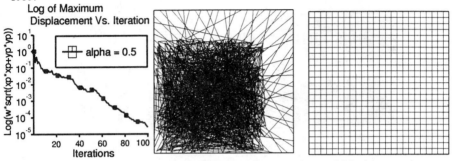

Figure 6: Randomised grid.   Optimised after 100 iterations.

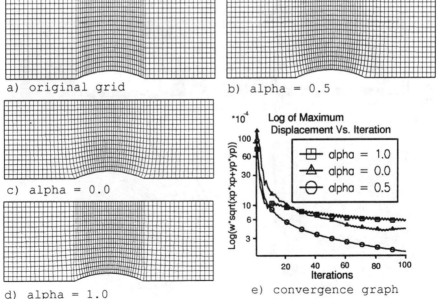

Figure 7: 10% Bump Case.

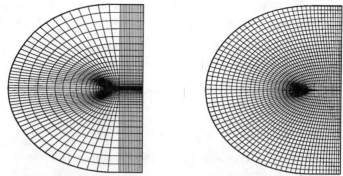

Figure 8: NACA0012 Optimised after 100 iterations

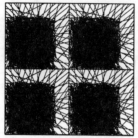 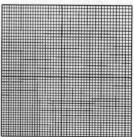

Figure 9: Randomised Multi-Block squares

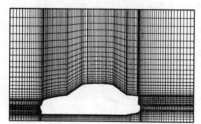 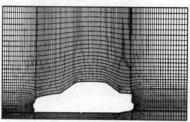

Figure 10: Multi-Blocked Car Grid. Using Eight Blocks

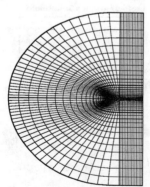 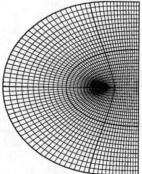

Figure 11: Multi-Blocked NACA0012 grid. Using Eight Blocks

LOCAL MESH ENRICHMENT FOR A BLOCK STRUCTURED 3D EULER SOLVER

Thilo Schönfeld

C.E.R.F.A.C.S.

European Centre for Research and Advanced Training in Scientific Computing
42 avenue Gustave Coriolis, F-31057 Toulouse Cedex, France

SUMMARY

This article summarizes the work completed at C.E.R.F.A.C.S. in sub-task 2.5 of the BRITE/EURAM EuroMesh project concerning the mesh embedding method. An unstructured multi-block data environment is introduced to manage point enrichment on multiple levels for three-dimensional structured meshes. The local grid refinement technique has been incorporated into a finite volume program that solves the Euler equations. Flow adaption is utilized to cluster points automatically in domains of interest.

1 INTRODUCTION

One essential challenge in computational fluid dynamics is to resolve the conflicting requirements of accuracy and efficiency. The solution quality of flow calculations is strongly related to the number of mesh points of which an increase means an improved resolution of flow features; a benefit which, however, is achieved at the expense of higher computational costs. In general dominating salient flow features like shocks, cover only a relatively small part of a typical flow field, whereas large regions are of low activity. One promising way to improve the resolution locally is to add grid points exclusively to domains of physical interest. This method, called *local grid refinement*, is of particular importance for three-dimensional problems for which limiting the number of grid points is crucial.

Over time, two principal strategies have evolved for the local refining of grids. In the more "classical" grid embedding approach the overall physical domain is covered by nested fine grids. Often the zonal fine grids are modular and stored independently from the global mesh with its own solution vectors. Early publications dealing with this type of varying grid resolution for two-dimensional problems are presented in e.g. [1]. However, this approach suffers from its inflexibility to align divided cells e.g. along a shock. This disadvantage is overcome in the alternative unstructured grid refinement (or point enrichment) technique. With this approach the rigidity of the ordered (i, j, k)-index system is broken and replaced by a system of pointers (see amongst others [2], [3]).

We propose a method that offers high flexibility for the local refinement of the grid and permits the refinement of arbitrarily shaped three-dimensional regions. Since in general one does not know a-priori where to refine the grid, flow adaption is used. Written in multi-block formulation the program permits the application to any given configuration.

2 THE BASIC PROGRAM

The grid refinement technique has been incorporated into an unstructured finite volume program which solves the three-dimensional Euler equations on regular grids of hexahedral cells.

For unsteady compressible flows in three space dimensions, the integral form of these equations is

$$\iiint_\Omega \frac{\partial}{\partial t} \vec{W} \, d\mathcal{V} = - \iint_{\partial \Omega} \overline{\overline{F}} \cdot \vec{n} \, dS \qquad (1)$$

where $\vec{W}$ is the vector of conservative variables, and $\overline{\overline{F}}$ is the convective flux tensor

$$\vec{W} = (\rho, \rho u, \rho v, \rho w, \rho E)^T \quad , \quad \overline{\overline{F}} = (\rho \vec{q}, \; \rho u \vec{q} + p_x, \; \rho v \vec{q} + p_y, \; \rho w \vec{q} + p_z, \; \rho H \vec{q})^T.$$

Here $\vec{q}$ denotes the vector of the Cartesian velocity components $u, v,$ and w, ρ the density, E represents the total energy and $H = E + p/\rho$ the total enthalpy. An arbitrary control volume is described by the volume Ω, its surface $\partial \Omega$, and the outward pointing vectors $\vec{n}$ normal to the boundary. The system is closed by the equation of state that relates pressure p to the primitive variables.

2.1 Finite Volume Technique

This integral form of the conservation law is the starting point of the finite volume method for the quantities describing the fluid flow. Applying the mean value theorem to Eq.(1) and introducing vector $\vec{Q}_i$ of the fluxes through the surface of cell $\mathcal{V}_i$ yields

$$(\frac{\partial}{\partial t} \vec{W})_i = - \frac{1}{\mathcal{V}_i} \sum_{k=1}^{6} \vec{Q}_{i_k} \qquad (2)$$

which indicates that the surface integral can be approximated by the sum over the six faces of the control volume. In analogy to the common (i, j, k)-description of the structured counterpart here discretized points, or mesh nodes, are denoted by a single index. Using a forward difference quotient in the time direction, Eq.(2) can be expressed by the following semi-discrete approximation:

$$W_i^{n+1} = W_i^n - \frac{\Delta t}{\mathcal{V}_i} P(W_i^n) \qquad (3)$$

where $P(W)$ represents the spatial discretization operator. The resulting coupled system of nonlinear ordinary differential equations is integrated separately in spatial and temporal direction. To advance solution in time an explicit *Runge-Kutta* three-stage scheme is employed in connection with the local time stepping convergence acceleration technique.

2.2 Cell-Vertex Space Discretization

For the evaluation of the fluxes we choose the *cell-vertex* space discretization with the primitive variables associated with the mesh nodes. The flux vectors located at the four vertices of each cell face are averaged to obtain a value at the midpoint of the face. Summing up the fluxes across the six cell faces yields the mean value of the rate of change at the cell center. Next this cell-centered value is related to the cell vertex. For this let us define the right hand side of Eq.(3) as the change or flux-residual $\mathcal{R}_i$ of cell i. The residuals of the eight cells that surround node N are averaged. In our work we follow the stencil proposed by Hall [4] who adopted a scheme using weighted averages

$$(\frac{\partial}{\partial t} \vec{W})_N = \frac{\sum_{k=1}^{8} d_k \mathcal{V}_k \mathcal{R}_k}{\sum_{k=1}^{8} \mathcal{V}_k} \qquad (4)$$

with d_k denoting distribution coefficients. Schemes for which $\sum_{k=1}^{8} d_k = c$ with the same constant c for each cell are said to be conservative [5]. Conveniently c is scaled to one and each of the eight surrounding cells contributes to the node residual by $d_k = 1/8$.

2.3 Formulation of Artificial Viscosity

In order to preserve numerical stability the well-known dissipation model consisting of blended second and fourth differences is modified into an unstructured formulation [6]. The diffusive Laplacian D^2 is obtained by summing up the m differences between the conservative variables at node N and its nearest neighbors k

$$D_N^2 = \sum_{k=1}^{m} \varepsilon_N^2 \, \Delta p_N \, (q_k - q_N) \frac{(a_k + a_N)}{2}, \quad D_N^4 = \sum_{k=1}^{m} \varepsilon_N^4 (D_k^2 - D_N^2) \frac{(a_k + a_N)}{2}. \quad (5)$$

While D^2 is applied only to domains of high gradients, the bi-harmonic background smoother D^4 acts as a overall smoother in the regions of low flow activity.

2.4 Boundary Conditions

For cell-vertex discretization, boundary values are located directly on surfaces. To impose the *solid wall* boundary condition, semi-control volumes are utilized with the node to be approximated on one face. The slip condition $\vec{V} \cdot \vec{n} = 0$ for inviscid flow is met by first setting explicitly to zero the normal components of mass and momentum fluxes (except the pressure terms), and secondly by forcing the total velocity to be tangential to the wall. At the *far field* boundary, the reliable method of Riemann invariants is used. A pointer system and auxiliary cells are applied to keep track of the information across the artificial interior *block interfaces*.

3 THE LOCAL GRID REFINEMENT TECHNIQUE

The refined zones are obtained by sub-dividing each global grid cell into eight hexahedra. The coordinates and the starting solution at the new nodes are linearly interpolated from the existing nodes of the initial mesh. Each cell is divided independently from its neighbors, a procedure that requires an extensive bookkeeping of the node connectivity.

3.1 Coarse/Fine Cell Interface Formulation

The topological similarity of coarse and fine mesh domains permits the application of the flow algorithm to the entire flow field except at nodes in the vicinity of the refinement interfaces. Usually, these latter nodes need a special discretization formulation which poses a number of problems for three-dimensional fields. One of the highest objectives when designing our flow solver was the uniform treatment of all interfaces, hereby avoiding the time-consuming fragmentation of the discretized domain. By avoiding the extra treatment of transition nodes optimal vectorization of long arrays becomes feasible once the problem of indirect addressing is solved.

Interface nodes are included in both the spatial and time integration process. No interpolation, which conveys no new information about the flow field, is necessary. This procedure is motivated by the unstructured approach which can result in a quite inhomogeneous grid and, consequently, a disadvantageously high ratio between nodes at interfaces and those lying inside a coarse cell. This effect may even be intensified in a cell-vertex scheme with interface nodes assigned directly to the junction of fine and coarse domains. On the other hand, combining sub-divided cells to a subset of refined nodes and aligning the interface with the global mesh lines yields regular shaped embedded domains. This approach maintains a structured data environment, as chosen in e.g. [7].

3.2 Interface Fluxes

The artificial interfaces between coarse and fine cells introduce an abrupt change in cell size which adversely effects the accuracy. In local refinement, one clusters nodes in domains of physical interest. Hence, interfaces generally are close to strong gradients, domains where flow conservation plays an important role. Thus conservation becomes the primary property to enforce at interfaces. As a second major consideration of mesh refinement so-called hanging nodes are produced, which are not fully connected to surrounding points.

Various discretization schemes for the interface nodes have been envisaged and are reported in a previous paper [8]. Most of them had to be disregarded as they turned out to be non-conservative. The most promising method we finally decided to implement is to modify the distribution coefficients of Eq.(4). By symmetry the condition on these coefficients becomes: $4A + 4B + C = 1/2$. This equation is satisfied by an infinite number of solutions. We choose $A = 1/32$, $B = 1/16$, and $C = 1/8$ which corresponds to the portion of the entire cell volume which is allocated the each node (dashed lines in Fig.1). The scheme preserves conservation since the coefficients sum to one for all cells.

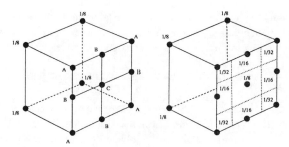

Figure 1: Modified distribution coefficients at coarse/fine interfaces.

3.3 Boundary Conditions

Refined nodes located at the outer boundaries are essentially treated in the same manner as described above for the global mesh nodes. A special register accounts for refined cells facing each other at block interfaces.

3.4 Interface Smoothing

The unstructured artificial dissipation model of Eq.(5) is well-suited for application to interface nodes. For hanging nodes summation is done over a number of differences with the surrounding points (less than six). In Fig.2, an example is given in which the Laplacian for node A is obtained by adding the differences with contiguous nodes B to E, a procedure which preserves conservation at these critical coarse/fine junctions.

4 DATA STRUCTURE

For the sake of efficiency the unstructured data environment is restricted to finite volume meshes with regular hexahedral cells. This implies that each three-dimensional element consists of six faces, each of them formed by four vertices. Further, each node usually is surrounded by eight cells, except nodes at boundaries and "hanging nodes". Five major integer registers are essentially necessary to provide connectivity information:

1. Node-neighbor register containing the indices of the adjacent nodes in each direction. This vector also provides information about the location of the node in the domain.

2. Cell-face register containing the six faces that form the corresponding cell.

3. Face-node register containing the indices of face vertices.

4. Node-cell register containing the indices of the eight cells that surround each node. The distribution coefficients are given by a similar register of type real.

5. "Parents-kids" register containing the indices of four small faces (kids) that form a big cell face. The "parent" of each small face is stored in the fifth vector.

These global registers are completed by "two-dimensional" registers for the interface and boundary nodes. With the help of "dummy" values nodes located at a block interface are connected to their neighbors in the adjacent block. Special care has to be taken to ensure the correct handling at facing blocks with different local coordinate directions.

The programming of a three-dimensional unstructured code is quite time consuming, especially when the data is complicated by local grid refinement facilities in a multi-block environment. Code debugging becomes substantially more frustrating, particularly because of the lack of regular (i, j, k)-indices to locate errors. The program is written in standard Fortran77 and has about 3900 lines of code. Approximately 30% of them are necessary for the solver part. The compact way of programming is one positive effect of the unstructured nature of the data. Often loops over the six faces per cell can replace the separate extra treatment of different faces/boundaries necessary in a structured environment. The remaining large part of the code are subroutines that are called either only once (setting up of pointers, etc.) or a few times (adaptive refinement) during calculation. Unfortunately, Fortran77 lacks the possibility to allocate storage dynamically. The arrays must be dimensionalized large enough in order to permit the storage of the additional elements of the local refinement procedure. It is envisaged to circumvent this inflexibility by the use of C-language routines.

Vector processing of unstructured data using indirect addressing is a more difficult matter than for a traditional structured system. Information about contiguous nodes is no longer given implicitly by an $(i, i+1)$ indexing but has to be provided by a set of pointers and registers. This results in longer access times since nodes adjacent in the flow domain are not necessarily stored next to each other. This is the penalty to be paid for the enhanced flexibility of cell-by-cell refining as compared to a structured data.

5 COMPUTATIONAL RESULTS

The test configuration chosen for the evaluation of the suggested refinement technique is the transonic flow past an isolated delta wing with a $65°$ leading edge sweep at moderate incidence.

Though the geometry of a single delta wing does not require a multi-block mesh, a structured O–O type single-block mesh has been sub-divided into eight initially equal-sized blocks in order to demonstrate the application of the program to a split flow domain.

Different views of example meshes for an adaptive multi-level refinement are given in Fig.4. The upper plot shows the grid with three levels of refinement (based on the total pressure loss sensor) at a spanwise cut and the upper surface. Note that the resolution of the very fine domains corresponds to that of a global fine mesh with the strikingly high number of 11 796 480 cells !

Fig.5 shows isoline contours of static pressure and total pressure loss at different chordlength cuts x/c = const. and at the upper wing surface for the wing with a round leading edge at $M_\infty = 0.85$ under $\alpha = 20°$ angle of attack. The solution obtained on a global coarse mesh (port side) is compared to the results of a local refinement computation (starboard side). The larger number and the higher density of isolines in the refinement plots indicate an improved resolution of the primary vortices coming close to a global fine solution which uses a total of 205 128 points (Fig.5c). Although only 12% of the global coarse cells are sub-divided, the number of nodes rises by a factor of 1.78. This demonstrates that three-dimensional grid embedding has to be treated very carefully. Except for some unavoidable effects due to post-processing, it is a common feature of all these isoline figures that no irregularities of the contours occur at coarse/fine interfaces.

The test case of the locally refined grid with a total of about 50 500 nodes results in an increase by 65% of CPU-time per iteration compared to the global coarse grid (28 424 nodes). It is evident that the global refining of the mesh (CPU-time increase by 620%) is less efficient than local refinement. However, the additional costs necessary due to the unstructured nature of data (compared to a structured system) are disregarded.

6 CONCLUSIONS

In the present work, the three-dimensional Euler equations have been solved on locally refined structured multi-block grids. The application of an unstructured data system, employed in conjunction with flow adaption, has appeared to be highly flexible for multi-level refinement of arbitrary shaped domains. Promising results have been obtained for transonic vortex flow. Work is under way with the main objective of the application to complex configurations and the local enrichment along shock waves for hypersonic flow.

REFERENCES

[1] Berger, M.J., Jameson, A.: "Automatic Adaptive Grid Refinement for the Euler Equations", AIAA Journal, Vol.23, No.4, pp. 561-568, April 1985.

[2] Kallinderis, Y.G., Baron, J.R.: "Adaptation Methods for a New Navier-Stokes Algorithm", AIAA Paper 87-1167, June 1987.

[3] Szmelter, J. et al.: "Solution of the Two-Dimensional Compressible Navier-Stokes Equations on Embedded Structured Multiblock Meshes", Proceedings of the *Third International Conference on Numerical Grid Generation in CFD*, pp. 287-299, Barcelona, June 1991.

[4] Hall, M.G.: "Cell-Vertex Multigrid Schemes for Solution of the Euler Equations", Proceeding of the *Conference on Numerical Methods for Fluid Dynamics*, Reading/U.K., 1985.

[5] Rudgyard, M.A.: "Cell-Vertex Methods for Compressible Gas Flows", Ph.D. Thesis, University of Oxford/U.K., 1990.

[6] Mavriplis, D.J.: "Accurate Multigrid Solution of the Euler Equations on Unstructured and Adaptive Meshes", ICASE Report 88-40, June 1988.

[7] Becker, K.: "Mesh Enrichment within an Embedded Block Structured Grid", Final report of Deutsche Airbus on BRITE/EURAM sub-task No.2.5 , 1992, to appear.

[8] Schönfeld, T., Rizzi, A.: "Transonic Vortex Flow Computations over Delta Wings Using an Unstructured Grid Enrichment Technique", Proceedings of the *4th International Symposium on Computational Fluid Dynamics*, pp. 1029-1034, Davis/CA, September 1991.

FIGURES

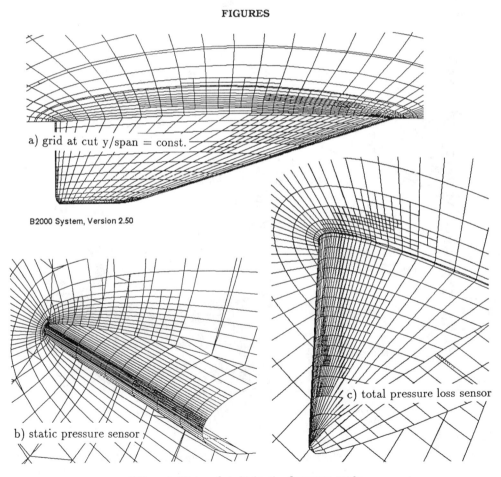

Figure 4: Views of multi-level refinement grids.

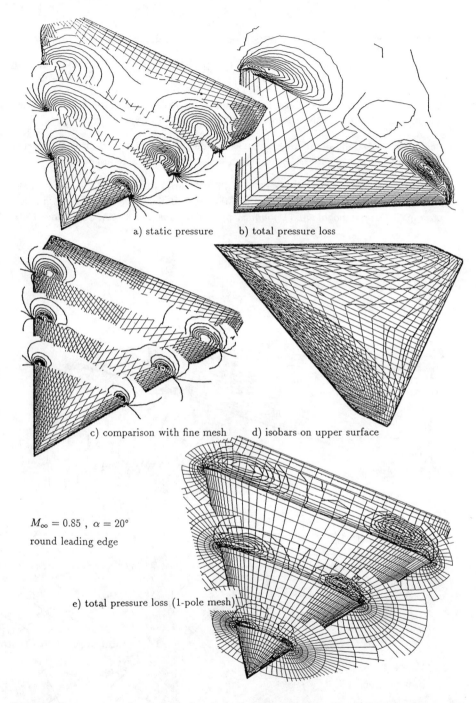

Figure 5: Comparison of isolines for locally refined grids (port side) with global coarse grid.

THE ADAPTATION OF TWO-DIMENSIONAL MULTIBLOCK STRUCTURED GRIDS USING A PDE-BASED METHOD

D Catherall
Defence Research Agency
Aerodynamics Department
Royal Aerospace Establishment
Farnborough
Hants.
GU14 6TD
UK

Summary

In previous work it has been shown that solution-adapted grids may be obtained by solving Poisson equations for the node ordinates with the control functions evaluated from a solution obtained on the original grid. However, this can lead to a situation where both the elliptic terms in the grid equations, and the equidistribution mechanism implied in the evaluation of the control functions, attract grid nodes to certain regions, so producing an 'overkill'.

In the work described here a new, but related, method, termed the LPE method, is introduced to overcome this limitation. In this approach each grid equation is formed from

1. an inverted Laplace equation (the L in LPE) which, if used in isolation, maximises smoothness and orthogonality.

2. an inverted Poisson equation (the P in LPE) with control functions evaluated from the original grid - if used in isolation the original grid is regenerated.

3. an Equidistribution equation (the E in LPE) which, if used in isolation, distributes nodes along each grid line so that node separations are inversely proportional to the local value of a sensor function formed from solution gradients.

Control is effected through choosing the relative weights to be placed on these three constituents of the grid equations. Grid control in the region of trailing edges is obtained separately through the use of source terms placed there.

Other features of the method described here include enforced grid orthogonality at grid boundaries and implementation on multiblock structured grids.

1 Introduction

A method is described here for adapting a given grid to be more optimum for the particular numerical problem being solved on it. The nodes of the grid are moved towards regions of high 'solution activity', characterised by high solution gradients, while, at the same time, smoothness and orthogonality are promoted. The adaptivity element appears to be much simpler to implement than these latter requirements, and relies on the principle of equidistribution:

If w is some measure of solution activity (we will call it the sensor function - see section 2) then we aim to equate, along each grid line, the product of w and the local grid interval, ie

$$w_i(x_{i+1} - x_i) = \text{constant}$$

where x_i is the distance along a grid line to the node i and w_i is the mean value of w within the interval (i,i+1). If ξ is the computational variable which varies along the grid line ($\xi = 1, 2, 3, \ldots$ at nodes) then the above equation is a discretisation of the equation

$$w.(dx/d\xi) = \text{constant} . \tag{1}$$

In a one-dimensional problem equidistribution is achieved by moving the nodes until the above equation is satisfied. In two or more dimensions we could follow this procedure along each grid line in turn. This would achieve one of the goals of adaptivity in that nodes would cluster where w is high, but the resulting grid is likely to be highly skewed. We thus seek a way of controlling such grid properties as smoothness and orthogonality within an adaptive procedure.

Three options have been fairly extensively investigated. These are the variational approach [1,2], the spring-analogy approach [3,4] and the method based on modifying an elliptic grid generator [5]. Although the third method has some good features it was found to have some basic deficiencies [5]. The good features included the ability to promote smoothness and orthogonality in the interior of the grid, while providing a good response to adaptivity demands. The bad features included a lack of orthogonality at grid boundaries (boundary nodes were adapted first in a one-dimensional manner and then their positions used as Dirichlet boundary conditions for the solution of the grid equations at internal nodes) and an 'overkill' in node bunching in some regions. This latter deficiency was due to the nature of the grid equations. If the adaptive terms are left out the grid equations are inverted Laplace equations - a grid produced by satisfying these equations will attract nodes towards convex regions of high surface curvature [6] such as the leading edge of an aerofoil. This is generally desirable because solution gradients tend to be high in such regions. However, when the adaptive terms are included in the grid equations they also attract nodes to these regions, and 'overkill' results. The LPE method (section 3) is designed to overcome this problem.

2 Sensor functions

Three alternative forms of sensor function have been used with the current method. These are

$$w = \left|\frac{dq}{dx}\right| \tag{2a}$$

$$w = \sqrt{|d^2q/dx^2|} \tag{2b}$$

$$w = \left|\frac{dq}{dx}\right| + \beta\left|\frac{d^2q}{dx^2}\right| \tag{2c}$$

where q is a chosen solution quantity (eg Mach number). The first two forms are related to solution truncation errors for, respectively, a first-order and a second-order-accurate solution method for a first order equation. The third form (which is used in the examples shown in section 5 with a value of 0.05 for β) is a compromise, which has been found to be generally more useful. The inclusion of both first and second derivatives is suggested in reference 7.

Provision is made in the method for several physical quantities to be used for q. If n physical quantities are used (each scaled with respect to free-stream values) then, within each grid interval, w_n is evaluated from (2) using each of the n quantities in turn. Whichever of these evaluations results in the largest value for w is the one accepted. w is then constrained to lie within chosen minimum and maximum values w_{min} and w_{max}, which correspond, for the line in question, to chosen maximum and minimum grid spacings. Formally, then, we take in each grid interval

$$w = \max\{w_{min}; \min\{w_{max}; \max_n(w_n)\}\} . \tag{3}$$

For an inviscid flow it is normally only necessary to use one physical quantity (n=1), such as the Mach number. However, for a viscous flow at least two quantities (n=2) are needed. This is because a quantity such as the Mach number adequately reflects solution gradients normal to a solid surface (through the boundary layer), but is unsuitable along the surface (where the Mach number is zero). The pressure, on the other hand, reflects solution gradients along the surface but is fairly constant through the thickness of the boundary layer. The form (3) will automatically select whichever of the quantities locally reflects the highest gradients.

3 LPE method

With a surface-conforming grid one family of grid lines can be labelled ξ=constant ($\xi=1,2,3,\ldots$ at nodes) and the other family η=constant ($\eta=1,2,3,\ldots$ at nodes), with solid surfaces and far-field boundaries corresponding to lines of constant ξ or η. In an elliptic grid generation method the grid is produced by numerically solving Poisson equations:

$$\frac{\partial^2 \xi}{\partial x^2} + \frac{\partial^2 \xi}{\partial y^2} = \overline{P} \; ; \quad \frac{\partial^2 \eta}{\partial x^2} + \frac{\partial^2 \eta}{\partial y^2} = \overline{Q} \; . \tag{4}$$

x, y are Cartesian coordinates, and $\overline{P}$, $\overline{Q}$ are functions used to control grid stretchings. If equations (4) are solved in discrete form, with zero right hand sides, then [6] smoothness (in the sense of uniformity of grid interval) and orthogonality are maximised (in some sense) and, in addition, the extremum principle applies. This ensures that extrema of solutions cannot occur within the field (ie away from the boundary), so that grid cross-over is prohibited. These desirable features are not necessarily strictly maintained with non-zero control functions, but in practice it is usually found that there is a strong bias towards them.

Equations (4) are not in a suitable form for numerical solution and are normally inverted so that x and y become the dependent variables:

$$g_{22}\left(\frac{\partial^2 x}{\partial \xi^2} + \frac{\partial x}{\partial \xi}P\right) + g_{11}\left(\frac{\partial^2 x}{\partial \eta^2} + \frac{\partial x}{\partial \eta}Q\right) - 2g_{12}\frac{\partial^2 x}{\partial \xi \partial \eta} = 0$$

$$g_{22}\left(\frac{\partial^2 y}{\partial \xi^2} + \frac{\partial y}{\partial \xi}P\right) + g_{11}\left(\frac{\partial^2 y}{\partial \eta^2} + \frac{\partial y}{\partial \eta}Q\right) - 2g_{12}\frac{\partial^2 y}{\partial \xi \partial \eta} = 0 \; . \tag{5}$$

Here the metrics of the transformation between (x,y) and (ξ,η) space are:

$$g_{11} = \left(\frac{\partial x}{\partial \xi}\right)^2 + \left(\frac{\partial y}{\partial \xi}\right)^2 \; ; \quad g_{22} = \left(\frac{\partial x}{\partial \eta}\right)^2 + \left(\frac{\partial y}{\partial \eta}\right)^2 \; ;$$

$$g_{12} = \frac{\partial x}{\partial \xi}\frac{\partial x}{\partial \eta} + \frac{\partial y}{\partial \xi}\frac{\partial y}{\partial \eta} \; ; \quad J = \frac{\partial x}{\partial \xi}\frac{\partial y}{\partial \eta} - \frac{\partial x}{\partial \eta}\frac{\partial y}{\partial \xi} \; , \tag{6}$$

while P and Q are related to $\overline{P}$ and $\overline{Q}$ via

$$P = \frac{J^2}{g_{22}} \overline{P} \; ; \quad Q = \frac{J^2}{g_{11}} \overline{Q} \; .$$

In the LPE method a combination of <u>L</u>aplace, <u>P</u>oisson and <u>E</u>quidistribution equations are solved to obtain a new grid. If r represents x or y, then inverted Laplace equations may be obtained by setting P = Q = 0 in equations (5):

$$g_{22}\frac{\partial^2 r}{\partial \xi^2} + g_{11}\frac{\partial^2 r}{\partial \eta^2} - 2g_{12}\frac{\partial^2 r}{\partial \xi \partial \eta} = 0 \; . \tag{7}$$

Rewriting (5) in this form results in inverted Poisson equations:

$$g_{22}\left(\frac{\partial^2 r}{\partial \xi^2} + \frac{\partial r}{\partial \xi}P\right) + g_{11}\left(\frac{\partial^2 r}{\partial \eta^2} + \frac{\partial r}{\partial \eta}Q\right) - 2g_{12}\frac{\partial^2 r}{\partial \xi \partial \eta} = 0 \quad . \tag{8}$$

If the initial grid was formed using an elliptic grid generator, then the control functions, P and Q, for it will be known. Alternatively, if equations (8) are discretised, and the node positions from the initial grid are substituted for x and y, then equations (8) may be used to evaluate P and Q, for the initial grid, at each node. P and Q are thus fuctions of ξ and η, rather than of x and y.

The equidistribution equation (1) may be written in the form:

$$wi\frac{\partial si}{\partial \xi} = \text{constant} \quad ; \quad wj\frac{\partial sj}{\partial \eta} = \text{constant}$$

where Si and Sj are the arc lengths measured along i-varying and j-varying grid lines respectively. These two equations may be differentiated along their respective grid lines, making use of the identity $dS^2 = dx^2 + dy^2$, to give

$$\frac{1}{g_{11}}\left(\frac{\partial x}{\partial \xi}\frac{\partial^2 x}{\partial \xi^2} + \frac{\partial y}{\partial \xi}\frac{\partial^2 y}{\partial \xi^2}\right) + \frac{1}{wi}\frac{\partial wi}{\partial \xi} = 0 \tag{9a}$$

$$\frac{1}{g_{22}}\left(\frac{\partial x}{\partial \eta}\frac{\partial^2 x}{\partial \eta^2} + \frac{\partial y}{\partial \eta}\frac{\partial^2 y}{\partial \eta^2}\right) + \frac{1}{wj}\frac{\partial wj}{\partial \eta} = 0 \quad . \tag{9b}$$

The LPE method involves solving a combination of these equations:

$$\lambda_L(7) + \lambda_P(8) + \lambda_E\left[g_{22}\frac{\partial r}{\partial \xi}(9a) + g_{11}\frac{\partial r}{\partial \eta}(9b)\right] = 0 \tag{10}$$

where λ_L, λ_P and λ_E are weights at the user's disposal. The first term in equation (10) promotes smoothness and orthogonality, the second term (because of the above choice of P and Q) promotes the initial grid, while the third term promotes equidistribution. Experience will tell what values to place on the three weights in equation (10). With a very poor initial grid, λ_P will probably need to be small, or even zero. On the other hand, if the initial grid is very good, and merely requires some small adaptation in response to the solution (for example some bunching of nodes in the shock region) then λ_L could be small or zero.

The overkill in bunching near the leading edge, which occurs with the method referred to in the previous section [5], should not be present in the LPE method, since each of the equations (7), (8) and (9) could be satisfied simultaneously when equation (10) is satisfied. Because they are combined in parallel, rather than in series, there should be no additive effect on the bunching.

The adaptive procedure is to produce a flow solution on the initial grid, adapt this grid by numerically solving equations (10), interpolate the flow solution onto this grid to use as initial conditions for the flow solver, run the flow solver on this new grid, and alternate adaptation with the flow solver until a satisfactory grid and solution are obtained. Usually two to three adaptations are sufficient.

4 Other features of the method

The features briefly described in this section are more fully covered in reference 8.

Equations (10) are solved iteratively using Point Relaxation. After each iterative cycle the physical quantities q (section 2) are interpolated onto the current grid and used to re-evaluate the sensor functions (equation (2)). In order to test the method in complex two-dimensional situations it has been implemented for multiblock grids. Internal block boundaries (where adjacent blocks meet) are treated as though they were transparent. External boundaries (solid surfaces or far-field boundaries) can have either a Dirichlet boundary condition (where nodes remain fixed), or a Neumann boundary condition, imposed. In the latter situation the grid equations are satisfied on the boundary and are augmented by an orthogonality condition ($g_{12} = 0$). This is the condition utilized with all the examples in section 5.

There are two problems associated with grid adaptation in the region of trailing edges, and we will consider first the one that is easier to overcome. This arises when we use an inviscid flow solution to drive the adaptation. Theoretically there should be a stagnation point there (though perhaps only on one surface), but numerical solutions of the Euler equations do not normally capture this feature, so that gradients in this region are under-predicted. This leads to a comparative scarcity of points in the region after adaptation, leading in turn to an even larger under-prediction of solution gradients there after running the flow code on the adapted grid, since Euler flow solvers are often rather sensitive to grid point distribution in the trailing edge region. A remedy for this problem is to adjust the numerical solution at and near the trailing edge before applying the adaptation procedure. One way of adjusting the solution is to subtract from each of the n quantities q used for the adaptatation (see section 2) an amount $(q_{TE} - q_{targ}) \cdot 0.01^{r/r_{max}}$, where q_{TE} is the value of q at the trailing edge, q_{targ} the target value, r is the distance from the trailing edge and r_{max} is the distance over which it is desired to spread this reduction (down to 1% of its value at the trailing edge).

A much more serious problem in the trailing edge region occurs because the grid equations are not satisfied at the trailing edge itself. This is because the trailing edge is fixed and so a point situated there is fixed also, and cannot move in order to satisfy the grid equations. Figure 1 illustrates the effect this can have - the grid is obviously unacceptable in this region.

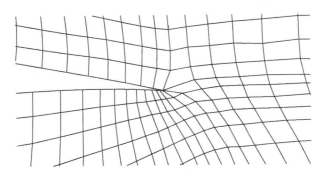

Figure 1: Grid in region of trailing edge if no control applied.

Before considering how to improve the situation here it may be instructive to review the procedure followed by modern elliptic grid generators. In these systems inverted Poisson equations (5) are solved for the ordinates of grid nodes. Interior grid orthogonality and smoothness are controlled by the second derivatives in these equations (equivalent to an inverted Laplace equation), but control of grid stretching and of orthogonality at grid boundaries is exercised [10] through the control functions P and Q. To obtain orthogonality at boundaries the usual procedure is to obtain, by some means, the node distribution on the boundaries, and to evaluate the control functions there on the assumption of grid orthogonality. The control functions in the interior of the field are interpolated from these boundary values (with some added stretching control) and then the grid equations are solved for interior nodes with the boundary values as Dirichlet conditions. Because of the choice of control functions at the boundary (including the trailing edge) approximate boundary grid orthogonality is ensured, including the trailing edge region.

In the LPE approach to grid adaptation we utilize the control function distribution of the initial grid through equation (8), but, because of the presence in the grid equations (10) of the other terms, this no longer ensures orthogonality at the boundaries, and we have to employ other means as described above (setting g_{12} to zero at the boundary). However, we do not satisfy the grid equations at the trailing edge itself, as explained earlier, and, in fact, the values of the control functions evaluated there are not utilized. We can, perhaps, imagine that if we satisfied the grid equations at the trailing edge and allowed the trailing edge point to move accordingly, then the Laplace element would tend to move the point towards the centroid of the surrounding points and orthogonality would be promoted. We do not, of course, have the freedom to allow this, but an alternative is proposed:

> Since the problem arises from non-satisfaction of the grid equations at the trailing edge, we will ensure that they are satisfied there, without moving the trailing edge point, by modifying the control functions appropriately in the region.

We do this by adding 'source' terms to P and Q, centred on the trailing edge. The terms 'source', 'doublet' and 'vortex' are used here because of their effect on the grid, which is similar to the effect on the flow of imposing such 'point sources' in classical hydrodynamic potential theory. We apply a 'doublet' term by adding to P and Q the terms

$$P_D = \alpha_P e^{-\gamma_D r} \; ; \; Q_D = \alpha_Q e^{-\gamma_D r} \qquad (11)$$

respectively with $r=\sqrt{((\xi - \xi_{TE})^2 + (\eta - \eta_{TE})^2)}$. γ_D is used to restrict the extent of the region surrounding the trailing edge which is affected (for example, if $\gamma_D = 0.75$ the source term will have decreased to about 1% of its value at the trailing edge by the time we are 6 grid intervals away). We satisfy the grid equations at the trailing edge without allowing the trailing edge node to move, but instead use them to compute α_P and α_Q in equation (11). After evaluating α_P and α_Q, P_D and Q_D (11) are added to P and Q and the grid equations solved normally at other nodes. The effect is that the trailing edge remains fixed and the grid around it moves to align the trailing edge (roughly) with the centroid of the surrounding nodes.

A 'vortex' term is applied by adding to P the term

$$P_V = \mu r e^{-\gamma_V r} \qquad (12)$$

in the region above the aerofoil and wake cut, with a similar term (but a different value of μ) in the region below them. This allows us to control the slope of the grid line $\xi = \xi_{TE}$ (assuming that the aerofoil is a curve of constant η). Thus we could force this line to be normal to the aerofoil surface at the trailing edge - note that we have forced orthogonality of grid lines elsewhere on the surface but have not, so far, been able to do so at the trailing edge. We implement this by modifying the way we satisfy the grid equations at the node above the trailing edge (on $\xi = \xi_{TE}$). We now have three unknowns: the values of x and y at the node and the value of μ from equation (12) (r = 1 at this position). These are determined from the two grid equations and the relationship between x and y which will produce the desired slope of the line $\xi = \xi_{TE}$. A similar procedure is applied below the trailing edge. If it is desired to control the slope of the wake cut at the trailing edge this can be achieved in a similar manner by including a 'vortex' term in Q.

A 'sink' term may be used to control grid spacings at the trailing edge. In the examples which follow (section 5) it has not been included; we have chosen instead to allow these spacings to evolve naturally (though with modifications to the solution as described at the beginning of this section). Figure 2 shows the effect of including the 'doublet' and 'vortex' terms in the example shown in figure 1.

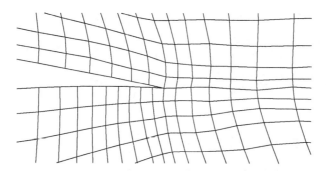

Figure 2: Effect of 'source' terms on trailing edge region.

In the far-field solution gradients are vanishingly small and the Equidistribution element in equation (10) has little influence on the grid. The Laplace element will promote uniformity in the absense of boundary curvature, whereas the Poisson element will promote the original grid. It is evident that the node distribution from the original grid is preferable, in the far-field, to a uniform one, but this will only be achieved if λ_p is relatively large. We would thus like λ_p to be larger in the far-field than in the aerofoil region. A crude way of arranging this is to multiply the values of P and Q, computed from the original grid at the start of the computation, by an amount dependant on the distance from the centre of the field, linearly interpolated between the values of λ_p chosen for the centre and for the far-field.

5 Examples

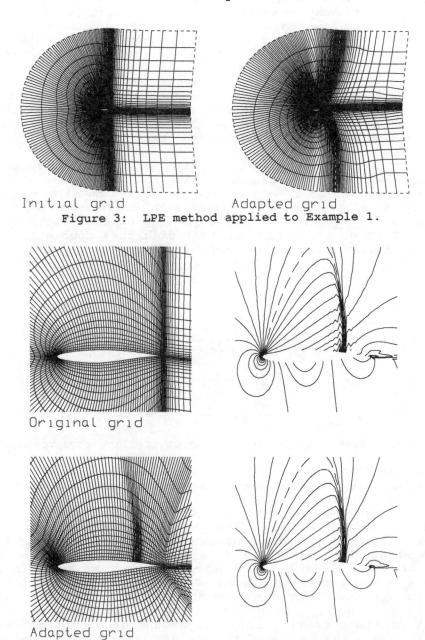

Figure 3: LPE method applied to Example 1.

Figure 4: LPE Example 1. Grid and Mach contours in vicinity of aerofoil.

The first example chosen to demonstrate the LPE method is of inviscid flow around a RAE2822 aerofoil at a Mach number of

0.75 and an incidence of 3°. The initial and adapted grids are shown in figure 3.

The flow solver [11] was run on the initial grid, the Mach number distribution was used to adapt the grid using the LPE method and flow variables were interpolated onto the adapted grid. The flow solution was continued on this new grid, followed by a further adaptation and another run of the flow solver. Values of unity were used for each of the λ's in equation (10), although λ_p was increased to 2.5 in the far-field. We can see that the LPE method does not suffer from the 'overkill' inherent in the method based on modifying an elliptic grid generator [5].

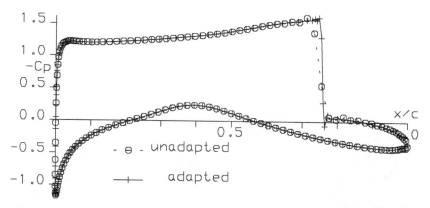

Figure 5: LPE Example 1. Surface pressure distributions.

The grid and Mach contours in the neighbourhood of the aerofoil are shown in figure 4, and surface pressure distributions are shown in figure 5. The major improvement to the original grid can be seen to be the concentration of grid nodes in the shock region, so that on the adapted grid the shock is much sharper. To achieve this without adapting the grid would necessitate making the original grid much finer globally, thus adding considerably to computer storage and computing time requirements. In a three dimensional situation the related increase in computing cost would be even larger.

Nevertheless, the changes we have made to the grid appear to be relatively minor, and it is tempting to question whether the improvement achieved justifies the effort expended. We note that, in this example, the initial grid was of fairly high quality, despite some odd-looking features. Quite a lot of effort went into the production of this grid, in particular to achieve a clustering of grid nodes in the leading and trailing edge regions, and to seek cells of unit aspect ratio around the aerofoil surface. With a poor initial grid (see the second example below) more dramatic improvements may be expected. It may be noted that the major part of the effort has gone into the mesh quality aspects (eg, the promotion of smoothness and orthogonality) rather than into the adaptive aspects, which are reasonably simple to implement.

Before considering other examples of the LPE method we will look at the role of the three λ's in equation (10). If λ_P is set to unity, and the other two values set to zero we obtain, not surprisingly in view of our choice for the control functions P and Q (section 3), the original grid as displayed in figures 3 and 4. If, instead, we set λ_L to unity and the other two λ's to zero we obtain the grid displayed in figure 6a (we have, of course, implemented the boundary condition and trailing edge treatments described in section 4). As may be seen, nodes are clustered where the surface is convex and of high curvature, as predicted [6].

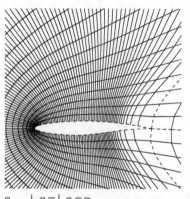 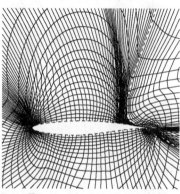

a. Laplace   b. Equidistribution
Figure 6:   LPE Example 1.   Effect of accentuating λ_L or λ_E.

Figure 6b shows the grid produced when λ_E is set to unity and the other two λ's to 0.25 (it was, unfortunately, not possible to obtain a solution of the grid equations with both λ_L and λ_P set to zero). Although well adapted to solution gradients this grid is very deficient in other qualities. We can see from these three figures that it is important to include, to an adequate degree, all three elements in the grid equations.

The second example chosen to demonstrate the LPE method is of inviscid flow around a NACA0012 aerofoil at a Mach number of 0.85 and an incidence of 1°. For this example the initial grid has deliberately been selected to be of poor quality, particularly around the leading edge (figure 8). The grid is, in fact, so coarse in the leading edge region that the flow solver [11] failed to produce a converged solution, the one taken leading to a lift coefficient which is about 40% too small, with high levels of spurious entropy being produced in the leading edge region.

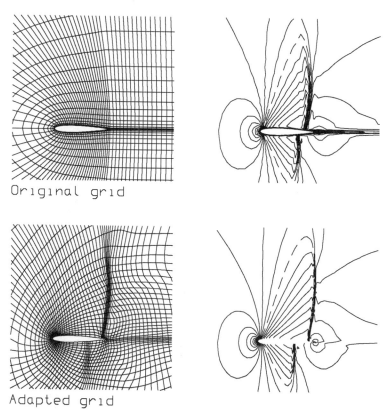

Figure 7: LPE method applied to Example 2.

Figures 7 and 8 show the effect of grid adaptation, starting with this poor initial grid and solution. Values of unity were used for λ_L and λ_E, and of 0.25 (2.0 in the far-field) for λ_P. After 3 adaptations (alternating with runs of the flow solver) the grids and solutions (in the form of Mach contours) shown in the figures were obtained. The lift coefficient was 0.4126, which compares well with those from the currently accepted 'best' solutions to this problem [12,13] of 0.4154 on a 256x32 grid with a shock fitting method [12] and 0.3938 on a 560x64 grid with shock capturing [13]. The grid used here has 128x32 intervals with the far-field boundary placed at 20 chord lengths from the aerofoil.

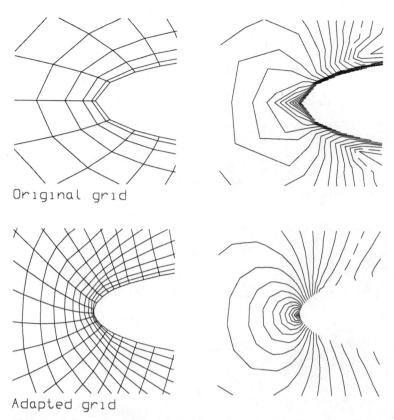

Figure 8: LPE Example 2. Detail of leading edge region.

The third example chosen to demonstrate the LPE method uses a 37-block grid around a NLR7301 two-element aerofoil. A Navier-Stokes flow solver was used to compute the flow at a Mach number of 0.185, an incidence of 13.1° and a Reynolds number of 2.51 million. Mach number and pressure were utilized for adapting the grid, and all three λ's were set to unity (λ_p was set to 3.0 in the far-field). Figures 9 and 10 show the result of a single adaptation (only the grid and solution, not the flow solver, were available to the author). Block boundaries are indicated by broken lines. Figure 9 shows the grid in the vicinity of the second aerofoil (in the interests of clarity only every other grid line is displayed in figure 9), while figure 10 shows the region of the gap between the two. These figures are not very dramatic - the Mach number is too low for shock waves - and are included merely to demonstrate that the LPE method works with viscous flows and in a multiblock environment.

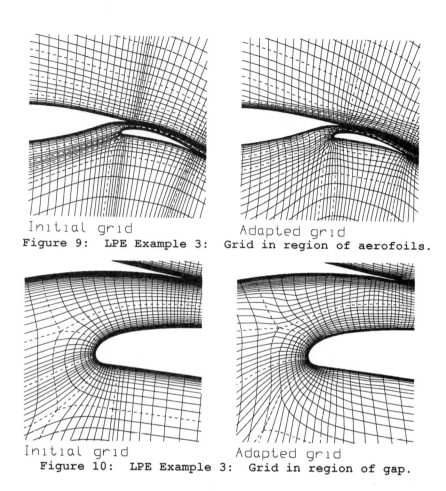

Figure 9: LPE Example 3: Grid in region of aerofoils.

Figure 10: LPE Example 3: Grid in region of gap.

6 Final remarks

The mesh movement approach to solution adaptivity permits the use of such a method with any appropriate flow solver - a simple interface to pass relevant information between the flow solver and the adapter is the only additional requirement. This contrasts with methods based on grid embedding, in that these latter methods require the flow solver to be modified to accommodate the new data structure.

The major problem with methods based on mesh movement is to promote such mesh quality features as smoothness and orthogonality, while implementing the adaptivity criteria. However, once we have a method which exhibits these properties, we can be fairly certain of obtaining grids of good quality, even when the grid with which we started is of poor quality. Methods based on mesh embedding do not have the same advantage, particularly if the starting mesh is highly skewed.

The potential benefits to be derived from mesh adaptation are easy to imagine. Even a good mesh (as in Example 1 of section 5) is open to improvement once a solution on it is available, so that we may in turn obtain a solution of a quality only to be matched by using a globally much finer mesh - an option which is much more expensive computationally. The benefits are, however, much more evident when the mesh available is only of poor quality (as in Example 2 of section 5), since we are able to utilize this to construct a high-quality mesh on which a high-quality solution may be obtained. This is likely to be of particular benefit when we have a three-dimensional multiblock grid around a complex configuration, such as a complete aeroplane. The construction of such a grid inevitably results in poor mesh quality somewhere in the field (such regions are often extensive) - a situation which can be rectified if a good mesh adapter is available.

It is believed that the LPE method is the most promising line to pursue - the major complications that have been experienced in developing this method arise, not from the LPE method itself, but from other sources, such as the need to control orthogonality at grid boundaries and the application to multiblock grids.

Further research is needed into the choice of sensor function to control the adaptation - a closer link with solution errors might be desirable. It would also be desirable to establish the connection between solution error and other grid qualities such as smoothness and orthogonality, so that the importance of such qualities may be judged - this will, in turn, influence the emphasis to be placed on them during the adaptation process.

Acknowledgement

The grid and Navier-Stokes solution for the NLR7301 (Example 3 of section 5) were provided by Mr J.J. Benton of British Aerospace (CAL), Woodford.

References

[1] Brackbill, J.U. and Saltzman, J.S.: "Adaptive zoning for singular problems in two dimensions", Journal of Computational Physics, 46, p342, 1982.

[2] Jacquotte, O.-P. and Cabello, J.: "A variational method for the optimisation and adaptation of grids in Computational Fluid Dynamics", in Numerical Grid

Generation in CFD '88 (Ed. Sengupta, S., Haüser, J., Eiseman, P.R., and Thompson, J.F.), pp 405 to 414, Pineridge Press, Swansea, U.K., 1988.

[3] Nakahashi, K. and Deiwert, G.S.: "A self-adaptive-grid method with application to airfoil flows", AIAA Paper 85-1525, 1985.

[4] Catherall, D.: "Solution-adaptive grids for transonic flows", in Numerical Grid Generation in CFD '88 (Ed. Sengupta, S., Haüser, J., Eiseman, P.R., and Thompson, J.F.), pp 329 to 338, Pineridge Press, Swansea, U.K., 1988.

[5] Catherall, D.: "The adaptation of structured grids to numerical solutions for transonic flow", International Journal for Numerical Methods in Engineering. Vol 32, No 4, pp 921 to 937, 1991.

[6] Thompson, Joe F., Warsi, Z.U.A. and Mastin, C. Wayne: "Numerical grid generation - foundations and applications", North-Holland, 1985.

[7] Hagmeijer, R.: "Adaptive grid generation development, a starting point", NLR CR 91138 L, 1991.

[8] Catherall, D.: "The adaptation of two-dimensional multiblock structured grids using a pde-based method - final report for Euromesh project", 1992.

[9] Shaw, J.A., Forsey, C.R., Weatherill, N.P. and Rose, K.E.: "A block structured mesh generation technique for aerodynamic geometries", in Proceedings of the first International Conference on Numerical Grid Generation in CFD (Ed Haüser, J. and Taylor, C.), Pineridge Press, 1986.

[10] Thomas, P.D. and Middlecoff, J.F.: "Direct control of the grid point distribution in meshes generated by elliptic equations", AIAA Journal, Vol 18, p652, 1980.

[11] Hall, M.G.: "Cell-vertex multigrid schemes for solution of the Euler equations", in Numerical methods for Fluid Dynamics II (Ed Morton, K.W. and Baines, M.J.), Oxford University Press, 1986.

[12] Paisley, M.F.: "Developments in shock fitting with the Euler equations", RAE TR 88075, 1988.

[13] Pulliam, T.H. and Barton, J.T.: "Euler computations of AGARD Working Group 07 airfoil test cases", AIAA Paper 85-0018, 1985.

CONTRIBUTION TO THE DEVELOPMENT OF A MULTIBLOCK GRID OPTIMIZATION AND ADAPTION CODE

Olivier-Pierre JACQUOTTE (*), Grégory COUSSEMENT (*)
Fabienne DESBOIS (*) and Christophe GAILLET (* and **)

* ONERA - Aerodynamics Department
BP 72 - 92322 CHATILLON CEDEX, FRANCE

** Aérospatiale - Missile Division
91370 VERRIERE LE BUISSON, FRANCE

ABSTRACT

This paper presents the work done at ONERA - Aerodynamics Department and Aérospatiale - Missile Division, from February 1990 to January 1992 in the framework of the "Euromesh" project, CEC Brite/Euram Aeronautics Project AERO0018. The work done has consisted in the development of a 3D optimization/adaption code based on a variational principle, and surface parametrization routines based on bi-cubic patches. Other works related to the applications of surface parametrization (displacement of nodes on surfaces, surface mesh optimization) are presented.

INTRODUCTION

It is now a well-known fact that the mesh generation and adaption are very important subjects in modern computational fluid mechanics (CFD) and that the smart use of good mesh generators can have a significant impact on the efficiency of numerical simulation. Although finite element or finite volume methods based on unstructured meshes are obtaining more and more successful theoretical results [3, 9, 10], industries still feel the need to have methods and codes for the generation of "good" structured grids. The notion of quality for these grids (this sentence also applies to unstructured meshes) is however very difficult to quantify: methods for their generation are usually based on mostly intuitive principles that produce the best possible grid with respect to a given criterion, hoping that the last judge, the CFD code, will first, accept the grid and then give an accurate solution. However, three types of constraint must generally be taken into account when generating a grid: the physical problem to be solved, the numerical scheme to be used for the discretization of the equations and the geometry of the domain; the handling of these constraints is different whether one belongs to the structured or to the unstructured world. For structured grids for instance, the scheme and the code are not usually considered directly, only a priori known geometric requirements (regularity, smoothness, orthogonality, cell deformation) are used in the construction of the generation method: for example, methods using elliptic partial equations rely on the observation that Laplace operator has smoothing and regularizing properties [8, 22]. Again for structured grids, the adaption usually handles the physics of the problem by the introduction in the basic algorithm of weights or control functions, whose action is switched off in an initial generation, and that have the ability to move the nodes towards (or away from) given areas in the domain for the grid adaption. The result of these considerations is that structured grids are judged through their generation methods, from which one expects the following qualities: their ability to respect the given geometric criteria, or to treat locally severe cases (concavity close to high curvatures for instance), their robustness in preventing the cells from overlapping, their possibility to obtain strong refinements, and their "tuning-free" feature, meaning that a correct grid can be obtained without too many tries or parameter adjustments.

Among the various difficulties encountered in the generation of grids for CFD, anyone who needs to construct a grid around a realistic configuration will also confront the problems linked to the treatment of surfaces. These problems include surface definition, construction of grids conforming to the surface, mesh quality control. In particular, construction of 3D grids by means of partial differential equation solution or optimization often requires this moving node feature, so does 3D grid adaption : for example the accurate capture of a shock attached to a body needs adaption of the mesh at the shock stem on the surface. Most CAD systems offer routines conceived for this surface handling. However it is never possible to extract these routines and to use them outside their interactive environment. Consequently, it appeared necessary to us to develop a method that can be implemented in a standard way and used in any grid algorithm and in particular in our 3D variational optimization/adaption methods [12-18].

With the motivation of fulfilling these requirements, we have been proposing for several years a method for the optimization and adaption of structured grids. This method is based on a clever use of principles of continuum mechanics and on several mathematical properties that ensure the good behavior of the method, the robustness of the optimization algorithm and code, and the quality of the resulting grids.

In the first part, we will present a formalism that enable a comprehensive unified description of the method for mesh optimization with respect to the cell deformation and adaption with respect to a physical solution. Then more practical features concerning the implementation of the method are presented: optimization algorithm, multiblock approach, respect of any type of boundary condition (free node, fixed node, node moving along a curve or a surface). In the second part, some aspects of the numerical implementation are presented: computation of the element contribution, the computation of the functional and its derivatives. The last part of this paper is devoted to the study of a surface construction and parametrization method that operates outside CAD system and its application for the optimization and adaption of mesh on continuously differentiable (C^1) interpolated surfaces. To illustrate the concepts developed in this paper, we will give several results of multi-block grid optimization, three-dimensional adaption and surface mesh optimization.

PART A
VARIATIONAL FORMULATIONS
FOR THE MESH GENERATION AND ADAPTION

In this part, we review an optimization method that has been thoroughly presented and studied in previous publications [5, 12, 16]. We introduce here a new formalism and new notations that enable a constructive and comprehensive presentation of the adaption. For further use, we refer to Fig.1 for the node numbering in a cell in 2 and 3 dimensions.

I. MESH DEFORMATION AND OPTIMIZATION

I.1 General Mesh Optimization

We consider a three-dimensional structured grid, formed by $imax \times jmax \times kmax$ nodes x_{ijk}, and we note:
- the elements (or cells) with the subscripts IJK,

$$(I, J, K) \in [1, imax-1] \times [1, jmax-1] \times [1, kmax-1]$$

- X_{IJK} the vector formed by the coordinates of the nodes of the element IJK

$$X_{IJK} = [\ x_{ijk}\ ]_{i=I, I+1;\ j=J, J+1;\ k=K, K+1}$$

- X the vector formed by the coordinates of the nodes in the whole mesh:

$$X = [\ x_{ijk}\ ]_{i=1,\ imax;\ j=1,\ jmax;\ k=1,\ kmax}$$

If one knows a mesh optimization criterion, $\sigma(X)$, the simplest algorithm taking it into account writes:

ALG0	1) Construct an initial mesh X_0 2) Initialize X by X_0 3) Find $\tilde{X} = Arg.Min._X\ \sigma(X)$

I.2 Mesh Optimization Based on Minimization of the Cell Deformation

For the optimization method introduced several years ago, we define, for each element IJK, a reference element, which is a rectangular parallelepiped, characterized by its three side lengths a_{IJK}, b_{IJK} and c_{IJK}, also noted A_{IJK}. We note abc_{IJK} the volume of this reference element. Similarly to X, the vector A will denote the vector formed by the reference cell lengths over the whole mesh:

$$A = [\, A_{IJK} \,]_{I=1,imax-1;\ J=1,jmax-1;\ K=1,kmax-1} .$$

For each element IJK, it is possible to define its deformation σ_{IJK} with respect to the reference element:

$$\sigma_{IJK} = \sigma(\, A_{IJK}\, ,\, X_{IJK}\,).$$

This element deformation was brought to the fore in [5, 12]. This deformation can be expressed as a polynomial function of the matrix invariants constructed from the three-linear transformation, $x(\xi)$ (Fig .2), between the reference element and the current element:

$$\sigma_{IJK} = abc_{IJK}^{\gamma-1} \int_{[0,a_{IJK}]\times[0,b_{IJK}]\times[0,c_{IJK}]} \sigma(I_1(\xi), I_2(\xi), J(\xi))\, d\xi$$

with

$$I_1(\xi) = \mathrm{trace}\ \nabla^t x . \nabla x\,;\ I_2(\xi) = \mathrm{trace}\ \mathrm{Cof}\ \nabla^t x . \nabla x\,;\ J(\xi) = \det \nabla^t x . \nabla x\,.$$

The power γ enables various weightings of the elementary contributions. For $\gamma = 1$, σ_{IJK} has the dimension of a volume, and each cell is given a weight proportional to its volume; for $\gamma = 0$, σ_{IJK} is dimensionless, and each cell has the same weight in the global functional; with $\gamma < 0$, it is possible to give more importance to small cells if they are located in regions where high mesh quality is desired. When for example $\sigma(I_1, I_2, I_3)$ reduces to I_1, we have:

$$\sigma_{IJK} = abc_{IJK}^{\gamma-1} \int_{[0,a_{IJK}]\times[0,b_{IJK}]\times[0,c_{IJK}]} (x_\xi^2 + y_\xi^2 + z_\xi^2 + x_\eta^2 + y_\eta^2 + z_\eta^2 + x_\zeta^2 + y_\zeta^2 + z_\zeta^2)\, d\xi\,.$$

It is interesting to define the master element of lengths $[0, 1] \times [0, 1] \times [0, 1]$ and to introduce a change of variable between the reference element and the master element (Fig.3):

$$\xi = (\xi,\eta,\zeta) \rightarrow \hat{\xi} = (\hat{\xi},\hat{\eta},\hat{\zeta}) = \left(\frac{\xi}{a_{IJK}}, \frac{\eta}{b_{IJK}}, \frac{\zeta}{c_{IJK}}\right) .$$

Corresponding to the 8 nodes of a cell, we have 8 shape functions forming the vector $\hat{N}(\hat{\xi})$ defined for $\hat{\xi}$; we also introduce the vectors $\hat{D}_i$ (respectively D_i) ($i=1,3$), derivatives of $\hat{N}(\hat{\xi})$ with respect to $\hat{\xi}$, $\hat{\eta}$ and $\hat{\zeta}$ (resp. to ξ, η and ζ):

$$\hat{D}_1 = \frac{\partial \hat{N}}{\partial \hat{\xi}}\ ,\quad D_1 = \frac{\partial \hat{N}}{\partial \xi} = \frac{1}{a_{IJK}}\hat{D}_1\,;$$

$$\hat{D}_2 = \frac{\partial \hat{N}}{\partial \hat{\eta}}\ ,\quad D_2 = \frac{\partial \hat{N}}{\partial \eta} = \frac{1}{b_{IJK}}\hat{D}_2\,;$$

$$\hat{D}_3 = \frac{\partial \hat{N}}{\partial \hat{\zeta}}\ ,\quad D_3 = \frac{\partial \hat{N}}{\partial \zeta} = \frac{1}{c_{IJK}}\hat{D}_3\,.$$

Corresponding to the 8 nodes of the master cell, the 8 shape functions forming the vector $\hat{N}(\hat{\xi})$ are classically defined:

$$\hat{N}(\hat{\xi}) = [\,(1-\hat{\xi})(1-\hat{\eta})(1-\hat{\zeta})\,;\ \hat{\xi}(1-\hat{\eta})(1-\hat{\zeta})\,;\ \hat{\xi}\hat{\eta}(1-\hat{\zeta})\,;\ \ldots\,]^t .$$

The transformation x and the gradient can then be expressed for $\hat{\xi}$:

$$x_{IJK}(\xi) = X_{IJK} . \hat{N}(\hat{\xi})$$
$$\nabla x = [\,X_1, X_2, X_3\,]$$

with

$$X_i = X . D_i\,.$$

If $\sigma(I_1, I_2, I_3)$ reduces to I_1, the elementary contribution can be expressed as:

$$\sigma_{IJK} = abc_{IJK}^{\gamma} \int_{[0,1]^3} [\frac{x_\xi^2}{a_{IJK}^2} + \frac{y_\xi^2}{a_{IJK}^2} + \frac{z_\xi^2}{a_{IJK}^2} + \frac{x_\eta^2}{b_{IJK}^2} + ...] \, d\xi.$$

I.3 Useful Expression of a Cell Deformation

The general expression for the cell deformation measure lets appear a function of the invariant of 3×3 matrix $\nabla^t x . \nabla x$. Since we later want to optimize the global mesh deformation, it is relevant to select a function σ that leads to
- a well-posed optimization problem,
- a functional easy to optimize.

The answers to these questions has been drawn in [12, 16] and a family of polynomial functions satisfying convexity properties has been exhibited; they can be written as:

$$\sigma = C_1 (I_1 - I_3 - 2) + C_2 (I_2 - 2 I_3 - 1) + K (J - 1)^2$$

with $I_3 = J^2$ and where the convexity condition requires that:

$$3K > C_1 + C_2 > 0.$$

Several interpretations can be given in two and three dimensions to the terms in the expression of σ. For example, the term $(J - 1)^2$ appears to be a penalty term that acts as a volume control term: it tends to maintain the volume of the current cell to stay as close as possible to the reference cell volume. In two dimensions, the invariants are linked by:

$$I_2 = I_1 + I_3 - 1$$

and the functional reduces to:

$$\sigma_{2d} = C (I_1 - 2J) + K (J - 1)^2$$

with K and C positive. The term $I_1 - 2J$ can be interpreted as least-square formulation of Cauchy-Riemann relations ensuring the conformity of the cells.

In three dimensions, we can consider a sub-family of functional:

$$\sigma_{3d} = C (I_1 + I_2 - 6J) + K (J - 1)^2.$$

It can be shown that the term $I_1 + I_2 - 6J$ is a least square formulation of three-dimensional relations expressing that the mesh lines ξ =Cnst., η =Cnst., ζ =Cnst. are tri-orthogonal everywhere.

In the next part devoted to the numerical implementation of the method, we will consider the general polynomial expression:

$$\sigma = C_1 I_1 + C_2 I_2 + C_3 J + C_4 I_3 - K \tag{1}$$

where K is chosen in order to obtain $\sigma = 0$ for the rigid transformation $(x(\xi) = \xi$ and $(I_1, I_2, J) = (3, 3, 1))$:

$$K = 3 C_1 + 3 C_2 + C_3 + C_4.$$

By summation of the elementary contributions σ_{IJK}, one defines a normalized measure of the mesh deformation σ:

$$\sigma = \frac{\sum_{IJK} \sigma(A_{IJK}, X_{IJK})}{\sum_{IJK} |abc_{IJK}|^{\gamma}}.$$

This measure of the mesh deformation is a function of the node coordinates, and of the reference element lengths:

$$\sigma = \sigma(A, X)$$

and is used as an optimization criterion to obtain a new mesh:

> **ALG1** 1) Construct an initial mesh X_0
> 2) Initialize X by X_0
> 3) Find $\tilde{X} = Arg.Min._X \; \sigma(A, X)$

The measure σ is entirely determined by the knowledge of the reference vector A, so will the optimized mesh.

In order to find the optimal mesh, we need to minimize a polynomial of the node coordinates X. This can be done with a Preconditioned Conjugate Gradient Algorithm that can be written as follows:

PCGA Initialization: Choose $X_0, \varepsilon, H_0 = 0$,
Iteration Loop:
1. $G_n = D\sigma(X_n)$,
2. $\tilde{G}_n = M^{-1} G_n$,
3. $\lambda_n = (\tilde{G}_n - \tilde{G}_{n-1}).G_n / \tilde{G}_{n-1}.G_{n-1}$,
4. $H_n = \tilde{G}_n + \lambda_n H_{n-1}$,
5. Descent Step: $\rho_n = Arg.Min._\rho \; \sigma(X_n + \rho H_n)$,
6. $X_n = X_{n-1} + \rho_n H_n$,
7. if $|\rho_n H_n| > \varepsilon$ set n to $n+1$ and go to 1.

This algorithm essentially requires the computation of the functional derivatives with respect to the node coordinates at the current configuration ($D\sigma(X_n)$), and the minimization of the polynomial:

$$P(\rho) = \sigma(X + \rho H).$$

However, the solution of this minimization problem is also solution of

$$P'(\rho) = D\sigma(X + \rho H).H = 0.$$

The functional σ is a polynomial of the node coordinates of degree 4 ($K = 0$) or 6 ($K \neq 0$); $P(\rho)$ has the same degree and therefore the degree of $P'(\rho)$ is 3 or 5. The convexity property on which the determination of the polynomial in (1) is based, ensures the uniqueness of the minimum of P, therefore out of the 3 or 5 roots of P', only one is real, the other 2 or 4 are complex conjugate. In preliminary studies, this feature has indeed been observed, proving the key role of the convexity condition. The descent step practically used in (PCGA) is:

Descent Step: Find ρ_n solution of $D\sigma(X_n + \rho H_n).H_n = 0$.

These remarks show that the calculation of the functional is in fact never required and may in practice never been carried out. This makes the method very efficient, especially with respect to other variational methods [6, 21] that effectively require this calculation.

In this algorithm, M is a preconditioning matrix that is meant to approach as much as possible the matrix of the second partial derivative of σ (Hessian matrix). For quadratic functionals, if M is precisely chosen to be this Hessian matrix, the algorithm converges in one iteration. For the type of functional encountered in this study, we choose M to be an approximation of the **first invariant**, that happens to be quadratic. The simplest choice is a diagonal scaling:

$$M = Diag \; I_1$$

with

$$X^t . I_1 . X = \sum_{IJK} abc_{IJK}{}^{\gamma-1} \int_{[0,a_{IJK}] \times [0,b_{IJK}] \times [0,c_{IJK}]} I_1(\xi) \, d\xi.$$

For each node, the corresponding term is the sum of the 8 surrounding cell contributions, whose values can be written:

$$M_{IJK} = abc_{IJK}{}^{\gamma} \left(\frac{1}{a_{IJK}{}^2} + \frac{1}{b_{IJK}{}^2} + \frac{1}{c_{IJK}{}^2} \right).$$

We have practically verified the benefit of this preconditioner in terms of convergence efficiency: it divides the number of iterations by 2, for a memory cost of one array and almost no cpu cost.

I.4 Multidomain Algorithm

The multidomain approach consists in decomposing the global computational domain in sub-domains, each of them meshed with a structured grid. Several decomposition topologies can be adopted: with or without sub-domain overlapping, with or without node coincidence at sub-domain interfaces. The approach chosen here is a decomposition without overlapping and with node coincidence at sub-domain interfaces; efforts have been made to enforce this last condition. However, it is still possible to use the method and the code for other cases (overlapping or no coincidence): in that case, data related to interface nodes should be omitted.

As mentioned in Section I.2, the global functional associated to one sub-domain is obtained by the summation of elementary contributions. For each node, the assembly process consists in adding the gradients associated to this node in each element. In the multidomain approach, the assembly process can be extended to the case where a node belongs to several sub-domains. The gradient component associated with an interface node will be calculated by summation of the gradients computed in all sub-domains which it belongs to. For the single domain algorithm, the topology of the assembly is rather obvious: a node (i, j, k) belongs to 8 cells that are usually labeled $(I-1, J-1, K-1)$, $(I, J-1, K-1)$, $(I, J, K-1)$, $(I-1, J, K-1)$, $(I-1, J-1, K)$, $(I, J-1, K)$, (I, J, K) and $(I-1, J, K)$. The information needed is more complex in the multidomain case: for each interface node, it is necessary to know the number of sub-domains which it belongs to, and for each of them, the indices of the node in this sub-domain.

Remark: this general assembly process enables the handling of so-called multiple nodes, that are, for example, the nodes on the wake in a C-mesh around a profile. The multidomain algorithm is necessary to treat this kind of topology if one wants to optimize the position of such nodes.

I.5 Boundary Conditions

One considers now what happens on the boundaries of the domain (**Remark**: the boundary conditions can be prescribed to nodes which do not belong to the domain boundaries but we will still refer them as boundary conditions since they will usually be prescribed on the boundaries). First note that the conjugate gradient algorithm that is used to minimize the functional can be interpreted as the iterative calculation of displacements that pull the nodes from their position at one step to their position at the next step. Nodes that are interior to the domain are free to move and are driven by the computed gradient. On the boundaries, one may also want to let the nodes be entirely free: in that case, the boundary shape will change and the domain will deform as iterations go. On the other hand, if we want to fix a set of boundary nodes, it is sufficient to cancel the corresponding displacement vectors : this will enforce a fixed-node boundary condition. Between these two types of conditions, later referred as free or fixed, it may be interesting to consider an intermediate condition referred as constrained-free: the nodes are able to move on a given surface or a given curve. In order to enforce this constraint, we compute the new position of the node, then project it on the constraint surface Σ (resp. curve Γ); the descent step and the updating step in the conjugate gradient algorithm (PCGA) can be written, for the corresponding nodes:

4.a $H_{n-\frac{1}{2}} = \tilde{G}_n + \lambda_n H_{n-1}$,
4.b $H_n = \Pi (H_{n-\frac{1}{2}})$ (resp. $H_n = \Lambda (H_{n-\frac{1}{2}})$),
6.a $X_{n-\frac{1}{2}} = X_{n-1} + \rho_n H_n$,
6.b $X_n = \Pi^\Sigma (X_{n-\frac{1}{2}})$ (resp. $X_n = \Lambda^\Gamma (X_{n-\frac{1}{2}})$)

where, in the step 4, the descent direction vector $H_{n-\frac{1}{2}}$ is projected on the plane tangent to the surface Σ (resp. on the vector tangent to the curve Γ) at the node and where, in step 6, the updating node position is projected on the surface Σ (resp. curve Γ). At equilibrium (convergence of the algorithm), the gradient will not be zero but will be orthogonal to the surface (resp. curve), therefore the node will not move anymore.

Practically, the projection can be difficult or expensive to compute; we modify this step in such way: suppose that we dispose of a parametric representation of the surface Σ: $X^\Sigma (u,v)$ (resp. of the curve Γ: $X^\Gamma(t)$); at iteration $n-1$, we know the position X_{n-1} through the parameters u_{n-1} and v_{n-1} (resp. t_{n-1}):

$$X_{n-1} = X^\Sigma (u_{n-1} , v_{n-1}) \quad (\text{resp. } X_{n-1} = X^\Gamma (t_{n-1}))$$

and we are interested in u_n and v_n (resp. t_n) defining X_n. We proceed as follows:
1) We evaluate the derivatives of the node position with respect to the parameters:

$$X_u = \frac{\partial X^\Sigma}{\partial u}(u_{n-1}, v_{n-1}) \; ; \; X_v = \frac{\partial X^\Sigma}{\partial v}(u_{n-1}, v_{n-1}) \; (\text{resp. } X_t = \frac{\partial X^\Gamma}{\partial t}(t_{n-1}))$$

this enables to obtain a vector normal to the surface in X_{n-1}:

$$N = X_u \times X_v \, .$$

2) We project the gradient on the tangent plane defined by X_u and X_v (resp. on the tangent vector defined by X_t):

$$H_n = H_{n-\frac{1}{2}} - \frac{H_{n-\frac{1}{2}} \cdot N}{\|N\|^2} N \; (\text{resp. } H_n = \frac{H_{n-\frac{1}{2}} \cdot X_t}{\|X_t\|^2} X_t) \, .$$

3) We determine the components of $\rho_n H_n$ in the tangent plane basis X_u and X_v (resp. tangent vector basis X_t):

$$\rho_n H_n = \Delta u \, X_u + \Delta v \, X_v \quad (\text{resp. } \rho_n H_n = \Delta t \, X_t) \, .$$

4) We update:

$$u_n = u_{n-1} + \Delta u \; ; \; v_n = v_{n-1} + \Delta v \quad (\text{resp. } t_n = t_{n-1} + \Delta t) \, .$$

5) We update the displacement component:

$$\rho_n H_n = X^\Sigma(u_n, v_n) - X^\Sigma(u_{n-1}, v_{n-1}) \quad (\text{resp. } \rho_n H_n = X^\Gamma(t_n) - X^\Gamma(t_{n-1})) \, .$$

These steps can be represented in Fig.5 (resp. Fig.6).

This displacement algorithm requires a parametrization of the surface considered. This subject is detailed in part C which is devoted to the Geometric Surface Modelling.
Remark: the displacement of nodes on a plane can however be easily implemented without these routines. Let N be a vector normal to the surface, then we have directly:

$$X_n = X_{n-1} + \rho_n \left(H_n - \frac{H_n \cdot N}{\|N\|^2} N \right) \, .$$

I.6 Geometric Optimization

If no indication about the physics of the problem to be solved is given, the algorithm will perform a geometrical optimization, and one of the difficulties lies in the determination of the reference lengths. Our objective in the development of this method was to construct tools to be used after classical algebraic structured mesh generation methods. These methods are typically able to fill in a mesh domain from the boundary meshes and enable the respect of given refinements and a control of the aspect ratio of the cells; however they fail in the control of the overall slope continuity of the mesh lines and the control of the orthogonality between them. Thus we suppose that we have been able to construct a first mesh X_0 that possesses some required density qualities. In order for the optimized mesh to respect (at least approximately) the mesh densities, we determine the element reference lengths A from the initial elements in 3 steps:

Step 1: we first approximate the initial element by a non rectangular parallelepiped by averaging the corresponding opposite sides; this defines three vectors a_{IJK}, b_{IJK} and c_{IJK}:

$$a_{IJK} = X_{IJK}{}^t \cdot (-1,+1,-1,+1,-1,+1,-1,+1) / 4$$
$$b_{IJK} = X_{IJK}{}^t \cdot (-1,-1,+1,+1,-1,-1,+1,+1) / 4$$
$$c_{IJK} = X_{IJK}{}^t \cdot (-1,-1,-1,-1,+1,+1,+1,+1) / 4$$

In the rest of the paragraph, we will omit the subscripts IJK.

Step 2: we then determine the reference lengths a b and c:

$$a^2 = \frac{|a \times b|}{|b|} \cdot \frac{|a \times c|}{|c|}$$

$$b^2 = \frac{|\mathbf{b} \times \mathbf{c}|}{|\mathbf{c}|} \cdot \frac{|\mathbf{b} \times \mathbf{a}|}{|\mathbf{a}|}$$

$$c^2 = \frac{|\mathbf{c} \times \mathbf{a}|}{|\mathbf{a}|} \cdot \frac{|\mathbf{c} \times \mathbf{b}|}{|\mathbf{b}|}.$$

These expressions have been shown to be compatible with the least-square formulation mentioned earlier. They can be re-written:

$$a = |\mathbf{a}| | \sin(\mathbf{a},\mathbf{b}) \sin(\mathbf{a},\mathbf{c}) |^{1/2}$$
$$b = |\mathbf{b}| | \sin(\mathbf{b},\mathbf{c}) \sin(\mathbf{b},\mathbf{a}) |^{1/2}$$
$$c = |\mathbf{c}| | \sin(\mathbf{c},\mathbf{a}) \sin(\mathbf{c},\mathbf{b}) |^{1/2}$$

where $(\mathbf{x},\mathbf{y})$ denotes the angle between the two vectors $\mathbf{x}$ and $\mathbf{y}$.

Step 3: the reference volume is then

$$abc = |\mathbf{a}| |\mathbf{b}| |\mathbf{c}| | \sin(\mathbf{a},\mathbf{b}) \sin(\mathbf{b},\mathbf{c}) \sin(\mathbf{c},\mathbf{a}) |$$

which is in general slightly different from the initial volume, $(\mathbf{a}, \mathbf{b}, \mathbf{c})$. In order to keep a global reference volume equal to the domain volume, the reference lengths are multiplied by a factor ρ such that:

$$\rho^3 = \frac{(\mathbf{a},\mathbf{b},\mathbf{c})}{|\mathbf{a}| |\mathbf{b}| |\mathbf{c}| | \sin(\mathbf{a},\mathbf{b}) \sin(\mathbf{b},\mathbf{c}) \sin(\mathbf{c},\mathbf{a}) |}.$$

These three steps are carried out for all the cells, they enable the determination of the reference vector A from a mesh X, we will call this transformation A:

$$A = A(X).$$

Remark: if the initial mesh is not satisfactory, it is also possible to define the reference cells with respect to the repartitions on the boundary of the domain. In that case, the values of a, b and c are first computed on the edges or on the faces of the initial mesh, and then interpolated inside the domain.

The overall geometrical optimization algorithm can be written as follows:

ALG2	1) Construct an initial mesh X_0 2) Compute the reference lengths $A_0 = A(X_0)$ 3) Initialize X by X_0 4) Find $\tilde{X} = Arg.Min._X \ \sigma(A_0, X)$

II. PHYSICAL OPTIMIZATION OR ADAPTION

II.1 Basic Algorithm

Within the framework presented above, mesh adaption can easily be handled: if there exists a way to extract information from a physical solution and assign new reference lengths to each cell, the optimization of the functional defined with these lengths will produce the adapted grid. The difficulty is once again shifted to the determination of the reference lengths.

It appears natural to retain some information from the initial mesh in the determination of these lengths, so we chose to determine them as modification of the reference lengths introduced for the geometrical optimization in the previous paragraph. To do so, we introduce for each element IJK three weighting factors, ω^a_{IJK}, ω^b_{IJK} and ω^c_{IJK}, that will act on the three element reference lengths (Fig. 4); so we note

$$\Omega_{IJK} = (\omega^a_{IJK}, \omega^b_{IJK}, \omega^c_{IJK})$$
$$\Omega_{IJK} \otimes A_{IJK} = (\omega^a_{IJK} a_{IJK}, \omega^b_{IJK} b_{IJK}, \omega^c_{IJK} c_{IJK})$$
$$\Omega \otimes A = [\Omega_{IJK} \otimes A_{IJK}]_{I=1,imax-1; J=1,jmax-1; K=1,kmax-1}.$$

Depending on the values of the coefficients ω^α_{IJK} ($\alpha = a,b,c$) with respect to 1, the adaptive optimization will tend to refine or impoverish the mesh around the cell IJK in the corresponding direction (for example on the mesh line "$I\ varying$" for $\alpha = a$).

$\omega \leq 1$: cell contraction

$\omega \geq 1$: cell expansion.

Practically, we consider the solution ϕ to be given at nodes of the initial mesh:

$$\Phi_0 = \phi(X_0) = [\phi(x_{0,ijk})]_{i=1,\ imax;\ j=1,\ jmax;\ k=1,\ kmax}$$

and we determine the weights from Φ_0:

$$\Omega = \Omega(\Phi_0). \tag{2}$$

Several ways enabling the transformation of the solution into weights will be presented in the last part. The overall physical optimization or adaption algorithm can be written as follows:

> **ALG3**
> 1) Construct an initial mesh X_0
> 2) Compute the reference lengths $A_0 = A(X_0)$
> 3) Compute the weights $\Omega = \Omega(\Phi_0)$
> 4) Initialize X by X_0
> 5) Find $\tilde{X} = Arg.Min._X\ \sigma(\Omega(\Phi_0) \otimes A_0, X)$

II.2 Motivation for a Field Interpolation

The minimization problem just introduced enables the construction of adapted grids presenting the desired refinements; however, in the third step of **ALG3**, the weights Ω are computed from the field ϕ, supposed to be known at the initial mesh nodes. In one dimension, this would be equivalent to finding a repartition $x(\xi)$ from the equation

$$\frac{dx}{d\xi} = \omega(\xi). \tag{3}$$

But during the adaption, cells move with respect to their original position. thus, the adaption moves away from the region where it should take place. In one dimension, equation (3) should be replaced by:

$$\frac{dx}{d\xi} = \omega(x). \tag{4}$$

A thorough study, theoretical in one dimension and numerical in two dimensions, of the differences and consequences in the adaption whether one uses (3) or (4) has been described in [17]. It showed the need to modify the basic adaption algorithm **ALG3** and to use an approach similar to (3-4) in order to obtain properly adapted meshes in 2 or 3 dimensions. The generalization of (3-4) to the mesh optimization algorithm introduced above would lead to a minimization where the control X would also have to appear in the weight computation

$$\Omega = \Omega(\Phi)\ \text{with}\ \Phi = \phi(X)$$

instead of (2); this makes the problem much more difficult to solve. An iterative fixed-point algorithm can however lead to the desired result; it consists in computing after each optimization the weights on the newly obtained mesh; it writes:

> **ALG4**
> 1) Construct an initial mesh X_0, set n to 0
> 2) Compute the reference lengths $A_0 = A(X_0)$
> 3) Interpolate $\Phi_n = \phi(X_n)$
> 4) Compute the weights $\Omega = \Omega(\Phi_n)$
> 5) Initialize X by X_n
> 6) Find $X_{n+1} = Arg.Min._X\ \sigma(\Omega(\Phi_n) \otimes A_0, X)$
> 6) If not converged, set n to $n+1$ and go to 3)

The cost of this algorithm modification is the iterative computation at current nodes (X_n) of the adaption field ϕ, known at initial mesh nodes X_0; practically, one has observed that the number of iterations needed in **ALG4**, and therefore the number of such interpolations, is very limited: not exceeding 3. This new algorithm turns out to furnish the desired results, in particular the gaps between physical solution and adaption disappear [17].

Remark: another conclusion obtained in [17] is that the interpolation process tends to produce refinements that are more spread apart than the ones obtained without it.

The problem that this last algorithm **ALG4** poses is the accuracy of the interpolation. In [17] several classical interpolation methods are reviewed. It emerges from this investigation that the tri-linear interpolation method chosen in our approach is sufficiently robust and accurate to treat the problem of interpolation in an adaption process.

PART B
NUMERICAL IMPLEMENTATION OF THE METHOD

In this part, we show how the variational method previously described has been implemented. We restrict the presentation to the computation of elementary contributions of the various terms of interest; the global values are then assembled in a standard finite element manner. A first implementation developed by CABELLO [5] is first briefly presented; this work ended up with a code that turned out to be difficult to work with and to be memory and cpu expensive. In order to overcome these difficulties, a new implementation was performed and some of its elements are presented: the computation of the functional, the gradient and the polynomial coefficients are described. The results obtained with the resulting code *OPTIM3D* [18] will illustrate the next part.

I. COMPUTATION OF ELEMENT CONTRIBUTIONS

I.1 Analytical Calculation

The integrals encountered in Section A.I are integrals over the reference cell or the unit master cell. The functions integrated are polynomials, so, provided that σ has the polynomial form (1), the integrals can be exactly and analytically evaluated. For example, for the first invariant we can obtain:

$$\begin{aligned} I_1 = [\ & r_1^2 + r_2^2 + r_3^2 + r_4^2 + r_5^2 + r_6^2 \\ & + r_7^2 + r_8^2 + r_9^2 + r_{10}^2 + r_{11}^2 + r_{12}^2 \\ & + r_1.r_3 + r_1.r_5 + r_3.r_7 + r_5.r_7 + r_2.r_4 + r_2.r_6 \\ & + r_4.r_8 + r_6.r_8 + r_9.r_{12} + r_9.r_{10} + r_{11}.r_{12} + r_{10}.r_{11} \\ & + r_4.r_8 + r_6.r_8 + r_9.r_{12} + r_9.r_{10} + r_{11}.r_{12} + r_{10}.r_{11} \\ & + \tfrac{1}{2} (r_1.r_7 + r_3.r_5 + r_4.r_6 + r_2.r_8 + r_9.r_{11} + r_{10}.r_{12}) \ ] / 9 \end{aligned}$$

where the r_i are the edges introduced in Fig.3. For the second and third invariants, this calculation is rather tedious; the results can be found in [5].

I.2 Numerical Calculation: Use of an Integration Rule

Even though we have demonstrated in Part A that the computation of the functional is not required for the minimization algorithm, it is useful to dispose of routines performing this calculation for verification purposes; we develop this calculation process when one uses numerical integration. It relies on the use of an integration rule: it is well known, in the finite element literature, that the integration required in the matrix are carried out using approximations as:

$$\int_\Omega f(\xi) \, d\xi = \sum_{l=1,L} W^l f(\xi^l)$$

where the ξ^l are the so-called integration points and W^l are weights associated with each point. For polynomials of degree 2, this integration is exact when the $L = 8$ Gauss points ($\xi^l = \tfrac{1}{2} (1 \pm 1/\sqrt{3}, 1 \pm 1/\sqrt{3}, 1 \pm 1/\sqrt{3})$) are chosen as integration points, with $W^l = 1/8$ for $l=1,8$. This integration is exact for I_1, not exact for I_2 and I_3 but sufficient.

We introduce several notation modifications: for

$$x_{IJK} = X_{IJK} \cdot N(\xi) \quad \text{and} \quad X_i|_{IJK} = X_{IJK} \cdot D_i(\xi) \quad (i=1,3)$$

the global functional is computed by the series of calculations:

$$X_1|_{IJK}^l = X_1|_{IJK} \; (\xi^l) \; ; \; X_2|_{IJK}^l = X_2|_{IJK} \; (\xi^l) \; ; \; X_3|_{IJK}^l = X_3|_{IJK} \; (\xi^l)$$

$$I_1|_{IJK}^l = (X_1|_{IJK}^l)^2 + (X_2|_{IJK}^l)^2 + (X_3|_{IJK}^l)^2$$

$$I_2|_{IJK}^l = (X_1|_{IJK}^l \times X_2|_{IJK}^l)^2 + (X_2|_{IJK}^l \times X_3|_{IJK}^l)^2 + (X_3|_{IJK}^l \times X_1|_{IJK}^l)^2$$

$$J_{IJK}^l = (X_1|_{IJK}^l , X_2|_{IJK}^l , X_3|_{IJK}^l)$$

$$I_3|_{IJK}^l = (J_{IJK}^l)^2$$

$$\sigma_{IJK}^l = abc_{IJK}^\gamma \; (\; C_1 \, I_1|_{IJK}^l + C_2 \, I_2|_{IJK}^l + C_3 \, I_3|_{IJK}^l + C_4 \, J_{IJK}^l - K \;)$$

$$\sigma = \frac{\sum\limits_{IJK} \sum\limits_{l=1,L} W^l \, \sigma_{IJK}^l}{\sum\limits_{IJK} | \, abc_{IJK} \, |^\gamma} \; .$$

From a practical point of view, the integration points are computed and stored at the beginning of the execution, as well as the shape function vectors and derivatives evaluated at these points. In the following, we will remove the indices *IJK* and *l*, and understand that a double summation over elements and integration points is carried out to obtain global quantities.

I.3 Influence of the Integration Rule

In order to increase the computational efficiency of the method, it may be interesting to consider the use of so-called under-integration: the numerical integration previously described relies on the computation of all quantities at 8 integration points, then on the summation of all 8 contributions. If one desires to reduce computational time, one may consider to use less than 8 points, say 6, 4, 2 or even 1, provided that there is no degradation of the solution quality. Indeed the difficulties related to under-integration has been throughoutly studied in the early 80's [4,11] for problems in Solid Mechanics. For Q_1-trilinear elements involved in this study, 8 Gauss points exactly integrate stiffness matrices associated to the Laplace operator in a rectangular reference cell and is sufficient (no loss in accuracy) for non rectangular reference cell [11] as well as for non-linear operator; indeed the key condition is that the 8-point rule preserves the rank and the kernel of the elementary Laplace operator (*rank* $A^e = 7$; *ker* $A^e = R$).

The Gauss rule of order just lower involves only one point (the centroid of the reference cell). This 1-point rule leads to a rank of 3, introducing 4 spurious modes known as hourglass modes; two classes of method overcome this difficulty. A priori-methods restore a rank-sufficient matrix by adding a stabilization matrix, constructed from the spurious modes:

$$\tilde{A}^e = A^e_{under.} + \Sigma_i \, \varepsilon_i \, H_i \, . \, H_i^t$$

where the ε_i are scaling coefficients. This kind of method can be extended to the optimization problem: a stabilization functional can be added after underintegration.

On the other hand, a posteriori-methods tries to solve the under-integrated problem, up to within arbitrary spurious modes, and then eliminate the instabilities in a post-processing operation; this can also be extended to optimization problem.

A simple way to evaluate performances of non standard integration rules, is to observe the process of construction of the elementary matrix; we can obtain an upper estimate of the rank of the numerically integrated matrix:

$$rank \; A^e_{num.} \leq \min (\, 3 \, L \, , \, 7 \,)$$

where L is the number of integration points. This inequality shows that at least 3 integration points are necessary, but may not be sufficient. Since we do not have theoretical results giving more accurate estimates on the rank of the under-integrated matrix, future works aimed at an improvement in the computational efficiency of the method will consist in trying 1-, 2-, 4- and 6- point rules to see if oscillations appear. For the 1-point rule, the point is the centroid of the reference cell, for the 2- (respectively 4-) point rule, the integration points will be 2 (resp. 4) out of the 8 Gauss points located in opposite corners of the cube. 4 (resp. 2) choices are possible: for example we will consider a 2-point rule involving P_1 and P_8, and a 4-point rule involving P_1, P_4, P_6 and P_7. For the 6-points rule, the points are the 6 Gauss points located in the opposite faces of the cube. These Gauss points are chosen such that:

$$\xi^{1,2} = \tfrac{1}{2}\,(1\pm 1/\sqrt{3}\,,\,0\,,\,0)\ ;\ \xi^{3,4} = \tfrac{1}{2}\,(0\,,\,1\pm 1/\sqrt{3}\,,\,0)\ ;\ \xi^{5,6} = \tfrac{1}{2}\,(0\,,\,0\,,\,1\pm 1/\sqrt{3}).$$

II. COMPUTATION OF THE FUNCTIONAL

In compact notation, we have, for the elementary contribution:

$$\sigma = (abc)^\gamma\,(\,C_1 I_1 + C_2 I_2 + C_3 J + C_4 I_3 - K\,)$$

with

$$X_i = X\,.\,D_i \quad (i=1,3) \tag{5}$$

$$I_1 = X_1^2 + X_2^2 + X_3^2 \tag{6}$$

$$I_2 = (X_1 \times X_2)^2 + (X_2 \times X_3)^2 + (X_3 \times X_1)^2 \tag{7}$$

$$J = (X_1, X_2, X_3) \tag{8}$$

$$I_3 = (X_1, X_2, X_3)^2. \tag{9}$$

In these expressions, the invariants are dimensionless. The vectors X_1, X_2 and X_3 respectively correspond to x_ξ, x_η and x_ζ, discretized and evaluated at the integration points. The functional is entirely determined by the knowledge of:
- the 5 constants C_1, C_2, C_3, C_4 and γ,
- the reference lengths a, b and c for each of the cells.

Remark: we have the remarquable identity:

$$I_1 + I_2 - 6J = (X_1 - X_2 \times X_3)^2 + (X_2 - X_3 \times X_1)^2 + (X_3 - X_1 \times X_2)^2.$$

Note also that, with the compact notation, we have:

$$\int_\Omega (\det \nabla^t x.\nabla x)^2\,d\xi = I_3 \neq J^2 = [\int_\Omega (\det \nabla^t x.\nabla x)\,d\xi\,]^2.$$

The computation of the functional requires for each element and each integration point the computation of:
- three vectors X_1, X_2, X_3,
- three vector products $X_1 \times X_2, X_2 \times X_3, X_3 \times X_1$,
- seven dot products, including $(X_1, X_2, X_3) = (X_1 \times X_2)\,.\,X_3$,

and the summation of all the contributions.

III. COMPUTATION OF THE FUNCTIONAL GRADIENT

At each iteration, we need to compute the gradient of the functional σ for the current configuration. We remind that σ is a polynomial of the node coordinates, hence, by gradient, we mean the derivative of this polynomial with respect of these node coordinates. The gradient is a field of vectors associated to each mesh node. As previously mentioned, the contributions from each cell are obtained by summation of the values computed at each integration point, then assembled to form the global vector. At the cell level, the gradient is a three-dimensional vector associated to the 8 cell nodes; it can be written as three 3×8 arrays. One can show that the usual derivation formulae work for the expressions (6-9) and that we have:

$$DI_1 = 2\,(X_1\,.\,D_1^t + X_2\,.\,D_2^t + X_3\,.\,D_3^t)$$

$$DI_2 = 2\,[\,(X_2^2 + X_3^2)X_1 - (X_1\,.\,X_2)X_2 - (X_1\,.\,X_3)X_3\,]\,.\,D_1^t$$
$$+ 2\,[\,(X_3^2 + X_1^2)X_2 - (X_2\,.\,X_3)X_3 - (X_2\,.\,X_1)X_1\,]\,.\,D_2^t$$
$$+ 2\,[\,(X_1^2 + X_2^2)X_3 - (X_3\,.\,X_1)X_1 - (X_3\,.\,X_2)X_2\,]\,.\,D_3^t$$

$$DJ = (X_2 \times X_3)\,.\,D_1^t + (X_3 \times X_1)\,.\,D_2^t + (X_1 \times X_2)\,.\,D_3^t$$

$$DI_3 = 2J\,[\,(X_2 \times X_3)\,.\,D_1^t + (X_3 \times X_1)\,.\,D_2^t + (X_1 \times X_3)\,.\,D_3^t\,].$$

The computation of the gradient essentially requires, for each element and each integration point, the computation of:
- three vectors X_1, X_2, X_3,
- three vector products $(X_i \times X_j)$,

- seven dot products, including $(X_1, X_2, X_3) = (X_1 \times X_2) . X_3$, and the assembling summation of all the contributions.

IV. COMPUTATION OF THE POLYNOMIAL COEFFICIENTS

More interesting is the computation of the coefficients of the polynomial to minimize in the descent step of the algorithm. At a current iteration (X and H given), we write this polynomial:

$$P'(\rho) = D\sigma(X+\rho H) = a_0 + a_1 \rho + \frac{a_2}{2}\rho^2 + \frac{a_3}{3!}\rho^3 + \frac{a_4}{4!}\rho^4 + \frac{a_5}{5!}\rho^5$$

and we evaluate the coefficients as derivatives of P at 0. We remind that σ is a scalar function of the node coordinates (vector X), therefore the i th derivative of σ is a tensor of order i that must be applied to i vectors in order to obtain the real $D^i\sigma(X).(H_1...,H_i)$. After simplifications due to the respective order of the invariants with respect to x, we obtain:

$$a_0 = D\sigma(X).H$$
$$= C_1 DI_1(X).H + C_2 DI_2(X).H + C_3 DJ(X).H + C_4 DI_3(X).H$$
$$a_1 = D^2\sigma(X).(H,H)$$
$$= C_1 D^2I_1(X).(H,H) + C_2 D^2I_2(X).(H,H)$$
$$+ C_3 D^2J(X).(H,H) + C_4 D^2I_3(X).(H,H)$$
$$a_2 = D^3\sigma(X).(H,H,H)$$
$$= C_2 D^3I_2(X).(H,H,H) + C_3 D^3J(X).(H,H,H)$$
$$+ C_4 D^3I_3(X).(H,H,H)$$
$$a_3 = D^4\sigma(X).(H,H,H,H)$$
$$= C_2 D^4I_2(X).(H,H,H,H) + C_4 D^4I_3(X).(H,H,H,H)$$
$$a_4 = D^5\sigma(X).(H,H,H,H,H) = C_4 D^5I_3(X).(H,H,H,H,H)$$
$$a_5 = D^6\sigma(X).(H,H,H,H,H,H) = C_4 D^6I_3(X).(H,H,H,H,H,H).$$

The quantities mentioned here are computed as follows. We first define the vectors H_i, for $i=1,3$, similarly to (5):

$$H_i = H . D_i$$

then the various quantities appearing in the coefficients can be expressed as:

for I_1:
$$DI_1(X).H = 2(X_1 . H_1 + X_2 . H_2 + X_3 . H_3)$$
$$D^2I_1(X).(H,H) = 2(H_1^2 + H_2^2 + H_3^2)$$

for I_2:
$$DI_2(X).H = 2[(X_1 \times X_2).(H_1 \times X_2 + X_1 + H_2)$$
$$+ (X_2 \times X_3).(H_2 \times X_3 + X_2 + H_3)$$
$$+ (X_3 \times X_1).(H_3 \times X_1 + X_3 + H_1)]$$
$$D^2I_2(X).(H,H) = 2[2(X_1 \times X_2).(H_1 \times H_2) + (H_1 \times X_2 + X_1 \times H_2)^2$$
$$+ 2(X_2 \times X_3).(H_2 \times H_3) + (H_2 \times X_3 + X_2 \times H_3)^2$$
$$+ 2(X_3 \times X_1).(H_3 \times H_1) + (H_3 \times X_1 + X_3 \times H_1)^2$$
$$D^3I_2(X).(H,H,H) = 12[(H_1 \times H_2).(H_1 \times X_2 + X_1 + H_2)$$
$$+ (H_2 \times H_3).(H_2 \times X_3 + X_2 + H_3)$$
$$+ (H_3 \times H_1).(H_3 \times X_1 + X_3 + H_1)]$$
$$D^4I_2(X).(H,H,H,H) = 24[(H_1 \times H_2)^2 + (H_2 \times H_3)^2 + (H_3 \times H_1)^2]$$

for J:
$$DJ(X).H = (H_1, X_2, X_3) + (X_1, H_2, X_3) + (X_1, X_2, H_3)$$

$$D^2 J(X).(H,H) = (H_1, H_2, X_3) + (H_1, X_2, H_3) + (X_1, H_2, H_3)$$
$$D^3 J(X).(H,H,H) = 3 (H_1, H_2, H_3)$$

for I_3:

$$DI_3(X).H = 2 J \, DJ(X).H$$
$$D^2 I_3(X).(H,H) = 2 [DJ(X).H]^2 + 2 J \, D^2 J(X).(H,H)$$
$$D^3 I_3(X).(H,H,H) = 2 [3 DJ(X).H \cdot D^2 J(X).(H,H) + J \, D^3 J(X).(H,H,H)]$$
$$D^4 I_3(X).(H,H,H,H) = 8 DJ(X).H \cdot D^3 J(X).(H,H,H) + 6 (D^2 J(X)(H,H))^2$$
$$D^5 I_3(X).(H,H,H,H,H) = 20 D^2 J(X).(H,H) \cdot D^3 J(X).(H,H,H)$$
$$D^6 I_3(X).(H,H,H,H,H,H) = 20 (D^3 J(X).(H,H,H))^2.$$

This computation requires, for each element and each integration point, the computation of:
- six vectors (X_i and H_i for $i=1, 3$),
- nine vector products ($X_i \times X_j$, $H_i \times H_j$ and $X_i \times H_j + H_i \times X_j$),
- 28 dot products, including $(X'_1, X'_2, X'_3) = (X_1 \times X_2) \cdot X_3$,

and the summation of all the contributions.

In order to avoid problems in the calculations of the roots of $P'(\rho)$, it can be useful to normalize it in a certain way; indeed we do not know a priori the order of magnitude of H, but we have the following estimates:

$$a_i = O(|H|^{i+1}) \text{ for } i=0,5.$$

Therefore we introduce a dimensionless root $\hat{\rho}$:

$$\hat{\rho} = \frac{a_1}{a_0} \rho$$

and the polynomial can be written:

$$Q(\rho) = 1 + \rho + \frac{a_2 a_0}{2 a_1^2} \rho^2 + \frac{a_3 a_0^2}{3! a_1^3} \rho^3 + \frac{a_4 a_0^3}{4! a_1^4} \rho^4 + \frac{a_5 a_0^4}{5! a_1^5} \rho^5$$

where all the new coefficients are of order 1.

PART C
RESULTS

The method presented in the previous parts has been tested for a few years at ONERA and we refer to [5, 16-18] for the first results. A new and more efficient programming of the method has been undertaken and is detailed in part B. The first section is devoted to the multi-block grid optimization, showing the method capability to handle complex three-dimensional topologies. The next section shows in details the sequence of procedures that lead to the adapted grids.

I. MULTIDOMAIN RESULT

We first present grids obtained with the multiblock code. The grid considered was constructed for the study of the interaction between a nacelle and a wing fixed between two plane. Figure 7 shows a three-dimensional view of the surface grids of the two bodies and the grid on one of the planes (root plane). This configuration is simplified and does not include the pylon supporting the nacelle; neverless, it is close to the geometry of the second engine under a long-distance carrier wing and it is considered difficult due to the relative high sweep angle, the small chord of the wing and the close position of the nacelle under the wing. The domain has been decomposed into four sub-domains: one above the wing, one between the wing and the nacelle, one under the nacelle and one inside the nacelle. The boundary conditions are the following: the nodes on the side planes are allowed to slide, the node on the wing and nacelle surfaces are fixed and the other boundary nodes are free. The dimensions of the grid are respectively 79× 31×18, 79×31× 14, 79×31×18 and 26×7×7.

Figures 8 to 10 show the results of the single-block optimization for the subdomains and their global deformation. We can remark that the single-block optimization gives a better volume smoothness and a better orthogonal properties. However the solution strongly depends on the initial mesh due to the computation of the reference cells with respect to the initial mesh cells.

The global deformation of the domains is large, but multi-block tests have shown that the freedom of the block-interface boundaries improves the mesh quality. Figures 9 and 10 show the grids before and after single- or multi-block optimization: in particular one can see the deformation of the block interfaces obtained when the multi-block code is used. For this case (4 blocks, 125 000 nodes), the running time on CRAY-XMP-416 is about half an hour.

II. ADAPTION RESULTS

II.1. NACA0012 Airfoil

At the present time, the most commonly used strategy in structured grid adaption is based on the use of error indicators, rough information about the possible location of the errors; they are generally determined as gradients of a variable characteristic of the flow solution and, in that case, the adaption will refine the grid where large variations of the solution occur. The variable driving the adaption can be, for instance in Euler calculations, the pressure, the Mach number or the density; in the following, we will consider the gradient of the static pressure, and we will use the example shown on Fig.11 (Euler calculation, $M_\infty = 0.85$, $\alpha = 1.0^o$) to illustrate the functional construction. Within our framework, the weights are computed according to the following steps:

1) The variable gradient components are computed and normalized between 0 and 1. The gradient considered can either be the real physical gradient ($\nabla_x \phi = \lim_{\Delta x \to 0} \Delta\phi/\Delta x$) or a logical gradient where only differences are considered ($\nabla_i \phi = \phi_{i+1} - \phi_i$); the latter one corresponds to a variation per cell in each index direction, and turns out to be more suitable for the computation of the adaption weights (Fig.12.a).

2) In order to further limit the adaption areas, the gradient components are limited: a value between 0 and 1 is chosen; above (respectively below) this value, the gradient components are set to 1 (resp. to 0); this delimits the adaption areas. This lower value acts as a filter to ignore transient and minor values while the upper value extends the adaption zone. Alternatively, the number or the percentage of cells forming this area can be fixed and the adaption threshold can then be recovered.

3) The limited gradient components are smoothed in order to later obtain regular weights and reference lengths (Fig.12.b). A smoother based on a Laplacian operator proves to be efficient in removing any oscillation or sharp variation in the weights.

4) The weights in each of the three directions are calculated as a linear function of the values obtained in 3); this linear function is determined by the refinement ($\omega_0 \leq 1$) and the enlargement parameters ($\omega_1 \geq 1$). Practically, the ratio $\rho = \omega_1/\omega_0$ is given and the parameters are evaluated from a volume conservation equation.

5) These weights are then used to modify the reference lengths according to the presentation done in A.II.1. On Fig.13, we have plotted contours of one of the reference lengths before (initial mesh) and after multiplication by the weight. This kind of plot enables an easy control of the functional construction before its minimization.

6) By minimization of the functional obtained with these new reference cells, we obtain, after 3 cycles of interpolation and 20 iterations of conjugate gradient algorithm in each cycle, the adapted mesh shown on Fig.14.a. A solution of the Euler equation on this new mesh shows the benefit of the adaption (Fig.14.b).

II.2. ONERA M6 Wing

The adaption of a C-H mesh around the ONERA M6 (Fig.15) wing was performed. The grid considered is formed by one domain of 92906 nodes (103×41×22). As above the error indicator is based on the pressure gradient of an Euler flow solution ($M_\infty = 0.835$ and $\alpha = 3.06^o$) obtained around this geometry (Fig.15.a). In the computation of the transonic flow [7], a λ shock appears along the wing surface (Fig. 15.b). The objective of the adaption is to refine the mesh along the pressure shock. Within the framework used in the previous examples, based on the pressure field, the weight factors were computed and the resulting adapted mesh is presented (Fig. 16, 17). We notice the nice behavior of the adaption along the upper surface of the wing where the mesh lines are correctly aligned with the λ shock. The node motion along the wing surface were allowed by the use of three-dimensional surface interpolation based on bi-cubic spline (Part D). Practically the node motion on surfaces is obtained by projection of the node motion onto the interpolating surface as described in section A.I.5.

III. CONCLUSION OF PARTS A-B-C

From the tests presented before, and from other runs of the code, it appears that several choices can be made:
- The functional $I_1 + I_2 - 6J$ leads to good results and should be adopted for optimization;
- Eventhough it is implemented, the term $(J-1)^2$ does not seem to be useful for mesh optimization; it rather slows the convergence. This term was introduced for the mesh adaption.
- The four-point rule is sufficient: it preserves the quality of solutions and divides the computational time by a factor of two. Neverless the 6 point rule give a symmetric contribution in the integration rule for a computational time reduced by 1.5.
- The best choice for the parameter γ is 0.
- The program can be considered as well vectorized: the option *NOVECTOR* on CRAY multiplies the computational time by 15.
- The program is also suitably written to work with autotasking on a parallel machine: on an ALLIANT FX2800 the waiting time is nearly reduced by a factor proportional to the number of processor used.
- The multiblock code is able to handle arbitrary topologies.
- The adaption results show the good capability to obtain non-isotropic mesh adaption. Neverless the tests performed show that the crucial point is the determination of a good adaption criterion.
- The capability to accept any kind of boundary condition and specially surface node motion enables the handling by the code of any practical mesh optimization or adaption problem.

PART D
SURFACE MESH GENERATION AND OPTIMIZATION

I. SURFACE INTERPOLATION

Although realistic surfaces are always defined by CAD systems in great complexity (large number of patches, high order polynomials), it is often sufficient, for CFD purposes to simplify and redefine surfaces from a limited number of data. In this part, we will only consider interpolated surfaces, constructed from a set of control points forming a grid, and passing through these points. We present a method for the creation of such an interpolated surface in this section; more precisely, we define a continuously differentiable (C^1) map from the unit square in R^2 into R^3. We suppose that a set of control points is given by their coordinates $\{X_{ij}^{\Sigma} \ i=1, imax; \ j=1, jmax\}$; The construction of the map is based on the following steps:

I.1 Construction of a Curve Network and Parametrization

We first build two families (one for each index direction) of C^1 piecewise cubic curves that respect the given points and slopes evaluated by either a 3-point [19] or a 5-point [20] scheme; each control point is located at the intersection of two curves, one of each family. The use of a quasi-intrinsic parametrization [19] of these curves (s^j in the i-direction and t^i in the j-direction) provides a good approximation of the arc length on these curves. This parametrization furnishes unit vectors tangent to each curve at the control points:

$$S_{ij} = \frac{\partial X^{\Sigma}}{\partial s^j} |_{ij} \text{ and } T_{ij} = \frac{\partial X^{\Sigma}}{\partial t^i} |_{ij} .$$

We also calculate second derivatives at the control points by the same 3- or 5-point scheme applied to the first derivatives.

The computation of $\frac{\partial^2 X^\Sigma}{\partial s^2}|_{ij}$ and $\frac{\partial^2 X^\Sigma}{\partial t^2}|_{ij}$ (respectively $\frac{\partial^2 X^\Sigma}{\partial s \partial t}|_{ij}$) uses the coordinates of 5 control points (resp. 9 points) for the 3-point scheme and 9 control points (resp. 25 points) for the 5-point scheme.
Remark: special off-centered calculations are done to obtain these quantities close to the boundaries.

I.2 Parameter Scaling

The lengths of these $imax + jmax$ curves (s^j_{max} and t^i_{max}) are used as scaling factors to normalize these arc lengths:

$$u_{ij} = s^j(X^\Sigma_{ij}) / s^j_{max} \quad \text{and} \quad v_{ij} = t^i(X^\Sigma_{ij}) / t^i_{max}.$$

For each control point X^Σ_{ij}, we obtain two values, (u_{ij}, v_{ij}) which define a point U_{ij} in the unit square. The tangent vectors become:

$$\frac{\partial X^\Sigma}{\partial u}|_{ij} = s^j_{max} \frac{\partial X^\Sigma}{\partial s^j}|_{ij} \quad \text{and} \quad \frac{\partial X^\Sigma}{\partial v}|_{ij} = t^i_{max} \frac{\partial X^\Sigma}{\partial t^i}|_{ij}.$$

Similarly, second derivatives with respect to u and v become:

$$\frac{\partial^2 X^\Sigma}{\partial u^2}|_{ij} = s^j_{max}{}^2 \frac{\partial^2 X^\Sigma}{\partial s^{j2}}|_{ij} \quad ; \quad \frac{\partial^2 X^\Sigma}{\partial v^2}|_{ij} = t^i_{max}{}^2 \frac{\partial^2 X^\Sigma}{\partial t^{i2}}|_{ij}$$

and

$$\frac{\partial^2 X^\Sigma}{\partial u \partial v}|_{ij} = s^j_{max} t^i_{max} \frac{\partial^2 X^\Sigma}{\partial s^j \partial t^i}|_{ij}.$$

This process defines a discrete application Θ^{-1} between the control point set $\{X^\Sigma_{ij}\}$ and the point set $\{U_{ij}\}$ in the unit square:

$$U_{ij} = \Theta^{-1}(X^\Sigma_{ij}) \quad \text{for } i=1, imax; j=1, jmax.$$

The construction of the surface will consist in defining a C^1 function Θ, mapping the entire unit square onto a surface in R^3.

I.3 Map onto the Parameter Patch

For a point U (coordinates (u, v) in the unit square) belonging to a parameter patch $U_{IJ} = [U_{ij}, U_{i+1j}, U_{i+1j+1}, U_{ij+1}]$, one defines its local coordinates (α, β) with respect to these four points by the relation:

$$U = P_{IJ}(\alpha, \beta)$$
$$= N_0(\alpha, \beta) U_{IJ} + N_1(\alpha, \beta) \frac{\partial U}{\partial \alpha}|_{IJ} + N_2(\alpha, \beta) \frac{\partial U}{\partial \beta}|_{IJ} + N_3(\alpha, \beta) \frac{\partial^2 U}{\partial \alpha \partial \beta}|_{IJ}$$

where the shape functions N_0, N_1, N_2, N_3 are given by:

$N_0(\alpha,\beta) = [\ \mu_0(\alpha)\mu_0(\beta),\quad \mu_0(\alpha)\mu_0(1-\beta),\quad \mu_0(1-\alpha)\mu_0(1-\beta),\quad \mu_0(1-\alpha)\mu_0(\beta)]^t$
$N_1(\alpha,\beta) = [\ \mu_0(\alpha)\mu_1(\beta),\ -\mu_0(\alpha)\mu_1(1-\beta),\ -\mu_0(1-\alpha)\mu_1(1-\beta),\ \mu_0(1-\alpha)\mu_1(\beta)]^t$
$N_2(\alpha,\beta) = [\ \mu_1(\alpha)\mu_0(\beta),\quad \mu_1(\alpha)\mu_0(1-\beta),\ -\mu_1(1-\alpha)\mu_0(1-\beta),\ -\mu_1(1-\alpha)\mu_0(\beta)]^t$
$N_3(\alpha,\beta) = [\ \mu_1(\alpha)\mu_1(\beta),\ -\mu_1(\alpha)\mu_1(1-\beta),\quad \mu_1(1-\alpha)\mu_1(1-\beta),\ -\mu_1(1-\alpha)\mu_1(\beta)]^t$

where the functions μ_i ($i = 0, 1$) are the classical Hermite blending functions:

$$\mu_0(\alpha) = (1+2\alpha)(1-\alpha)^2 \quad ; \quad \mu_1(\alpha) = \alpha(1-\alpha)^2.$$

In the expression of P_{IJ}, the derivatives of U with respect to α and β at the control points, are evaluated by centered finite differences; for instance:

$$\frac{\partial U}{\partial \alpha}|_{ij} = \tfrac{1}{2}(U_{i+1j} - U_{i-1j}) \quad , \quad \frac{\partial U}{\partial \beta}|_{ij} = \tfrac{1}{2}(U_{ij+1} - U_{ij-1})$$

$$\frac{\partial^2 U}{\partial \alpha \partial \beta}|_{ij} = \tfrac{1}{4}(U_{i+1j+1} - U_{i-1j+1} - U_{i+1j-1} + U_{i-1j-1}).$$

Note that, according to the definition of P_{IJ}, the unit square is transformed into a parameter patch U_{IJ} that is curvilinear.

I.4 Map onto the Surface Patch

Similarly, we also define maps Q_{IJ} from the unit square to R^3:

$$X^\Sigma = Q_{IJ}(\alpha, \beta)$$
$$= N_0(\alpha, \beta)X_{IJ}^\Sigma + N_1(\alpha, \beta)\frac{\partial X^\Sigma}{\partial \alpha}\Big|_{IJ} + N_2(\alpha, \beta)\frac{\partial X^\Sigma}{\partial \beta}\Big|_{IJ} + N_3(\alpha, \beta)\frac{\partial^2 X^\Sigma}{\partial \alpha \partial \beta}\Big|_{IJ}.$$

In this expression, the derivatives of X^Σ with respect to α and β are computed from the values obtained in I.2) (derivatives of X^Σ w.r.t. u and v) by chain rule derivation of u and v (given in I.3)) w.r.t. α and β. Note that the applications P_{IJ} and Q_{IJ} are defined by similar expressions, but they map the unit square onto R^2 and R^3 respectively.

I.5 Construction of the Global Map

We define Θ by:

$$X^\Sigma = \Theta(U) = \sum_{IJ} \chi_{IJ}(U) \cdot Q_{IJ} \circ P_{IJ}^{-1}(U)$$

where χ_{IJ} is the characteristic function of the parameter patch U_{IJ}:

$$\chi_{IJ}(U) = \begin{cases} 1 & \text{if } U \in U_{IJ} \\ 0 & \text{otherwise} \end{cases}.$$

The choices made for P_{IJ} and Q_{IJ} in I.3 and I.4 ensure the C^1 continuity of Θ across the control points and the patch interfaces.

Remark: a parametrization from $[0, imax] \times [0, jmax]$ onto the surface, simply obtained by assembly of the functions Q_{IJ} would not be C^1.

II. MESH GENERATION

A usual way to construct a mesh on a surface consists in first constructing a mesh in the parameter domain, then in using the map from the parameter space onto the surface to obtain the node coordinates by transformation of the parameter nodes. This can be done for any kind of mesh, structured or unstructured, the meshes in the parameter space and on the surface keeping the same topology. We use this approach for the construction of structured grids on the surface, interpolated as explained in section I from a given grid of control points. We proceed as follows:

1. we construct an algebraic grid in $[0,1] \times [0,1]$ by bilinear interpolation from the distribution on the edges. This step leads to a set of parameter nodes U_{kl} with $k=1$, k_{max} and $l=1$, l_{max}.

2. For each node U_{kl} we compute $X_{kl}^\Sigma = \Theta(U_{kl})$ by the sequence:

   a) determination of the patch U_{IJ} to which U_{kl} belongs. This determination is carried out by a Bunny-Hop search method based on vector product sign considerations. This step finds U_{IJ} such that $\chi_{IJ}(U_{kl}) = 1$.

   b) computation of $(\alpha_{kl}, \beta_{kl})$ satisfying: $U_{kl} = P_{IJ}(\alpha_{kl}, \beta_{kl})$
   This inversion of P_{IJ} is carried out by a Newton algorithm.

   c) computation of the coordinates of the surface node X_{kl}^Σ using Q_{IJ}:

$$X_{kl}^\Sigma = Q_{IJ}(\alpha_{kl}, \beta_{kl}).$$

Remark: the Bunny-Hop search method supposes first that the parameter patches are quadrilaterals and gives a first guess of the indices of the patch; these indices are eventually updated if the local coordinates (α, β) are not found in $[0,1] \times [0,1]$.

This approach enables the generation of grids on the surface from the edge node repartitions that can be prescribed on the parameter unit square; these repartitions are then automatically transferred to the four sides of the surface. The choice made for the parametrization as being proportional to curvilinear arc length on the two-curve family network provides a uniformly spaced surface mesh whenever the edge distributions have the same property.

We illustrate the mesh construction process on the following example (Fig.18): we consider a 11×11 control point grid (Fig.18.a) presenting a severe discontinuity (jump). The parameter grid is first constructed in the unit square (Fig.18.b) and a uniform mesh is constructed in this square, then each parameter node is mapped onto the surface by the map Θ to obtain the surface mesh (Fig.18.c). The difference between the 3- or 5-point scheme for the evaluation of slopes can be perceived when looking closely at the discontinuity (Fig. 18.d and e). Whereas the 3-point scheme creates oscillations around the jump, the 5-point scheme constructs a sharper discontinuity without any overshoot. This algebraic method was used for the modification of grids given on both sides of an airplane wing. The initial nodes (Fig. 19.a and c) are used as control points for the definition of a surface, a mesh in the parameter space is constructed such that the control parameter points (U_{ij}) and the parameter nodes (U_{kl}) coincide on the boundaries of the unit square ($U_{ij} = U_{kl}$ on $\partial[0,1]\times[0,1]$). The results show that this algebraic reconstruction process improves the quality of the grids, in particular the node repartitions on the surface are more regular, and the mesh lines are more orthogonal to the boundaries, especially on the trailing edges.

III. SURFACE MESH OPTIMIZATION

The optimization method, presented in the previous parts, can also be applied for quality control of surface grids. Here, a surface measure is obtained by summation of elementary contributions σ_{KL} measuring the deformation of the current cells KL with respect to reference cells, that are chosen to be rectangles of side lengths a_{KL}, b_{KL}; these elementary contributions are expressed as function of the invariants of the matrices ∇x_{KL} and $\nabla x_{KL}^t \cdot \nabla x_{KL}$:

$$x_{KL}(\xi, \eta) = (1-\hat{\xi})(1-\hat{\eta}) x_{kl} + \hat{\xi}(1-\hat{\eta}) x_{k+1l} + \hat{\xi}\hat{\eta} x_{k+1l+1} + (1-\hat{\xi})\hat{\eta} x_{kl+1}$$

with

$$\xi = a_{KL} \hat{\xi} \text{ and } \eta = b_{KL} \hat{\eta} \; ; \; (\hat{\xi}, \hat{\eta}) \in [0, 1]\times[0, 1] \, .$$

The expression of the 2D functional (σ_{2d}) can be used with a slight difference: now ∇x_{KL} ix a 3×2 matrix and the equivalent of its determinant must be defined. This is done by:

$$J_{KL} = det \left(\frac{\partial x_{KL}}{\partial \xi}, \frac{\partial x_{KL}}{\partial \eta}, N \right) = det \left(\nabla x_{KL}, N \right)$$

where N is the unit vector normal to the surface. As before the first invariant is:

$$I_{KL} = Trace \; \nabla x_{KL}^t \cdot \nabla x_{KL} \, .$$

Finally, the elementary contribution used is:

$$\sigma_{KL} = (ab_{KL})^{\gamma-1} \int_{[0,a_{KL}]\times[0,b_{KL}]} \sigma_{2d} \left(I_{KL}(\xi, \eta), J_{KL}(\xi, \eta) \right) d\xi \, d\eta \, .$$

IV. RESULTS

The procedure described in the second section then enables the computation of the new node coordinates $X^{n+1} = \Theta(U^{n+1})$. The map between the unit square and the surface is not used elsewhere: this means in particular that the node gradients are directly evaluated from the node coordinates, which makes the method more physical and indeed independent of the parametrization. At convergence of the optimization algorithm, the gradient and descent direction vectors are different from zero but are normal to the surface, thus their projection vanishes and the nodes have reached their final position.

The Figures 20 and 21 illustrate the behavior of the algorithm and the results the method can provide. The first example was designed to verify the good coupling between the parametrization routines and the optimization algorithm. A 11×11 mesh has been constructed on an analytical surface ($z = xy$) around its saddle point; a perturbed mesh has then been obtained by small displacement of the nodes around their initial position, the mesh has been further distorted by displacement of another node in order to create overlapping cells (Fig.20.a). This mesh is used as initial mesh for the optimization method; one can observe how it transforms after 1 iteration (Fig.20.b), 3 iterations (Fig.20.c) when the mesh starts to unravel, and after 10 iterations, when the mesh has recovered its regular aspect (Fig.20.d).

On this second example, we considered a more irregular surface (not C^1) presenting a local singularity (a cell with a 180 degree angle). An initial 11×6 grid (Fig.21.a) is constructed on the double ellipsoid used in Hypersonics Workshops for the modelization of flows around space vehicles. It has two characteristics: a grid line coincides with the C^1 discontinuity line and a corner cell (on the cockpit) presents a 180 degree angle. This grid is used for the definition of a new surface, its corresponding parameter square is shown in Fig.21.b. In spite of the care taken in the method development, the 41×21 initial algebraic mesh obtained by the procedure described in the section 2 presents slight oscillations on the surface (Fig.21.c); they disappear after 200 iterations of optimization (Fig.21.d). Unfortunately, this optimization moves the nodes away from discontinuity line located at the intersection of the two ellipsoids; still, the surface remains well respected.

V. CONCLUSION OF PART D

A method for the representation of surfaces only defined by a grid of control points has been presented. One of the main objectives in its development was to obtain a method robust enough to be able to work with data as irregular as possible; first results are encouraging. Compared to other techniques, this method is quite general, quality control is obtained by application of simple basic ideas, uniform repartition can be easily achieved even in high curvature areas. In severe cases however, this algebraic procedure is not sufficient for the construction of high quality meshes; an optimization method that enables to achieve regularity and orthogonality has also been presented. In part C, result of adapted grid were presented where the surface interpolation method was incorporated to allow node motion on surfaces.

REFERENCES

[1] **Numerical Grid Generation in Computational Fluid Mechanics '88**, Ed. S. Sengupta, J. Haüser, P.R. Eiseman, J.F. Thompson, Pineridge Press, 1988; Proceedings of the Second Conference on Grid Generation in CFD, Miami-Beach, U.S.A., Dec. 5-9, 1988.

[2] **Numerical Grid Generation in Computational Fluid Dynamics and Related Areas**, Ed. A. S.-Arcilla, J. Häuser, P.R. Eiseman et J.F. Thompson, North Holland, 1991; Proceedings of the Third International Conference on Numerical Grid Generation in CFD and Related Fields, Barcelone, Spain, June 3-7, 1991.

[3] BABUSKA, I. and RHEINBOLD, W.C., "Error Estimates for Adaptive Finite Element Computations", **SIAM J. Muner. Anal.**, Vol.15, pp.736-755, 1978.

[4] BELYTSCHO, T. and TSAY, C.S., "A Stabilization Procedure for the Quadrilateral Plate Element with One Point Quadrature", **Comp. Meth. Appl. Mech. Engrg.**, Vol.42, pp 225-251,1984.

[5] CABELLO, J., "Méthode Variationnelle d'Optimisation et d'Adaptation de Maillages Structurés Tridimensionnels", **Doctoral Thesis**, Université de Paris-Sud, France, February 1990.

[6] CARCAILLET, R, "Optimization of Three-Dimensional Grids and Generation of Flow Adaptive Computational Grids", **AIAA Paper** N⁰ 86-0156, AIAA 24th Aerospace Sciences Meeting, Reno, NEV, U.S.A., 1986.

[7] COUAILLIER, V., "Multigrid Method for Solving Euler and Navier-Stokes Equations in Two and Three Dimensions", **Proceedings of the Eighth GAMM Conference on Numerical Methods in Fluid Dynamics**, Delft, The Netherlands, September 27-29,1989.

[8] COUSSEMENT, G., "Grid Generation Around Single and Multiple Element Aerofoils by the Use of Elliptic Partial Differential Equations", **VKI Report**, PR 1989-16, June, 1989.

[9] DEMKOWICZ, L., ODEN, J.T., and RACHOWICZ, W., "A New Finite Element Method for Solving Compressible Navier-Stokes Equations Based on an Operator Splitting Method and h-p Adaptivity", **Comp. Meth. Appl. Mech. Engrg.**, Vol.84, No.3, pp.275-326, 1990.

[10] DEMKOWICZ, L., ODEN, J.T., RACHOWICZ, W. and HARDY, O., "An h-p Taylor-Galerkin Finite Element Method for Compressible Euler Equations", **Comp. Meth. Appl. Mech. Engrg.**, Vol.88, No.3, pp.363-396, 1991.

[11] JACQUOTTE, O.-P., "Analysis and Control of Instabilities in Mixed and Underintegrated Finite Element Methods in Solid and Fluid Mechanics" **Ph.D. Dissertation**, University of Texas at Austin, U.S.A., August 1985.

[12] JACQUOTTE, O.-P., "A Mechanical Model for a New Mesh Generation Method in Computational Fluid Dynamics", **Comp. Meth. Appl. Mech. Engrg.**, Vol.66, No.3, pp.323-338, 1988.

[13] JACQUOTTE, O.-P., "Recent Progress on Mesh Optimization", in [2], pp.581-596.
[14] JACQUOTTE, O.-P., "Generation, Optimization and Adaptation of Multiblock Grids around Complex Configurations in Computational Fluid Dynamics", **Int. J. Num. Methods Eng.**, Vol.34,pp.443-454, 1992.
[15] JACQUOTTE, O.-P. and CABELLO, J., "A Variational Method for the Optimization and Adaptation of Grids in Computational Fluid Dynamics", in [1], pp. 405-414.
[16] JACQUOTTE, O.-P. and CABELLO, J., "Three-Dimensional Grid Generation Method based on a Variational Principle", **La Recherche Aérospatiale**, English Edition, 1990 N^0 4 (July-August), pp.7-19.
[17] JACQUOTTE, O.-P. and COUSSEMENT, G., "Structured Mesh Adaption: Space Accuracy and Interpolation Methods", **Proceedings of the 2nd Workshop on Reliability and Adaptive Methods in Computational Mechanics**, Krakow, Poland, Oct. 14-16, 1991, to appear in **Comp. Meth. Appl. Mech. Engrg.**, 1992.
[18] JACQUOTTE, O.-P. and GAILLET, Ch., "Construction de Maillages Multiblocs Tridimensionnels par une Méthode Variationnelle" (in english), **ONERA Report**, RT 7/3519 AY 424A, March 1991.
[19] McCONALOGUE, D.J., "The Quasi-Intrinsic Scheme for Passing a Smooth Curve through a Discrete Set of Points", **Computer Journal**, Vol. 13, No 4, Nov. 1970.
[20] PIEGL, L., "Hermite-and Coons-like Interpolants using Rational Bezier Approximation Form with Infinite Control Points ", **Computer-Aided Design**, Vol. 20, No 1, pp. 1-10, 1988 .
[21] SALTZMAN, J.S. and BRACKBILL, J.U., "Applications and Generalizations of Variational Methods for Generating Adaptive Meshes", **Numerical Grid Generation**, Ed. J.F. Thompson, North Holland, 1982.
[22] THOMPSON, J.F., "Elliptic Grid Generation", **Numerical Grid Generation**, Ed. J.F. Thompson, North Holland, 1982.

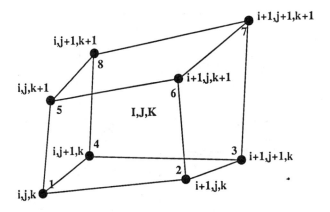

Fig. 1: Node and Cell Numbering.

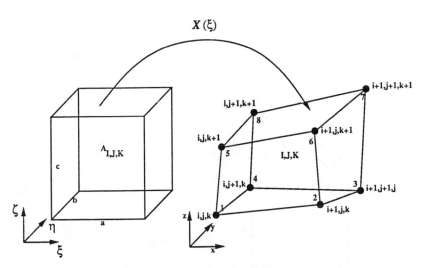

Fig. 2: Deformation of the Reference Cell.

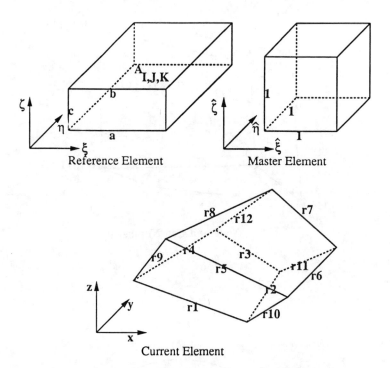

Fig. 3: Reference, Master and Current Cell.

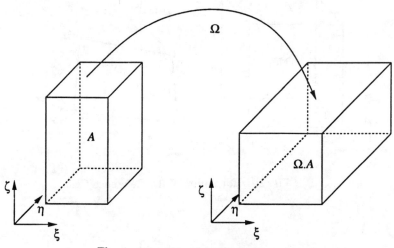

Fig. 4: Adaption of the Reference cell.

246

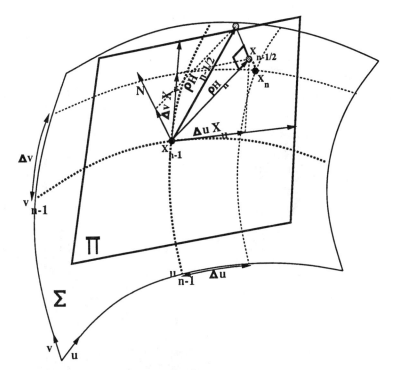

Fig. 5: Node Constrained to Move along a Surface.

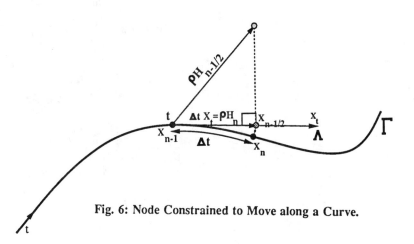

Fig. 6: Node Constrained to Move along a Curve.

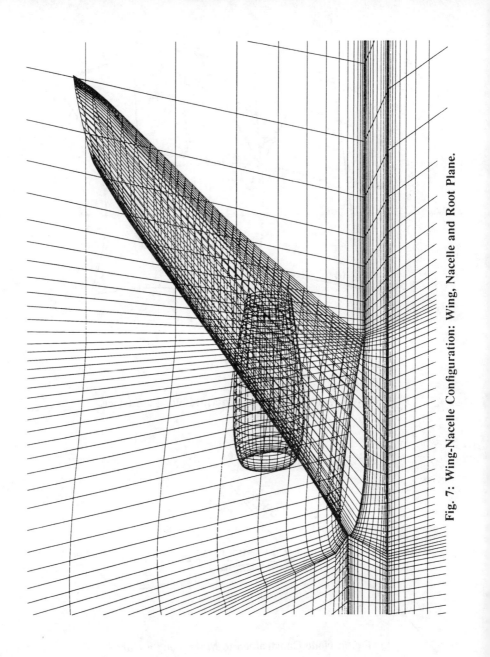

Fig. 7: Wing-Nacelle Configuration: Wing, Nacelle and Root Plane.

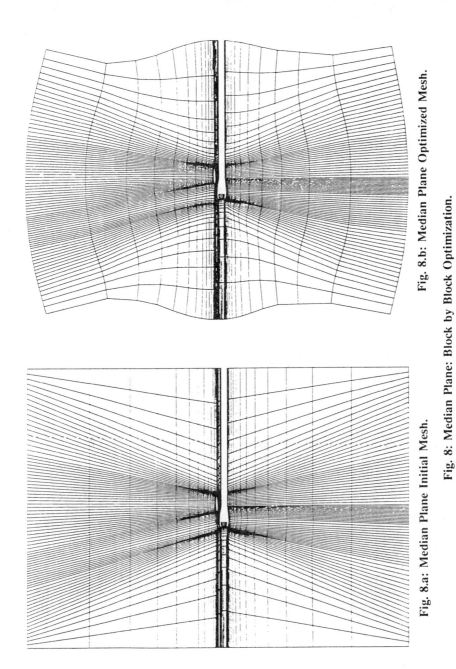

Fig. 8.a: Median Plane Initial Mesh.   Fig. 8.b: Median Plane Optimized Mesh.

Fig. 8: Median Plane: Block by Block Optimization.

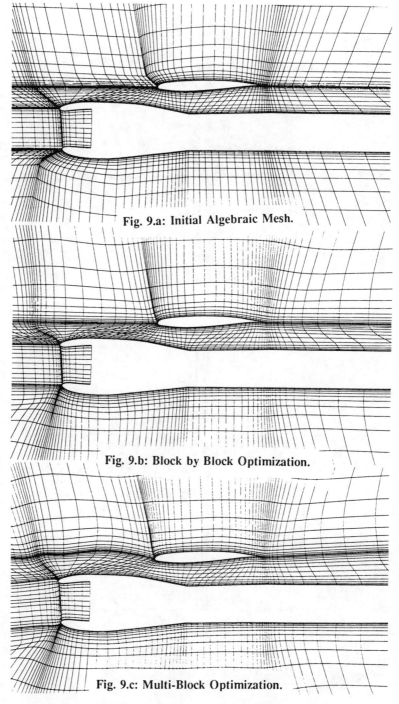

Fig. 9.a: Initial Algebraic Mesh.

Fig. 9.b: Block by Block Optimization.

Fig. 9.c: Multi-Block Optimization.

Fig. 9: Block by Block and Multi-Block Optimization : Side View.

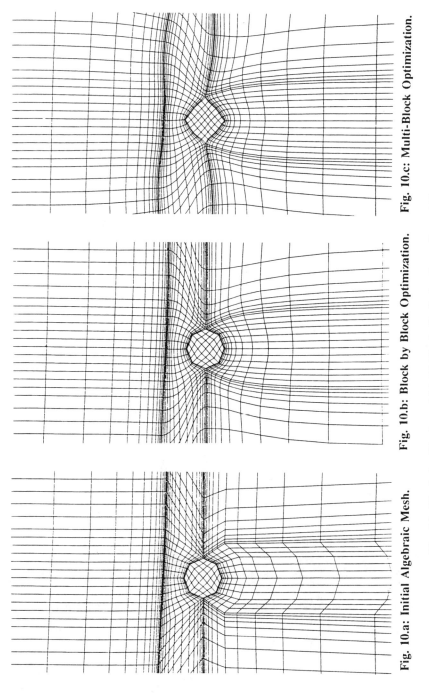

Fig. 10.a: Initial Algebraic Mesh. Fig. 10.b: Block by Block Optimization. Fig. 10.c: Multi-Block Optimization.

Fig. 10: Block by Block and Multi-Block Optimization : Front View.

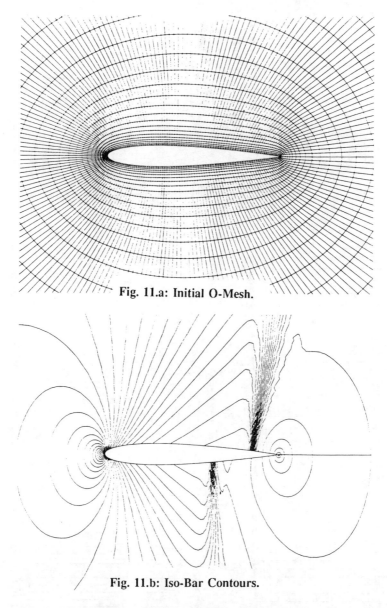

Fig. 11.a: Initial O-Mesh.

Fig. 11.b: Iso-Bar Contours.

Fig. 11: Initial Mesh and Pressure Field around a NACA0012 Airfoil.
(Euler Flow Calculation, $M_\infty = 0.85$, $\alpha = 1.0°$)

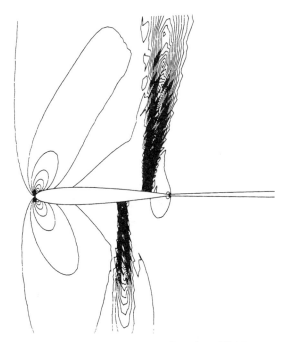

Fig. 12.a: Exact Pressure Gradient Field.

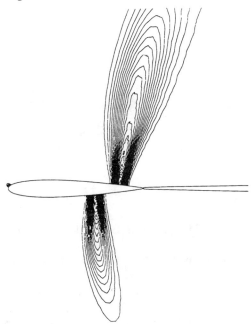

Fig. 12.b: Limited and Smoothed Pressure Gradient Field.
Fig. 12: Iso-Bar (Logical) Gradient Contours.

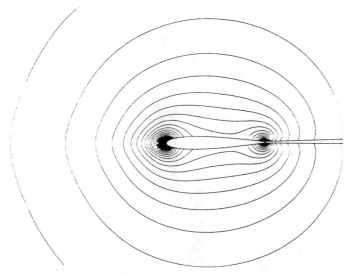

Fig. 13.a: Reference Length for the Initial Mesh.

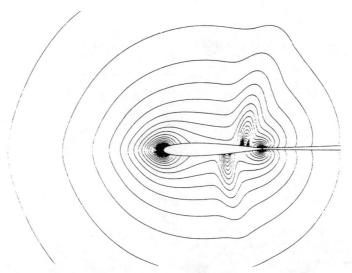

Fig. 13.b: Reference Length for the Adapted Mesh.

Fig. 13: Modification of the Reference Length for the Mesh Adaption.

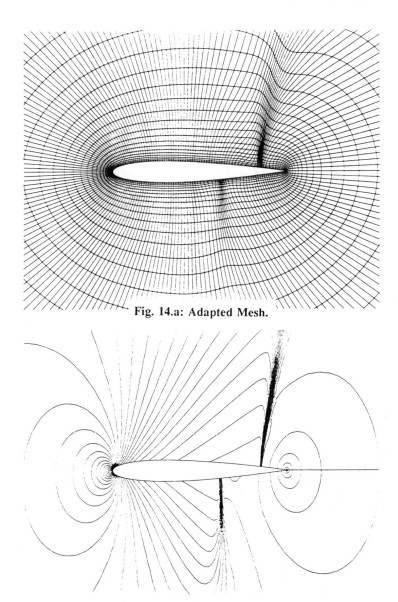

Fig. 14.a: Adapted Mesh.

Fig. 14.b: Pressure Solution on the Adapted Mesh.

Fig. 14: Results of the Mesh Adaption.
(Euler Flow Calculation, $M_\infty = 0.85$, $\alpha = 1.0°$)

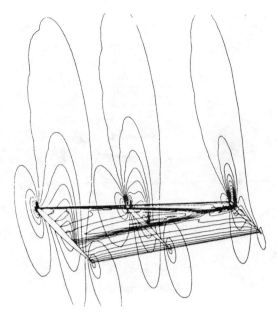

Fig. 15.a: Iso-Bar Contours in the Field.

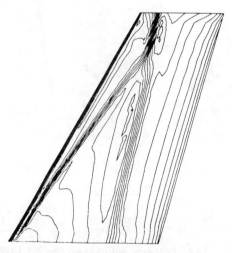

Fig. 15.b: Iso-Bar Contours along the Suction Face of the Wing.

Fig. 15: Pressure Field around ONERA M6 Wing.
(Euler Flow Calculation, $M_\infty = 0.835$, $\alpha = 3.06°$)

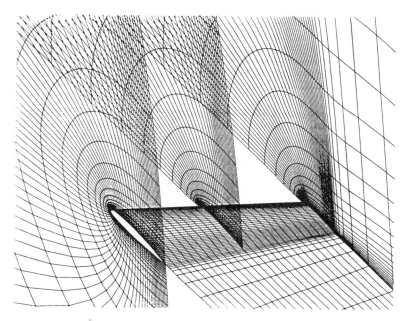

Fig. 16.a: Initial Mesh

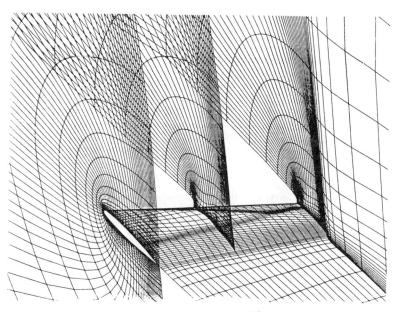

Fig. 16.b: Adapted Mesh

Fig. 16: Initial and Adapted Mesh around ONERA M6 Wing.
(103×41×22 Mesh Nodes)

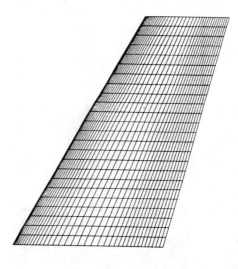

Fig. 17.a: Initial Mesh

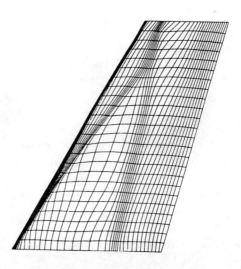

Fig. 17.b: Adapted Mesh

Fig. 17: Initial and Adapted Mesh around ONERA M6 Wing.
(Top View : Wing Mesh)

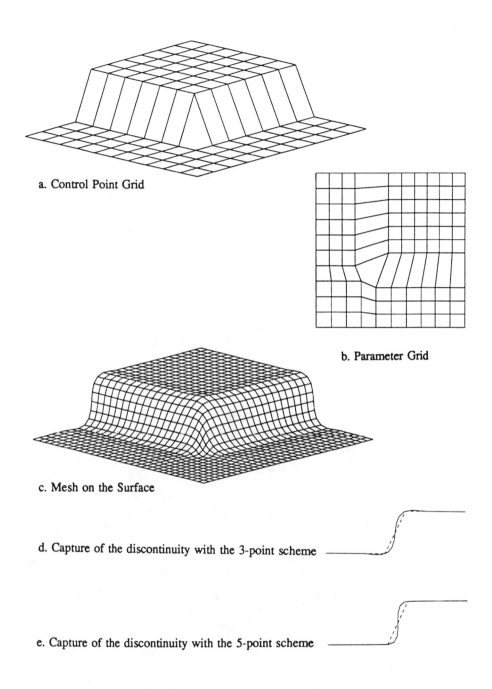

a. Control Point Grid

b. Parameter Grid

c. Mesh on the Surface

d. Capture of the discontinuity with the 3-point scheme

e. Capture of the discontinuity with the 5-point scheme

Fig. 18: Algebraic Mesh Generation.

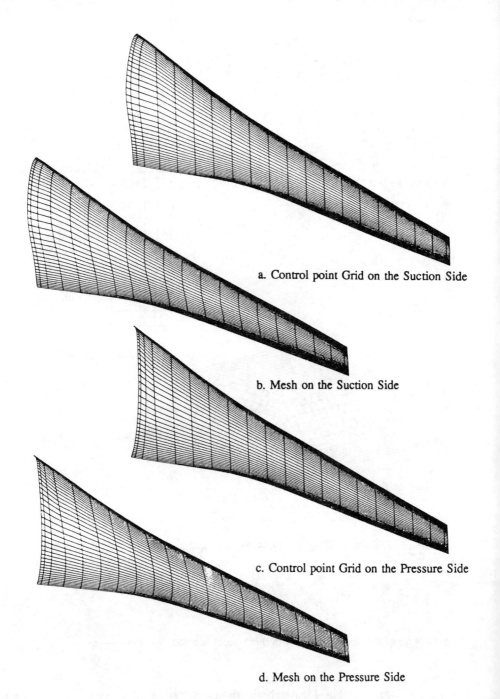

Fig. 19: Mesh Generation on a Wing.

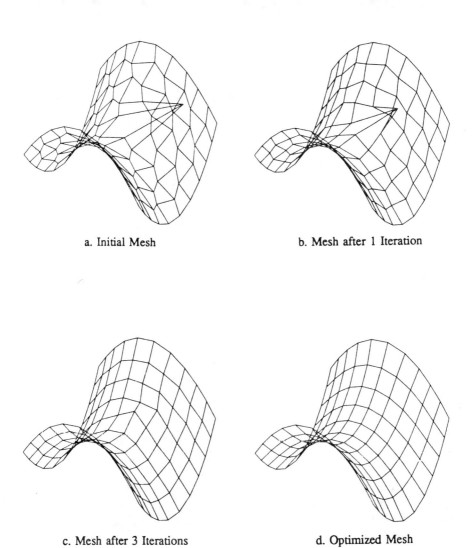

a. Initial Mesh

b. Mesh after 1 Iteration

c. Mesh after 3 Iterations

d. Optimized Mesh

Fig. 20: Mesh Optimization on a Saddle Surface.

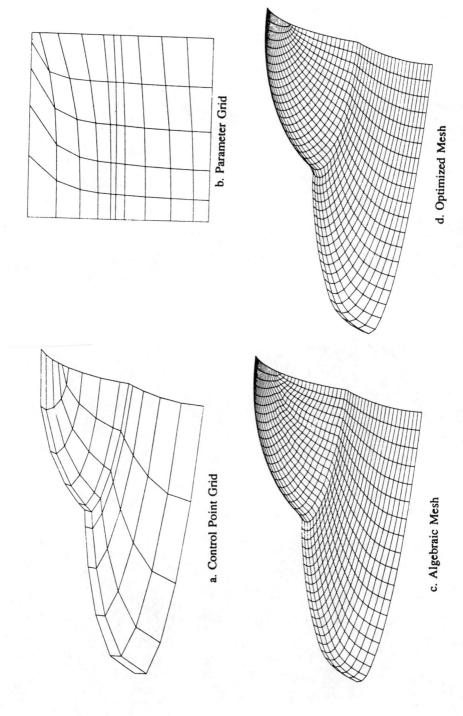

a. Control Point Grid

b. Parameter Grid

c. Algebraic Mesh

d. Optimized Mesh

Fig. 21: Mesh Generation and Optimization on a Double Ellipsoid.

GENERAL GRID ADAPTIVITY FOR FLOW SIMULATION

M. J. Marchant, N. P. Weatherill and J. Szmelter

Institute for Numerical Methods in Engineering,
Department of Civil Engineering,
University College of Swansea, Singleton Park,
Swansea, SA2 8PP, U.K.

Summary

Numerical solutions which simulate compressible flows can be greatly improved by mesh adaption. This paper describes the grid adaptivity techniques for mesh refinement, mesh derefinement and node movement. To achieve this flexibility the adaption must be implemented using a suitable data structure, details of which are given. For application to complicated geometries structured, unstructured and hybrid grids are used. Aspects of these mesh generation techniques will be described. Mesh adaptivity will be demonstrated on a variety of 2-dimensional aerospace applications for both inviscid and viscous flow simulation.

1. Introduction

The generation of computational meshes for complex aerodynamic geometries has occupied workers in the field of computational fluid dynamics (CFD) for many years. The methods used to discretise a domain bounded by a given set of points describing the geometry, all result in a network of points and their associated connectivities. Discretisation techniques can be categorised under two main headings; structured meshes, whose network of curvilinear lines naturally map into the rows and columns of a matrix, where neighbouring positions in a matrix map directly to adjacent points in the physical domain, and unstructured meshes, which require a connectivity matrix to define the arrangement of the assembled points. A third category exists which combines structured and unstructured mesh types to form, so-called, hybrid meshes.

Each mesh generation technique can, invariably, only produce an initial mesh point distribution within the domain which is based on the point distribution on the boundary, and therefore, mesh points will not reflect features which occur in the flowfield, such as shock waves and boundary layers. Once an initial flowfield solution has been produced, it becomes possible to perform adaptions and generate a mesh which reflects the flow conditions more appropriately. Most techniques for grid adaption are based upon an equidistribution principle, which states that, throughout the field, a variable w and the local mesh spacing ds at a point i, should satisfy the rule:

$$w_i ds_i = constant \qquad (1.1)$$

An explanation of this principle shows that if w_i, which can represent a measure of a physical flow variable such as pressure, or an error in the solution, is large then the local mesh spacing ds_i must be conversely small. Alternatively, if the value of w_i is small, the local mesh spacing ds_i should be large. A number of approaches to mesh adaption exist which attempt to satisfy the requirements of the equidistribution principle. Three of which are mesh enrichment, mesh derefinement and mesh point movement. Mesh point movement techniques reduce ds_i when w_i is detected to be large[1,2], whereas mesh enrichment introduces additional points to regions of high w_i, thus successively reducing the local mesh spacing ds_i with each level of refinement. Derefinement increases ds_i by removing points from regions of low w_i.

Point enrichment techniques can be applied to both structured and unstructured meshes. Within an unstructured mesh the application of point enrichment can result in local subdivision of triangular elements into further smaller fully connected triangular elements, and thus no alterations are required for the flow solver. This is unlike the situation which arises in structured meshes, where the addition of points breaks the curvilinear network of lines, and subsequently the efficiency of the data structure. Hence, so-called, 'hanging' nodes are formed which do not connect to all the surrounding points. As a consequence, local point enrichment on structured meshes requires either, special treatment within cell vertex flow solvers or the use of cell centred flow solvers. However, there are many advantages to structured meshes which make the effort of dealing with such difficulties worthwhile.

2. Data Structure for Adaptivity

In many branches of computational mechanics the choice of a suitable data structure with which to manipulate data and construct numerical operations is of key importance. The data structure has an important influence on the efficiency of an algorithm, and on the ability to make the computer software both readable and robust. Traditionally, structured grids have utilised an ordered point set (i,j) to describe a grid and as such proved highly efficient in the implementation of finite difference operators. However, such a data structure is found to be unsuitable when point enrichment is introduced. The regular pattern of points which map so neatly into an array matrix is lost and the additional points have to be inserted into a semi-ad-hoc data structure of pointers. Rather than construct such an approach, it proves more beneficial to abandon the indexing and resort to a more appropriate, well defined, data structure.

Since adaptivity in the form of point enrichment can result in the 'quadsecting' of a quadrilateral element, a data structure based upon a four branch tree structure would appear to be a logical choice. A quadtree data structure satisfies this criterion and has been shown to perform efficiently on enriched structured meshes [3] and with a few modifications also allows adaptivity techniques to be applied on general grid types.

2.1 Quadtree data structure

The tree structure provided by the quadtree data structure gives a natural environment for referencing information which has been produced in quadruples. With some modification to the original tree structure the approach proves highly flexible, in that it can also be applied to unstructured and hybrid grids.

Within the quadtree data structure the subdivision of a quadrilateral into four quadrilaterals is a natural process. For mesh adaption to different flowfield features, it is envisaged that two additional subdivision processes would be advantageous. Hence, three types of subdivision are used and these have been termed type 1, 2 and 3, shown in Figure 2.1. They represent possible subdivisions for a wide range of flow features.

For example, the type 2 subdivision may be particularly efficient for boundary layers in viscous calculations. These two additional types of subdivision also fit naturally into the quadtree data structure.

2.1.1 *Data arrays*

The implementation of the quadtree data structure to single and multiblock structured meshes requires six main arrays. These arrays contain :
1. List of neighbours for each cell [4 memory locations per cell]
2. List of the adjacent side number of each neighbouring cell
   [4 memory locations per cell]
3. The nodal connectivity of each cell [4 memory locations per cell]
4. The subdivision level of each cell [1 memory location per cell]
5. The number of the parent of a cell [1 memory location per cell]
6. Details of a cells subdivision [3 memory locations per cell].

The whole quadtree method revolves around an hierarchical or family tree, and array(4) is the generation or hierarchy pointer. It contains the number of subdivisions necessary to create a given quadrilateral from a quadrilateral contained in the original mesh. The quadtree, or family tree, is contained in arrays (5) and (6). Array (5) is a pointer to the parent cell, and array (6) utilises three storage locations, one stores the first cell produced by the subdivision, the second and third locations specify the number of subdivisions produced in local element coordinates, which enables the identification of cells subdivided by types 2 and 3. To reproduce a cells family tree these locations are searched recursively. The other arrays are self-explanatory for quadrilateral elements.

2.1.2 *Searching methods*

All potentially efficient data structures must allow data required for the flow solver or adaptivity process to be found simply, both in terms of programming effort and the computational time spent locating the information. In the implementation of the adaptivity and flow solving algorithms the search routines are required to perform three tasks; locate midside nodes, locate quadrilaterals at different levels of subdivision and identify three, five or more point singularities arising in the block connectivities of the multiblock meshes. Detailed information about the searching routines can be obtained from reference [3].

2.1.3 *Embedded interfaces*

A consequence of local mesh embedding is that nodes are produced which are not connected to four surrounding points. Instead, 'hanging' nodes are produced, which are only connected to three surrounding points. It is therefore necessary to formulate a technique which can deal with such meshes. Three approaches have been investigated. Two methods remove the hanging nodes by forming local transitional connections to surrounding points, whilst the third option leaves the nodes, but modifies the contour integrals within the cell vertex finite volume flow solver. When modifying the contour integral only two hanging nodes are allowed in each cell, these hanging nodes must also be on adjacent edges. All other possibilities are removed by smoothing using quadsection of cells. Within the cell centre flow solver no modifications are necessary to deal with hanging nodes.

The quadtree method is flexible enough to allow the removal of hanging nodes by triangular and quadrilateral transitional interfaces between refined and unrefined areas. To implement these interfaces it is firstly necessary to identify cells which contain one hanging node or two adjacent hanging nodes, and then to create the transitional interface. These are shown in Figures 2.2 and 2.3, for triangular and quadrilateral interfaces, respectively. The alternative technique for dealing with hanging nodes involves a modification of the contour integrals calculated within the cell vertex flow solver, so that cells containing five and six nodes can be computed. This involves the flow solver recognising cells with one hanging node as pentagonal and two hanging nodes as hexagonal. This technique will be described in detail in section 4.2.

2.1.4 *Smoothing and Clustering*

To minimise the discontinuities in cell volumes that occur at embedded interfaces it proves more efficient to locally smooth the levels of subdivision. This is performed such that there is only a difference of one subdivision level between adjacent cells. This avoids the complication of multiple hanging nodes on an edge when using a cell vertex flow solver. Smoothing is instigated by comparing the subdivision levels of each cell and its neighbours, if a difference of more than one level of subdivision is identified then the cell with the lower subdivision level is subdivided.

To further enhance the efficiency of the embedding process, it proves advantageous to cluster small refined regions into larger blocks. This statement may seem somewhat contradictory, considering that more elements will be subdivided as a result, but the process has the advantages of simplifying the mesh and also reducing the number of discontinuities in cell volumes. Particularly where cell volumes pass from small to large to small over a short distance. Clustering is performed by searching through the mesh identifying small regions bounded by elements at the same higher level of refinement. The extent of the search can be limited to a specific number of cells away from the cell being checked.

The quadtree data structure combined with the ability to investigate various embedding interfaces proved very successful for the cell vertex Euler and Navier-Stokes flow solvers. The implementation of adaptivity into cell centre flow solvers enabled a less complicated, less memory intensive data structure to be developed. The new data structure was developed using a subset of the data required by the quadtree data structure and allows meshes incorporating quadrilateral and triangular elements to be adapted either individually or simultaneously.

2.2 Modified quadtree data structure

To apply adaptivity techniques on general grid types and within a cell centre flow solver a modified quadtree data structure has been further developed. This implementation has enabled the number of storage locations per cell to be restricted to ten compared with the seventeen locations required for the previous data structure. The utilisation of these storage locations is broken down as follows:

1. List of neighbours for each cell [4 memory locations per cell]
2. The nodal connectivity of each cell [4 memory locations per cell]
3. Identification of the type of cell (triangle/quadrilateral)
    [1 memory location per cell]
4. The subdivision level of each cell [1 memory location per cell].

Figures 2.4 and 2.5 show the data ordering of neighbours (LNGBR) and connectivity (LSLST) for typical quadrilateral and triangular elements, respectively.

The hanging nodes formed in the buffer zone between refined and unrefined regions means that, since only one storage location is allocated to describe the neighbour of a cell along one edge, a method to allow the identification of all cells along a side is required. By comparing the subdivision levels of a cell and its neighbour, it is possible to ascertain whether there is more than one cell adjacent to an edge. If a neighbour has a higher subdivision level there is more than one cell along the edge. If the neighbour has the same subdivision level there is only one cell on the edge. Finally, when the neighbour has a lower subdivision level only part of the neighbouring cell lies along the edge. If the comparison of subdivision levels indicates that there is more than one cell on an edge, then the cell and its neighbour are used to search for the other neighbours. The implementation of the search is illustrated in the following example for the structured mesh shown in Figure 2.6. First, the side of the neighbouring cell which is connected to the cell of interest must be established. This allows a search to begin in each direction along the edge. This technique returns cells A, B and C as neighbours of cell D on side 1, as pictured in Figure 2.6, and can gather all information about the neighbouring cells for every cell in a structured grid. In unstructured and hybrid grids the data structure and the searches can be modified to cope with triangular elements, see Figure 2.7.

2.3 Refinement Strategy

The approximate number of cells to be refined is specified by the user as a percentage of the existing number of cells. This gives the user some control over the computational cost of the calculation. The target number of cells to be subdivided is achieved by iteratively adjusting the tolerance parameter of the adaptivity criteria described in section 5.1. Subsequent clustering of embedded regions usually results in a slight increase in the number of cells flagged for subdivision, but this is necessary to reduce the number of discontinuities in cell volumes. Mesh smoothing is not essential when using a cell centre flow solver because multiple hanging nodes on an edge do not introduce any additional complications.

2.4 Derefinement Strategy

The utilisation of point enrichment as an adaptivity technique has a well documented record for improving the resolution of flow features and consequently the overall accuracy of a flow simulation [3,4]. The obvious increase in computational cost can be offset by removing excess nodes from regions of flow inactivity, a technique referred to as derefinement. Another benefit of derefinement is that it allows more efficient satisfaction of the equidistribution principle because the value of ds_i can increase as well as decrease (the result of refinement). The technique of derefinement has only been applied on structured meshes because with unstructured meshes it is possible to generate a mesh without an unnecessary number of grid points in redundant flow regions. This is not always the case with structured meshes, mainly due to the constraints of grid line continuity.

Derefinement is employed at the same stage of the adaption process as point enrichment, that is, after the production of an initial solution. The mesh is examined on a cell by cell basis to identify groups of four cells, sharing a single common node, each at the same subdivision level. Once a group has been detected, values of a suitable criterion, such as gradient of pressure, within each of the cells are compared and if found to be less than a user prescribed tolerance, the cells can be derefined into a single cell with a subdivision level one lower than that of the four cells in the group. There are some situations in which derefinement is not allowed, this is where derefinement would cause an anomaly in the data structure. The check used to maintain the integrity of the data structure is explained with reference to Figures 2.8 and 2.9. If cells a,b,c and d, in Figure 2.8, are derefined into one cell, cell A in Figure 2.9, then cells A,B and C would be at the same level of subdivision, but would only share a common half edge. This would conflict with the neighbour searches described earlier and hence derefinements such as this, must be avoided. If this situation does not occur then the four cells can be derefined.

3. Grid Generation

To obtain accurate flow solutions around relatively complicated aerodynamic geometries it is necessary to have a flexible technique for mesh generation. For simple configurations, such as an aerofoil, it is possible to map a single set of curvilinear coordinates to the physical space to produce a mesh. However, such structured meshes have limited flexibility when applied to geometrically complicated domains, as it becomes extremely difficult to generate a single curvilinear system that appropriately discretises the whole domain. To relax this constraint, it is possible to subdivide the domain into a number of smaller, more manageable regions which are topologically equivalent to blocks, which each have there own local curvilinear coordinate system. This technique is known as multiblock mesh generation.

3.1 Elliptic Multiblock Mesh Generation

The method used to generate the mesh points in the physical domain is based on the elliptic partial differential equation approach. Laplace's equation produces an optimally smooth mesh point distribution. In two dimensions, the equations are

$$\xi_{xx} + \xi_{yy} = 0$$

$$\eta_{xx} + \eta_{yy} = 0 \qquad (3.1)$$

where ξ and η are the regular coordinates of the transformed plane, and x and y are the Cartesian coordinates for the real space. To utilise these equations for mesh generation, it is more convenient to perform a transformation so that x and y become the dependent variables. This leads to a system of coupled non-linear elliptic partial differential equations in x and y, namely,

$$\alpha\, r_{\xi\xi} - 2\beta\, r_{\xi\eta} + \gamma\, r_{\eta\eta} = 0 \qquad (3.2)$$

where $r = (x,y)^T$, and the metric coefficients are given by

$$\alpha = x_\eta^2 + y_\eta^2$$
$$\beta = x_\xi x_\eta + y_\xi y_\eta$$
$$\gamma = x_\xi^2 + y_\xi^2$$

where β represents orthogonality and for values of $\alpha/\gamma = 1$ and $\beta = 0$, the mapping is conformal.

To ensure adequate control of the grid point spacing equation (3.1) is augmented with source terms P and Q. The resulting transformed equations take the form

$$\alpha\left(r_{\xi\xi} + Pr_\xi\right) - 2\beta\, r_{\xi\eta} + \gamma\left(r_{\eta\eta} + Qr_\eta\right) = 0\,. \qquad (3.3)$$

With the source terms set to zero, the system of coupled non-linear equations maximise mesh smoothness and subsequently in the absence of boundary curvature the mesh points are evenly spaced. However, near convex boundaries the grid points will become more closely spaced, whilst near concave boundaries the mesh spacing will be more sparse.

The work of Thomas and Middlecoff [5] demonstrated that the boundary point distribution can be used to generate the control functions and thus control the field point distribution by the choice of appropriate boundary point spacing. The control functions defined by Thomas and Middlecoff are

$$P = -\frac{r_\xi . r_{\xi\xi}}{r_\xi^2} \tag{3.4}$$

and

$$Q = -\frac{r_\eta . r_{\eta\eta}}{r_\eta^2} . \tag{3.5}$$

The values of P and Q obtained for each boundary point are linearly interpolated along lines of constant ξ and η to give values to internal mesh points. However, this technique does not control grid line skewness at boundaries, which is known to be a desirable property for accurate solutions. Therefore, an additional term is included in the formulation of the control functions to produce grid lines that are orthogonal at the boundary. The boundary orthogonality terms are included in the control functions once an initial mesh has been obtained, since when the terms are evaluated it is necessary to use the coordinates of points in the field, adjacent to the boundary. The control functions are formulated as follows.

For orthogonality, $\beta = 0$ and thus equation (3.2) becomes

$$\alpha \, r_{\xi\xi} + \gamma \, r_{\eta\eta} = 0 . \tag{3.6}$$

Taking the scalar product of this equation with r_ξ and r_η and using the condition for orthogonality leads to

$$P = -\frac{r_\xi . r_{\xi\xi}}{r_\xi^2} - \frac{r_\xi . r_{\eta\eta}}{r_\eta^2} \tag{3.7}$$

and

$$Q = -\frac{r_\eta . r_{\eta\eta}}{r_\eta^2} - \frac{r_\eta . r_{\xi\xi}}{r_\xi^2} . \tag{3.8}$$

The first terms of equations (3.7) and (3.8) are equivalent to the control functions of Thomas and Middlecoff. The second terms are the corrective terms for orthogonality. This technique is an extension of the work of Thompson et al. [6], and further details can be obtained in reference [7].

To enhance the quality of the grids produced and particularly for meshes suitable for Navier-Stokes calculations, the grid generation incorporates point redistribution along lines radial to solid boundaries. This allows considerable control of the size of cells adjacent to boundaries. Figure 3.1 shows a mesh generated around a tandem configuration of two NACA0012 aerofoils with boundary orthogonality and radial line redistribution, suitable for a Navier-Stokes calculation. Figure 3.2 details the 23 block mesh generated for the Williams "B" landing configuration and the corresponding surface pressure coefficient for an inviscid calculation at a Mach number of 0.125 and zero incidence (the result is compared with the exact potential flow).

3.2 Unstructured mesh generation - Delaunay Triangulation

The generation of unstructured meshes using Delaunay triangulation was first proposed by Weatherill [8], and has subsequently lead to the development of robust algorithms suitable for both 2 and 3 dimensions [14,15].

The triangulation is based upon the concept of the in-circle criterion. The triangulation is the geometrical dual of the Voronoi diagram. This diagram is the construction of tiles in which regions are associated with points in such a way that a region is closer to a point than to any other point in the field. A more formal definition states that, if a set of points is denoted by $\{p_i\}$, then the Voronoi region $\{V_i\}$ can be defined as

$$\{V_i\} = \{p : \| p - p_i \| < \| p - p_j \|, \; for \; all \; j \neq i\} \tag{3.9}$$

i.e. the Voronoi region $\{V_i\}$ is the set of all points that are closer to p_i than to any other point. The sum of all points forms a Voronoi polygon. If points with common line segments are connected then the Delaunay triangulation is formed. The vertices of the Voronoi diagram are at the centres of the circles passing through three points which form the individual triangles. No other points can lie within a circle. In 2-Dimensions, this triangulation process ensures an optimally smooth triangulation of an arbitrary set of points. It is also possible to generate points consistent with the spatial distribution of points on the boundaries, or by controlling the point spacing by using a background mesh [14,15].

3.3 Hybrid mesh generation

To generate hybrid meshes requires the use of the basic techniques for both structured and unstructured mesh generation. Given a structured mesh, either single block or multiblock, another component can be introduced into the topology by deleting some of the original mesh points and then inserting the additional component with its own mesh. The two grids are then connected together using the Delaunay triangulation. Figure 3.3 demonstrates this process, Figure 3.3(a) shows the original mesh overlaid with an additional component. Figure 3.3(b) shows the mesh when the mesh points of the original mesh, which are under the component mesh, have been removed. The mesh resulting after removing mesh points of the component mesh that lie outside of the underlying cut out section is shown in Figure 3.3(c). The final mesh, in which the two meshes have been reconnected using the Delaunay triangulation, is shown in Figure 3.3(d).It is important to ensure that the boundaries of the two structured meshes, which form the inner and outer boundaries for the Delaunay triangulation, are retained in the triangulation. Much effort has been applied to the techniques which ensure the integrity of a given boundary in a Delaunay triangulation and these methods are applied in the triangulation procedure.

4. Cell based flow solvers

4.1 Governing Flow Equations

The flow of a viscous compressible fluid is governed by the Navier-Stokes equations. They represent conservation of mass, momentum and energy. For two dimensional unsteady flow the integral form is:

$$\frac{\partial}{\partial t} \int \int_\Omega \boldsymbol{w} dx dy + \int_{\partial \Omega} (\boldsymbol{F} dy - \boldsymbol{G} dx) = 0 \tag{4.1}$$

where x and y are the Cartesian coordinates, and the integrals are taken over a control volume Ω, bounded by the curve $\partial \Omega$. The flux tensor can be split into the inviscid 'I' and the viscous 'V' contributions, such that,

$$\boldsymbol{F} = \boldsymbol{F}^I + \boldsymbol{F}^V, \qquad \boldsymbol{G} = \boldsymbol{G}^I + \boldsymbol{G}^V. \tag{4.2}$$

The conserved variable $\boldsymbol{w}$ and the Cartesian flux function $\boldsymbol{F}$ and $\boldsymbol{G}$ are given by :

$$\boldsymbol{w} = \begin{bmatrix} \rho \\ \rho u \\ \rho v \\ \rho e \end{bmatrix}, \quad \boldsymbol{F}^I = \begin{bmatrix} \rho u \\ \rho u^2 + p \\ \rho u v \\ \rho u h_o \end{bmatrix}, \quad \boldsymbol{G}^I = \begin{bmatrix} \rho v \\ \rho u v \\ \rho v^2 + p \\ \rho v h_o \end{bmatrix},$$

$$\boldsymbol{F}^V = \begin{bmatrix} 0 \\ -\sigma_{xx} \\ -\sigma_{xy} \\ -u\sigma_{xx} - v\sigma_{xy} + q_x \end{bmatrix}, \quad \boldsymbol{G}^V = \begin{bmatrix} 0 \\ -\sigma_{xy} \\ -\sigma_{yy} \\ -u\sigma_{xy} - v\sigma_{yy} + q_y \end{bmatrix}. \tag{4.3}$$

The elements of the stress tensor and heat vector are as follows:

$$\begin{aligned} \sigma_{xx} &= \eta \left[\frac{4}{3} \frac{\partial u}{\partial x} - \frac{2}{3} \frac{\partial v}{\partial y} \right], \\ \sigma_{xy} &= \eta \left[\frac{\partial u}{\partial y} + \frac{\partial v}{\partial x} \right], \\ \sigma_{yy} &= \eta \left[\frac{4}{3} \frac{\partial v}{\partial y} - \frac{2}{3} \frac{\partial u}{\partial x} \right], \\ q_x &= -\gamma k \frac{\partial T}{\partial x}, \qquad q_y = -\gamma k \frac{\partial T}{\partial y}. \end{aligned} \tag{4.4}$$

η indicates laminar viscosity and is determined from Sutherland's law. k the coefficient of thermal conductivity, ρ is the density, u and v are the Cartesian velocity components, p is the pressure, σ is the stress, e and h_o are the total internal energy per unit volume and enthalpy, respectively. The temperature T is obtained from the equation of state which closes the system

$$p = (\gamma - 1)\rho^T \tag{4.5}$$

where γ is the ratio of specific heats. In this paper only laminar flows will be considered.

The boundary conditions for the Navier–Stokes equations at solid surfaces are that of no slip and no temperature gradient at the wall. A one-dimensional characteristic analysis is used to devise the boundary conditions at the farfield boundary. The Euler equations describing an inviscid compressible flow are obtained from equatons (4.1-4.5) by setting all components of the flux tensor and heat vector in equation 4.3 to zero. i.e. $\boldsymbol{F}^V = \boldsymbol{0}, \boldsymbol{G}^V = \boldsymbol{0}$. In this case, the zero normal flow boundary condition must be applied at a solid wall.

4.2 Numerical Formulation

The flow equations are solved by a finite volume scheme based on the work of Jameson [9,10]. The spatial and temporal integrals are performed seperately.

Spatial Discretization

The flux terms of the flow equations are approximated by defining a residual R_j as the net flux in each cell j, such that:

$$R_j = \sum_{i=1}^{m} [F(w_j)\Delta y - G(w_j)\Delta x] . \qquad (4.6)$$

Cell Vertex Formulation

For the cell vertex finite volume discretization the residual R_j in equation (4.6) is taken at point j placed at the grid point, as shown in Figure 4.1, together with the contour path for the evaluation of the flux balance. For the embedded grids, the treatment of so-called 'hanging nodes', i.e. nodes which are not fully connected to surrounding points in the mesh, is addressed differently. Transitional triangular or quadrilateral interfaces (see Figures 2.2 and 2.3) can be introduced, or alternatively such nodes can be treated directly. In the latter case, the contour integral for the fluxes within the solver is modified. In the adaptive procedure, smoothing and clustering routines are employed so that a refinement ratio of two between grids is obtained, and a maximum of two 'hanging nodes' in the cell is allowed. Hence, in practice the admissible subdivisions lead to two types of interfaces, as shown in Figure 4.2. Cells of type 'A' (no 'hanging nodes'), are treated in a standard way, both for the coarse and refined regions. Cells of type 'B' (one 'hanging node'), and type 'C' (two 'hanging nodes') are treated by the solver as pentagonal and hexagonal, respectively. For example the contribution of the residual, equation 4.6, to node 1 from cell 'B' (see Figure 4.2) is as follows:

$$\int_{\partial \Omega} (Fdy - Gdx) = \sum_{sides\, 4-3\,3-5\,5-2} [F\Delta y - G\Delta x] . \qquad (4.7)$$

Some examples of the control volumes taken for the 'hanging nodes' are shown in Figure 4.3.

To impose the flow boundary conditions, the flow variables are set within dummy cells. For every edge on the boundary, there is a separate dummy cell constructed, thus allowing for the treatment of unsmooth boundaries, such as corners etc. All components of the flow variables, within the nodes of the dummy cell, are determined from the values in the adjacent cell within the computational domain. The quadtree data structure, described in section 2.1 is used for the point enrichment procedure used for the cell vertex versions of the Euler and Navier-Stokes flow solvers.

Cell Centred Formulation

For the cell centred finite volume discretization the residual R_j in equation 4.6 is taken at point j placed at the centre of a cell. The same treatment is valid for both initial nonembedded grids, such as that shown in Figure 4.4, and adapted embedded grids, Figure 4.5, and it can be readily generalized for any combination of polygonal cells. To impose boundary conditions a set of dummy cells is constructed as before. The velocity components within a dummy cell are determined from the values in the adjacent cell within the computational domain. Pressure and density are set equal in both cells.

The modified quadtree data structure, described in section 2.2 is used for the point enrichment procedure used for the cell centred versions of the Euler and Navier-Stokes flow solvers.

Time Integration

Given the solution and residuals for a point at time level n, the solution at the new time $n + 1$ is obtained from a Runge-Kutta multi-stage scheme. For example, the three stage scheme is:

$$\boldsymbol{w}^1 = \boldsymbol{w}^n - \frac{0.6\Delta t[\boldsymbol{R}(\boldsymbol{w}^n)]}{A},$$
$$\boldsymbol{w}^2 = \boldsymbol{w}^1 - \frac{0.6\Delta t[\boldsymbol{R}(\boldsymbol{w}^1)]}{A}, \quad (4.7)$$
$$and \quad \boldsymbol{w}^{n+1} = \boldsymbol{w}^2 - \frac{\Delta t[\boldsymbol{R}(\boldsymbol{w}^2)]}{A},$$

where A is the area of the cell and Δt is the time step, which is chosen to be consistent with stability limitations due to both the inviscid and viscous characteristics of the Navier-Stokes equations.

4.3 Artificial Viscosity

To ensure stability it is necessary to augment the governing flow equations with terms which represent artificial dissipation. Two terms, $\boldsymbol{D}_o^1$ a diffusive Laplacian smoothing to capture the shock waves, and $\boldsymbol{D}_o^2$ a bi-harmonic diffusive smoothing acting as a low level background dissipation to reduce odd-even decoupling are introduced. A simple way to introduce these dissipation operators is to construct a Laplacian operator by summing over all the edges of a control volume the difference between the flow variables across each edge. This leads to a form for the bi-harmonic contribution, which for cell i is

$$\boldsymbol{D}_i^1 = \sum_{k=1}^{ne} \varepsilon_{ki}^1 (\boldsymbol{w}_k - \boldsymbol{w}_i),$$
$$\boldsymbol{D}_i^2 = \sum_{k=1}^{ne} \varepsilon_{ki}^2 (\boldsymbol{E}_k - \boldsymbol{E}_i), \quad (4.8)$$
$$where \quad \boldsymbol{E}_i = \sum_{k=1}^{ne} \varepsilon_{ki}^1 (\boldsymbol{w}_k - \boldsymbol{w}_i),$$

with the summations over the ne edges surrounding the cell i. The terms ε_{ki}^1 and ε_{ki}^2 are coefficients which incorporate pressure sensors. These two terms, $\boldsymbol{D}_i^1$ and $\boldsymbol{D}_i^2$, are then summed to produce a single dissipative term, which is added to equations (4.6).

Within the finite-volume flow solver, which is based on edge information, it is possible to vectorise every loop that passes over all edges by 'colouring' the edges. The edge data structure allows the calculation of a flux across an edge, this flux is then passed to each of the cells, or in the case of a cell vertex flow solver the nodes, that the edge separates. The vectorisation of the flow solver results in the reduction of computation time by a factor of approximately six.

5. Adaptivity

The adaptation of meshes to flow features has been undertaken by three techniques, point enrichment, point derefinement and node movement. Each of these approaches will be described in detail.

5.1 Adaption Criterion for Point Enrichment

To instigate point enrichment it is first necessary to identify cells which require subdivision. Techniques used to flag cells for subdivision can be generally divided into those based on the physics of the flowfield, and those based on the error in the solution of the flow equations. The physical criteria for adaptivity aim to detect features of the flowfield, typically the shock waves, stagnation points and boundary layers. Two such approaches have been investigated. The first criterion, which is more suitable for the cell vertex flow solver, is based on a comparison of the gradient of some variable Φ, for which average cell values are compared with neighbouring cells. Using the notation from Figure 5.1, the following procedure can be described: the average cell value Φ_A is taken as

$$\Phi_A = \frac{1}{4}(\Phi_a + \Phi_b + \Phi_c + \Phi_d) \tag{5.1}$$

and the value Φ_B in the neighbouring cell is given by

$$\Phi_B = \frac{1}{2}(\Phi_e + \Phi_f). \tag{5.2}$$

If

$$\frac{(\Phi_A - \Phi_B)}{(maximum\ field\ range\ of\ \Phi)} > tolerance \tag{5.3}$$

then the edge is flagged as requiring subdivision. The level of sensitivity can be controlled by applying expression 5.3 to one, two, three or all four direct neighbouring cells of the cell A. The choice of the type of subdivision is dictated by the flow feature which is to be detected. Thus, type 2 refinement, (see Figure 2.1), is typically used in cells in the boundary layer when solving the Navier Stokes equations. Type 3 subdivision is suitable for capturing grid aligned shock waves. Finally, type 1 subdivision is required to join regions which have been subdivided with types 2 and 3. It is also possible to use only type 1 subdivision in all flagged cells, although, this proves less efficient.

Many possibilities of the choice of the variable Φ, as a physical adaptivity criterion can be considered. For inviscid flows the performance of density, pressure, and Mach number have been studied and are compared in Figure 5.2, for the same sensitivity level. It is clear that the resulting meshes vary, particularly when pressure is used as the adaptivity criterion. The meshes produced by using Mach number and density as the adaptivity criteria are similar.

In addition to the adaptivity criteria, it is important to ensure that the proper tolerance of the chosen sensor is achieved to capture the flow field features adequately.

The influence of the sensitivity of the adaptive indicator on the flow solution is shown in Figure 5.3, for embedding by pressure gradients.

For inviscid flow calculations the gradient of pressure has proved to be the most appropriate adaptivity criterion for capturing most of the important flow features. Although none of the investigated physical criteria identifies all flow features.

For viscous flows, the choice is less obvious. Again, gradients of density, pressure, entropy and Mach number have been tested as well as their scalar combination and as expected, to properly model the rapid changes in the viscous boundary layers, the choice of Mach number seems to be the most suitable.

The second criterion used to perform adaptivity incorporates both the flow conditions within and locally surrounding each cell, as well as accounting for the size of the cell, and is more suitable for cell centre flow solvers. The adaptivity criterion is cell based, which disregards the number of connection to the cell. For each cell a maximum gradient of a user defined flow variable is stored, after calculating the gradient across each edge of the cell. The maximum gradient being defined as Φ, where

$$\Phi = \phi_i - \phi_o \tag{5.4}$$

and ϕ_i and ϕ_o are values of the flow parameter in cell i and its neighbouring cell o. The adaptivity criterion, Ψ_i, is thus defined as

$$\Psi_i = \Phi_i^2 * A_i^{1/2} \tag{5.5}$$

where A_i is the area of the cell i. A cell is flagged for subdivision after the value of Ψ has been compared with a tolerance level which is iteratively adjusted to achieve the desired percentage of subdivided cells. For inviscid calculations pressure has proven to be a suitable parameter on which to base the adaptivity, however, for viscous flow calculations Mach number is more applicable.

An alternative adaptivity indicator is based on the error in the solution of the governing equations. It follows the work of M.Berger *et al.* [11] and uses principles similar to Richardson extrapolation. The idea is to solve equations 4.1 to 4.4, first on a fine mesh, with a point spacing of h, and then to estimate the error on a coarser mesh with a point spacing of $2h$. The truncation error of the solution, in regions of smooth flow, derived from equation 4.6, can be written in the short form

$$\boldsymbol{Q}(h)\boldsymbol{u} = \boldsymbol{\tau} h^p \tag{5.6}$$

where p is the order of the numerical method, $\boldsymbol{\tau}$ represents derivatives of the solution $\boldsymbol{u}$ and $\boldsymbol{Q}$ incorporates the difference operators. After numerical discretization on the finer grid, the approximated values of the flow variables $\boldsymbol{U}$ are calculated from equation 4.6, written here in the short form:

$$\boldsymbol{Q}(h)\boldsymbol{U} = 0 \tag{5.7}$$

In the steady state the residuals $\boldsymbol{Q}(h)U$ on the finer mesh are assumed to converge to zero, while the residuals obtained from the coarser mesh $Q(2h)U$ give an error estimation, equation 5.6, at each point of the coarser grid:

$$error = \tau h^p = \frac{Q(2h)U}{2^p - 1}. \tag{5.8}$$

For the second order numerical method $p = 2$ and for the adaptive procedure used here it is necessary to calculate

$$error = \frac{Q(2h)U}{3}. \qquad (5.9)$$

A computation of the above expression is equivalent to the first stage of the three stage Runge-Kutta integration step performed on the coarser mesh.

In this work the cell vertex version of the Euler and Navier-Stokes flow solvers have been used with the Richardson extrapolation error indicator. The quadtree data structure enables the application of the error estimation procedure to be performed easily. In Figure 5.4, the initial mesh, without embedding, consists of cells at the the lowest subdivision level (in this case level zero). Each successive enrichment increments the subdivision level of a cell by one. The error estimation is obtained by comparing the values of the first stage of the Runge-Kutta scheme produced by a mesh at one subdivision level and the values given by a mesh at a subdivision level one higher, the values are compared only on nodes common to each mesh. The same treatment is valid for the further levels of subdivision. Figure 5.5 shows the enriched mesh produced by error estimation for an inviscid calculation around a NACA0012 aerofoil at Mach number 0.5 and zero incidence.

5.2 Adaption Criterion for Point Derefinement

The point derefinement described in section 2.4 is controlled by the choice of a derefinement criteria. In a similar way to point enrichment, many different approaches can be employed. The simplest technique is to use the values generated using the criteria for enrichment and instead of defining a limited above which cells are refined, a lower limit is defined below which derefinement can be performed. Therefore, once a group of cells suitable for derefinement have been identified, each cell is then checked against the derefinement tolerance level. If all four cells satisfy this condition, the group is then derefined into a single cell.

5.3 Adaption Criterion for Node Movement

The adaptivity technique described thus far involves the refinement of cells in regions of high solution activity. This embedding strategy has produced a marked improvement in the resolution of flow features, but at the expense of increased run times due to an increase in the total number of cells. The cost of computations increases rapidly with successive enrichments and a point of diminishing returns is quickly reached. An alternative approach to using embedding for mesh adaptivity is the use of node movement, where nodal concentration is increased in regions of high solution activity. This is achieved by the migration of nodes from regions of inactivity towards important flow features. Unfortunately, the node movement technique, when implemented alone, can result in areas suffering from node starvation, due to the limited resources available to the approach in the form of a fixed number of nodes in the mesh. This problem can be overcome by coupling the node movement and enrichment approaches to adaptivity. Since it has been possible to implement mesh embedding on general mesh types, it is necessary that any attempt at node movement must be equally applicable to all mesh types.

An approach which proves to be very successful is a modified Laplacian operator of the form

$$r_o^{n+1} = r_o^n + \omega \frac{\sum_{i=1}^{M} c_{io}(r_i^n - r_o^n)}{\sum_{i=1}^{M} c_{io}} \quad (5.3)$$

where r=(x,y), r_o^{n+1} is the position of node o at relaxation level n+1, c_{io} is the adaptive weight function between nodes i and o and ω is the relaxation parameter. The weight factor can be expressed as

$$c_{io} = k_1 + k_2 \left| \frac{\phi_i - \phi_o}{\phi_i + \phi_o} \right| \quad (5.4)$$

where ϕ can be taken as a measure of any flow parameter, such as pressure, density or a measure of the local error. The constants k_1 and k_2 provide a damping to background noise and amplification of gradients, respectively. The summation is taken over all edges connecting point o to i, where it is taken that there are M surrounding nodes. In practice, this relaxation is typically applied over 50 cycles with a relaxation parameter of 0.1. This approach proves trivial to implement on all mesh types, but yet its effects are impressive.

For viscous flows, Mach number has been chosen to represent ϕ, incorporated in the weighting factor (5.4). Such a choice reflects the high gradients in velocity, which are characteristic of viscous boundary layers.

6. Adaption Strategy

The possibility of employing two very different forms of adaptivity necessitates the development of an adaption strategy. Each of the adaption techniques results in a transformation of the previous grid. The consequence of point enrichment is that grid points are successively added to regions of high flow activity, this results in a rapid increase in the computational cost of the calculation. Alternatively, node movement does not increase the computational cost per solution iteration, but the migration of nodes to regions of high flow activity can culminate in areas of mesh which have an insufficient number of nodal points to resolve weaker flow features satisfactorily.

The adaption strategy proposed involves a combination of both node movement and mesh enrichment in such a way that it is possible to counter the adverse effects of each technique. To decide on an approach, it is necessary to assess the order of application, for example, node movement followed by point enrichment or vice versa. A number of combinations were attempted for different flow conditions and various configurations and it was concluded that a limited number of enrichments, say three, followed by an application of node movement produced the most efficient and accurate solution. The reasons for this become clear on explanation. By performing point enrichment several times on an initial mesh produces a grid which has sufficient node numbers in regions of both high and low solution activity, but due to the nature of the initial mesh some cell sizes may still not quite be ideal. To reduce or enlarge these cell sizes node movement can be implemented, and because an adequate number of nodes are already in position, node movement does not result in regions of node starvation and does allow suitable clustering of nodes around regions of high solution activity to resolve them more accurately.

To demonstrate the effect of point enrichment on the equidistribution function $w_i ds_i = constant$, Figure 6.1 shows the values of the function $w_i ds_i$ plotted against frequency of occurrence (normalised with respect to the number of cells in the mesh).

It should be noted that a cut-off value has been introduced and all cells with a value of $w_i ds_i$ greater than this value have been summed. Hence the plot shows a large frequency of occurrence of the highest value. The test problem chosen is that of the flow around a NACA0012 aerofoil at a Mach number of 0.85 and 1° of incidence. The initial mesh consists of 1600 cells and utilises an "O" topology. The figure shows how successive enrichments of the mesh (particularly levels 5 and 7) results in an increase in the frequency of a single $w_i ds_i$ value, which is, of course, a successful attempt to satisfy the equidistribution principle.

7. Results

The results presented aim to demonstrate the advantages and accuracy of a flexible approach to both inviscid and viscous compressible flow simulation using each of the adaptivity techniques described.

Inviscid Calculations

Figure 7.1 shows three meshes and their corresponding contours of pressure. The initial unstructured mesh and resultant inviscid flow solution for an RAE2822 aerofoil at Mach number 0.75 and 3° incidence is shown in Figure 7.1(a), the mesh consisted of 3172 cells and 1646 nodes. Figure 7.1(b) shows an improvement in the resolution of the shock wave formed on the upper surface of the aerofoil from the initial solution, by using three applications of point enrichment. Finally, Figure 7.1(c) is the mesh resulting after applying node movement. The final mesh utilised 7873 cells and 4274 nodes. The contours of pressure remain smooth throughout the field during each stage of the adaption.

The result shown in Figure 7.2 is for an inviscid flow of Mach number 0.85 at 1° incidence around a NACA0012 aerofoil. The adapted mesh has been produced from a structured "O" mesh using three level of enrichment. The mesh consists of 10122 cells and 10718 nodes, also pictured are the corresponding contours of pressure and surface pressure coefficient.

Two further results illustrate an adaption strategy which incorporates both point enrichment and node movement. The first of these, pictured in Figure 7.3, shows the structured mesh and Mach number contours for an inviscid flow of Mach 2.0 over a double incline ramp, each of which step up an angle of 10°. The solution was obtained by first enriching the mesh to gradients in pressure and derefining inactive regions of flow, this was followed by a series of node movements. The results can be compared directly with theoretical values and errors of less than 0.5 percent were obtained for each region of the flow.

The second example, shown in Figure 7.4, is a biplane configuration of two NACA0012 aerofoils. The mesh consists of two "O" meshes which have been overlapped and connected by a Delaunay triangulation to form an hybrid mesh. The flow conditions are of Mach number 0.85 and zero incidence. The solution has again been obtained by enriching the mesh twice and derefining regions of inactivity, followed by a number of applications of node movement. It is difficult to assess the accuracy of the solution because the case is non-standard, but it is apparent that the shocks are equally well defined in both the structured and unstructured regions of the mesh, and also the contour lines are smooth throughout the transitional regions.

The first example of a viscous calculation is for the Williams two component landing configuration at Mach number 0.5, zero incidence, with a Reynoulds number of 5000. The initial mesh was produced using the structured multiblock elliptic mesh generator utilising a 23 block topology. The solution was adapted to gradient of Mach number and the resulting enriched mesh and corresponding contours of Mach number are shown in Figure 7.5.

The second example is a viscous solution about a NACA0012 aerofoil at Mach number 0.5 and zero incidence with a Reynoulds number of 5000 using a structured mesh. The mesh refinement and derefinement was initiated from an "O" mesh with a grid line density more suitable for an inviscid calculation. The adaption has resulted in good resolution of the boundary layer, particularly in the wake region, which is usually more reliably captured by "C" type meshes. This was made possible by using an adaptivity criterion that accounts for the size of each cell and hence larger cells further away from regions of high gradients are flagged for subdivision. The computational mesh and corresponding contours of pressure are shown in Figure 7.6.

8. Conclusions

Mesh generation techniques for two-dimensional structured, unstructured and hybrid grids have been described. Details of two data structures suitable for each mesh type and with the flexibility to deal with a variety of adaption techniques have been given. The results produced by a combination of various meshes and both cell vertex and cell centred Euler and Navier-Stokes flow solvers, which can cope efficiently with meshes adapted using point enrichment, derefinement and node movement, clearly demonstrates the benefit of a flexible grid and adaption procedures.

9. Acknowledgements

The funding for this work has been provided by CEC BRITE/EURAM Area 5 Aeronautics Project Aero0018 Euromesh. The authors are grateful for this support.

10. References

[1] D. Catherall, 'The adaption of structured grids to numerical solutions for transonic flow', Int.J.Num.Methds.Engng, vol. 32, (1991), p921-937.

[2] J. Peraire, M. Vahdati, K. Morgan and O.C. Zienkiewicz, 'Adaptive remeshing for compressible flow computations', J.Comp.Phys., 72, No. 2, p449-466 (1987).

[3] A.Evans, M.J.Marchant, J.Szmelter and N.P.Weatherill, 'Adaptivity for compressible flow computations using point embedding on 2-D structured multiblock meshes', Int.J.Num.Methds.Engng, vol. 32, (1991), p896-919

[4] J.F. Dannenhoffer, 'A comparison of adaptive-grid redistribution and embedding for steady transonic flows', Int.J.Numer.Meth.Engng, 32, (1991) 653-663

[5] P.D. Thomas and J.F. Middlecoff, 'Direct control of grid point distribution in meshes generated by elliptic equations', AIAA J., 18, p652 (1980)

[6] J.F. Thompson, Z.U.A. Warsi and C. Mastin, 'Numerical Grid Generation, Foundations and Applications', Pub. North-Holland, 1985.

[7] M.J. Marchant and N.P. Weatherill, 'The Construction of Nearly Orthogonal Multi-block Grids for Compressible Flow Simulation', submitted to Comm. Applied Numerical Methods, Sept 1991.

[8] N. P. Weatherill, 'The generation of unstructured grids using Dirichlet tessellations', Dept. of Mechanical and Aerospace Engineering, Princeton University, USA. Report No. 1715, July 1985.

[9] A. Jameson, J. Baker and N.P.Weatherill, 'Calculation of transonic flow over a complete aircraft', AIAA Paper no. 86-0103, 1986.

[10] A. Jameson, W. Schmidt and E. Turkel, 'Numerical solution of the Euler equations by finite volume methods using Runge-Kutta timestepping schemes', AIAA Paper no. 81-1259, (1981)

[11] M.J. Berger and J. Oliger, 'Adaptive mesh refinement for hyperbolic partial differential equations', J. Comp. Physics, Vol 53 pp 484-512, 1984.

[12] M.J. Marchant and N.P. Weatherill, 'Adaptivity Techniques for Compressible Inviscid Flows', submitted to Computer Methods in App. Mech. and Eng.,Nov 1991.

[13] N.P. Weatherill and M.J. Marchant, 'Generalised Mesh Techniques for Computational Fluid Dynamics',4th Int. Symposium on CFD, Sept 9-12,1991, Davis, California, USA.

[14] N.P. Weatherill, 'Delaunay triangulation in Computational Fluid Dynamics', Computers Math. Applic, Vol 24, No 5/6, pp 129-150, 1992.

[15] N.P. Weatherill and O. Hassan, 'Efficient three-dimensional grid generation using the Delaunay triangulation', to be presented at the 1st European CFD Conference, Brussels, Sept, 1992. To be published by Elsevier, Ed. Ch. Hirsch.

Figure 2.1 Three types of subdivision

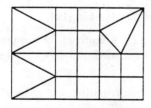

Figure 2.2 Triangular interface

for hanging nodes

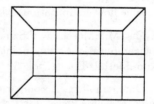

Figure 2.3 Quadrilateral interface

for hanging nodes

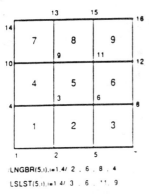

LNGBR(5,i),i=1,4/ 2 , 6 , 8 , 4
LSLST(5,i),i=1,4/ 3 , 6 , 11, 9

Figure 2.4
Data ordering for Quadrilateral cell

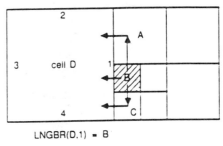

LNGBR(D,1) = B

Figure 2.6
Neighbours search for enriched structured mesh

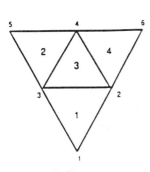

/LNGBR(3,i),i=1,3/ 1 , 4 , 2
/LSLST(3,i),i=1,3/ 3 , 2 , 4

Figure 2.5
Data ordering for Triangular cell

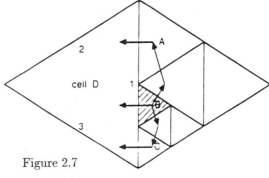

Figure 2.7
Neighbours search for enriched unstructured mesh

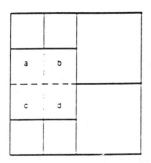

Figure 2.8 Four cells suitable for derefinement

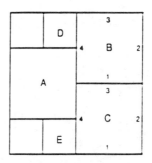

Figure 2.9
Data structure anomaly from derefinement

281

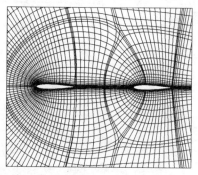

Figure 3.1 Navier-Stokes mesh for tandem configuration

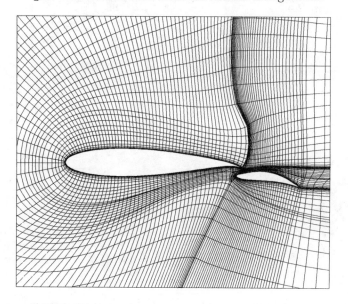

Figure 3.2 Landing configuration mesh and pressure coefficient

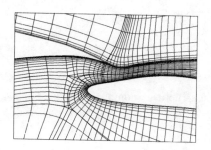
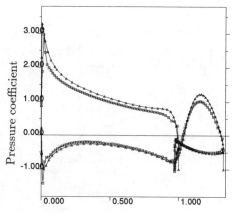

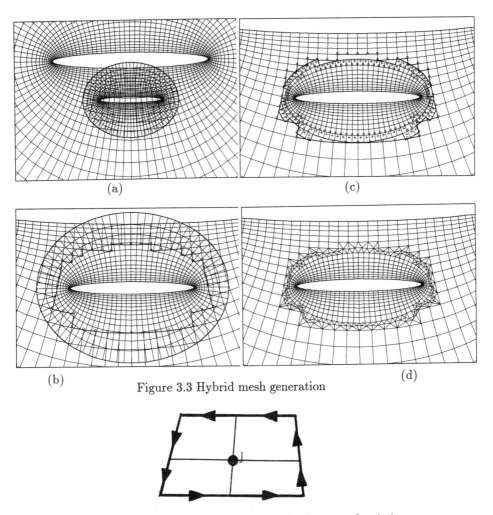

Figure 3.3 Hybrid mesh generation

Figure 4.1 Contour path for evaluation of cell vertex flux balance

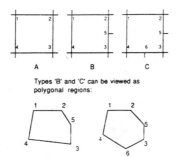

Figure 4.2 Treatment of hanging node interface using polygonal regions

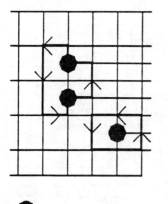

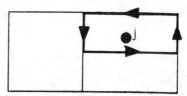

Figure 4.4 Contour path for evaluation of cell centre flux balance for a non-embedded mesh

● **Hanging node**

Figure 4.3 Example control volumes

Figure 4.5 Contour path for evaluation of cell centre flux balance for an embedded mesh

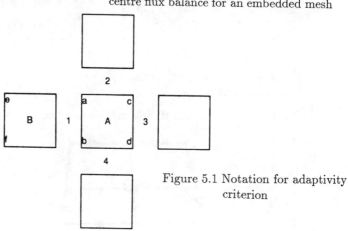

Figure 5.1 Notation for adaptivity criterion

Initial mesh and Pressure contours for example meshes given in Figures 5.2 and 5.3

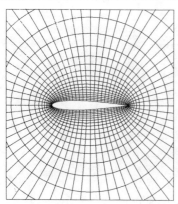
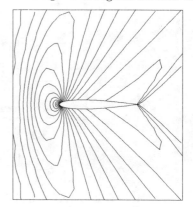

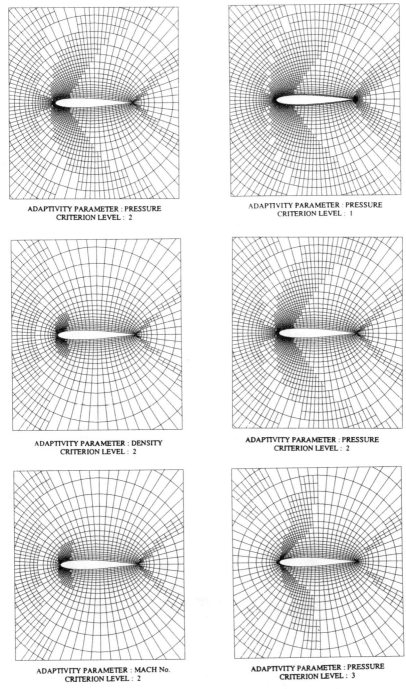

Figure 5.2 Enrichment using various flow parameters

Figure 5.3 Enrichment using different sensitivities of adaptivity criterion

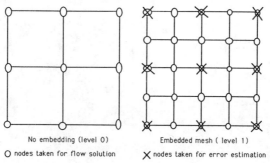

Figure 5.4 Implementation of error estimation based upon Richardson's Extrapolation

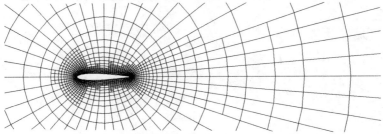

Figure 5.5 Mesh adapted using error estimation
Inviscid NACA0012 aerofoil, Mach number 0.5, zero incidence

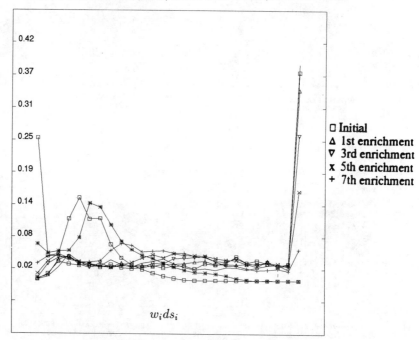

Figure 6.1 Frequency of occurrence
(normalised with respect to the total number of cells in the mesh)

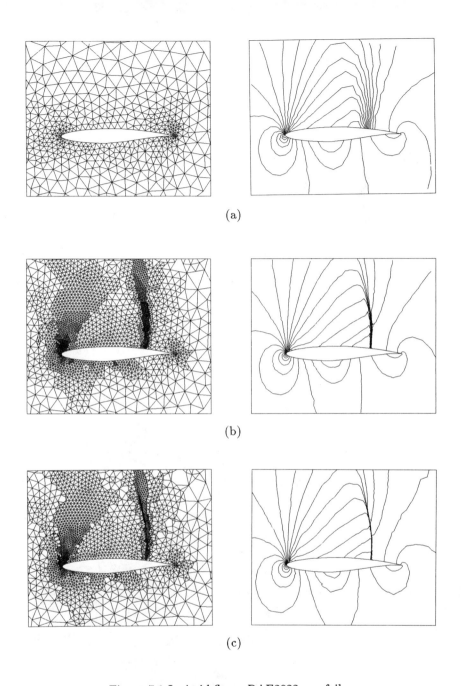

(a)

(b)

(c)

Figure 7.1 Inviscid flow - RAE2822 aerofoil
Mach number = 0.75 Incidence = 3°

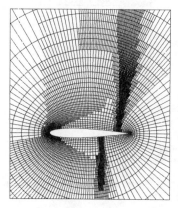
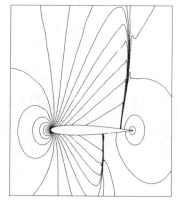

Figure 7.2 Inviscid flow - NACA0012 aerofoil   Mach number = 0.85 Incidence = 1°

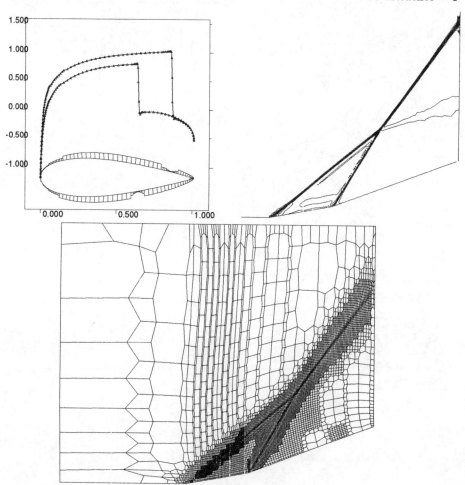

Figure 7.3 Inviscid flow - Double incline ramp Mach number = 2.0 Step angles = 10°

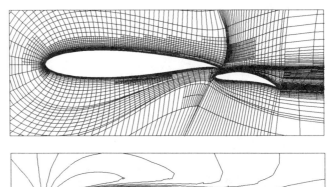
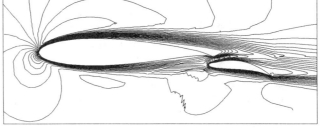

Figure 7.5 Viscous flow - Two component landing configuration
Mach number = 0.5 Incidence = $0°$ Re = 5000

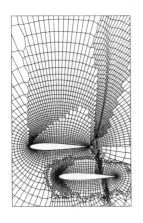

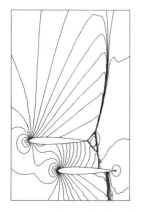

Figure 7.4 Inviscid flow - Biplane configuration
Mach number = 0.85
Incidence = $0°$

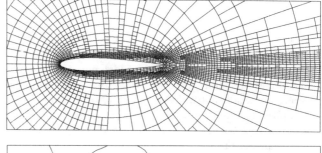
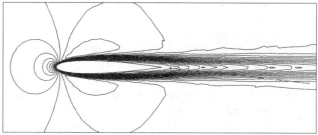

Figure 7.6 Viscous flow - NACA0012 aerofoil
Mach number = 0.5 Incidence = $0°$ Re = 5000

Error estimates and mesh adaption for a cell vertex finite volume scheme

J.A. Mackenzie, D.F. Mayers and A.J. Mayfield

Oxford University Computing Laboratory
Oxford OX1 3QD, England

Abstract

In this paper we investigate a posteriori error estimates for the cell vertex finite volume method and their applicability to adaptive mesh generation. We start from the cell vertex scheme reformulated as a Petrov-Galerkin finite element method and base the error estimate on the calculation of the finite element residual of the approximate solution. An analogy can be made between the residual in finite element methods and the truncation error of finite difference schemes. Numerical investigation of the performance of the error indicator for a model problem of two dimensional advection demonstrate its reliability and efficiency. The error indicator is then applied to a transonic Euler calculation where it is contrasted with an alternative procedure based on first and second differences of physical flow quantities which can be related to local interpolation error. Finally, we discuss the effective use of error estimates in adaptive mesh generation.

1 Introduction

Today there is increasing demand for high quality computational grids for fluid dynamic calculations. A "good" mesh will not only posess certain geometric characteristics such as grid smoothness but will also be well suited to the type of calculation being performed on it. This has led to the development of adaptive mesh generation techniques where the aim is to automatically produce such meshes.

Fundamental to the success of adaptive mesh generation – and adaptive methods in general – is the use of *reliable* and *efficient* error estimators. A reliable error estimator is one which can bound above and below the norm of the error. An efficient error estimator is one which does not dramatically over or underestimate the error.

Given the mathematical intractability of the governing equations of compressible gas flow, it is not surprising that there has been little work on error estimation for these equations. This is in contrast to the large literature on adaptive methods for solid mechanics problems and Stokes equations where of course the equations are quite different, usually being elliptic and linear. A literature survey of these techniques can be found in Jones [6]. Where there has been an attempt at estimating the error for the Euler and Navier-Stokes equations it has usually been at the level of estimating the local truncation error of finite difference and certain finite volume methods. Due to their inherent conservation properties, finite volume methods are the most popular discretisation schemes for these problems and are often interpreted as finite difference methods. This is often reflected in the techniques used to analyse such schemes. However, the cell vertex formulation of the finite volume method has a natural interpretation as a Petrov–Galerkin finite element method [13] [5]. The finite element framework then affords the possibility of applying some form of error estimation techniques, which are extensively used for elliptic problems in multidimensions, to the equations of compressible gas flow. In most of these techniques the finite element residual plays a crucial part in establishing local error bounds. The finite element residual is defined in the usual way as $r = Lw^h - f$ where L is the differential operator, w^h is the finite element solution and f is the data for the problem. The use of the residual of the cell vertex method was first proposed by Süli [13] for two dimensional linear advection but until now there has been no numerical evidence to suggest that the residual provides an efficient error estimator although it has been shown to be reliable for problems with smooth solutions.

The rest of this report is as follows; in section 2 we introduce some notation which will be used throughout the rest of the report. In section 3 we present the inviscid and viscous conservation laws of fluid dynamics and in section 4 we describe the cell vertex method used for their approximate solution. In section 5 we outline the basic theoretical background of the use of the finite element residual and remark on its similarity with the local truncation error of finite difference methods. In section 6 we examine the basis of using divided differences of the approximate solution as a local error indicator and in section 7 some numerical test cases are carried out where the residual error indicator is compared and contrasted with other methods. In section 8 we discuss the effective use of error estimates in adaptive mesh generation and finally we make some conclusions in section 9.

2 Notation

Let Ω denote an open subset of $\mathbb{R}^n$ with Lipschitz continuous boundary Γ. We will use $\alpha = (\alpha_1, \ldots, \alpha_n)$ to denote a multi-index with length

$$|\alpha| = \sum_{i=1}^{n} \alpha_i,$$

with the usual notation for derivatives:

$$\partial^\alpha u = \frac{\partial^{|\alpha|} u}{\partial x_1^{\alpha_1} \ldots \partial x_n^{\alpha_n}}.$$

For each integer $m \geq 0$ and real p with $1 \leq p \leq \infty$, we define the Sobolev space

$$W^{m,p}(\Omega) = \{v \in L^p(\Omega);\ \partial^\alpha v \in L^p(\Omega)\ \forall |\alpha| \leq m\},$$

which has a norm

$$\|u\|_{m,p,\Omega} = \left(\sum_{|\alpha| \leq m} \int_\Omega |\partial^\alpha u(\mathbf{x})|^p d\mathbf{x} \right)^{1/p} \qquad p < \infty$$

or

$$\|u\|_{m,p,\Omega} = \max_{|\alpha| \leq m} \left(ess\ sup_{x \in \Omega} |\partial^\alpha u(\mathbf{x})| \right), \qquad p = \infty.$$

We also provide $W^{m,p}(\Omega)$ with the following seminorm

$$|u|_{m,p,\Omega} = \left(\sum_{|\alpha|=m} \int_{\Omega} |\partial^{\alpha} u(\mathbf{x})|^p d\mathbf{x} \right)^{1/p} \quad p < \infty,$$

with the above modification when $p = \infty$. When $p = 2$, $W^{m,2}(\Omega)$ will usually be denoted by $H^m(\Omega)$.

3 The governing equations

Since the ultimate aim is to develop error indicators for the compressible Navier–Stokes equations it is appropriate here to state them in their conservation form. The steady two-dimensional, Navier–Stokes equations can be written in conservation form as

$$\nabla.(\mathbf{f}, \mathbf{g}) = \mathbf{0}, \quad (x_1, x_2) \in \Omega. \tag{1}$$

Here, $\mathbf{w}(x,y)$ is a vector of conserved variables and $\mathbf{f}(\mathbf{w}, \nabla \mathbf{w})$ and $\mathbf{g}(\mathbf{w}, \nabla \mathbf{w})$ are vector valued flux functions which can be split into inviscid and viscous components such that

$$\mathbf{f}(\mathbf{w}, \nabla \mathbf{w}) = \mathbf{f}^I(\mathbf{w}) + \mathbf{f}^V(\mathbf{w}, \nabla \mathbf{w}) \quad \text{and} \quad \mathbf{g}(\mathbf{w}, \nabla \mathbf{w}) = \mathbf{g}^I(\mathbf{w}) + \mathbf{g}^V(\mathbf{w}, \nabla \mathbf{w}). \tag{2}$$

For the Navier–Stokes equations the unknowns and the fluxes have the following non-dimensionalised form

$$\mathbf{w} = \begin{pmatrix} \rho \\ \rho u \\ \rho v \\ \rho E \end{pmatrix} \quad \mathbf{f}^I = \begin{pmatrix} \rho u \\ \rho u^2 + p \\ \rho u v \\ \rho u H \end{pmatrix} \quad \mathbf{g}^I = \begin{pmatrix} \rho v \\ \rho u v \\ \rho v^2 + p \\ \rho v H \end{pmatrix} \tag{3}$$

$$\mathbf{f}^V = \begin{pmatrix} 0 \\ -\tau_{x_1 x_1} \\ -\tau_{x_1 x_2} \\ -u\tau_{x_1 x_1} - v\tau_{x_1 x_2} + q_{x_1} \end{pmatrix} \quad \mathbf{g}^V = \begin{pmatrix} 0 \\ -\tau_{x_1 x_2} \\ -\tau_{x_2 x_2} \\ -u\tau_{x_1 x_2} - v\tau_{x_2 x_2} + q_{x_2} \end{pmatrix}, \tag{4}$$

where ρ, u, v, p, E and H denote the density, the two Cartesian components of velocity, the pressure, the total specific energy and the total specific enthalpy respectively. The deviatoric stress and heat conduction terms are given by

$$\tau_{x_1 x_1} = \frac{2\mu}{3Re}(2\frac{\partial u}{\partial x_1} - \frac{\partial v}{\partial x_2}), \quad \tau_{x_1 x_2} = \frac{\mu}{Re}(\frac{\partial u}{\partial x_2} + \frac{\partial v}{\partial x_1}), \quad \tau_{x_2 x_2} = \frac{2\mu}{3Re}(2\frac{\partial v}{\partial x_2} - \frac{\partial u}{\partial x_1}), \tag{5}$$

$$q_{x_1} = -\frac{\kappa}{(\gamma - 1)M_\infty^2 RePr} \frac{\partial T}{\partial x_1}, \quad q_{x_2} = -\frac{\kappa}{(\gamma - 1)M_\infty^2 RePr} \frac{\partial T}{\partial x_2}, \tag{6}$$

where γ, κ, Re, Pr and M_∞ denote the adiabatic constant, the coefficient of thermal conductivity, the Reynolds number, the Prandtl number and the freestream Mach number respectively. The viscosity μ is assumed to vary with temperature according to Sutherland's law [12].

4 The cell vertex method

The cell vertex finite volume method is a commonly used discretisation scheme for conservation laws expressing the equations of gas dynamics. The term cell vertex has been used to describe a number of quite different schemes but here we shall concentrate on one particular formulation. Most finite volume methods are thought of as conservative finite difference methods and we shall briefly give a derivation of the cell vertex method from a finite difference point of view. However, the cell vertex method can also be thought of as a finite element method and we also give its derivation in this setting as it is from the finite element formulation that we shall develop the error indicators.

4.1 Finite difference formulation

A finite volume method is a discretisation scheme based on the integral form of (1). We assume that Ω can be partitioned by a set of non-overlapping, convex quadrilateral cells such that

$$\bar{\Omega} = \bigcup_i \bar{\kappa}_i$$

and if the equations are integrated over any κ_i and Gauss' theorem is applied then weak solutions of the equations are obtained. Hence we replace (1) by

$$\oint_{\partial \kappa_i} (\mathbf{f} dx_2 - \mathbf{g} dx_1) = 0 \quad \forall \ \kappa_i, \tag{7}$$

and it is to this form that the cell vertex method is applied as an approximation procedure.

With the approximate solution at the cell vertex $\mathbf{x}_j = (x_1^j, x_2^j)$ denoted by $\mathbf{W}_j$, and defining $\mathbf{F}_j := \mathbf{f}(\mathbf{W}_j)$ and $\mathbf{G}_j := \mathbf{g}(\mathbf{W}_j)$, then with reference to Fig. 1 the numerical integration of the fluxes along each cell edge is given by

$$\oint_{\partial \kappa_i} (\mathbf{f} dx_2 - \mathbf{g} dx_1) \approx \frac{1}{2}[(\mathbf{F}_1 - \mathbf{F}_3)\delta x_2^{24} + (\mathbf{F}_2 - \mathbf{F}_4)\delta x_2^{31} \\ - (\mathbf{G}_1 - \mathbf{G}_3)\delta x_1^{24} - (\mathbf{G}_2 - \mathbf{G}_4)\delta x_1^{31}], \tag{8}$$

where $\delta x_i^{jk} = x_i^j - x_i^k$ for $i = 1, 2$.

This compact four-point approximation has many favourable characteristics, the most important being its accuracy on distorted grids. Morton and Paisley [11] have shown that the truncation error is second order accurate as long as the cells are parallelograms to within $O(h)$ where h is the diameter of the cell, i.e. the orientations of opposite sides differ by $O(h)$. However, it should be noted that analysis based on truncation errors may be misleading on non-uniform meshes. This is due to the phenomena of *supraconvergence* where a finite difference method may converge at a higher order than the order of the truncation error [10]. Stability is also very difficult to establish for multidimensional problems on general partitions. A mathematically more promising setting is a finite element approach.

4.2 Finite element formulation

The reformulation of the cell vertex method within a finite element framework was originally done in the work of Süli on two dimensional linear advection [13]. The finite element formulation necessitates the introduction of some more notation. First of all we shall assume that we have a partition $T^h = \{\kappa_i\}$ of Ω where each κ_i is a convex quadrilateral and that $\bar{\Omega} = \bigcup_{i=1}^{N} \bar{\kappa}_i$. We denote by S_j the subtriangle of κ_i with vertices $\mathbf{x}_{j-1}$, $\mathbf{x}_j$ and $\mathbf{x}_{j+1}$. The geometry of each κ_i is then characterised by

$$h_{\kappa_i} = \text{diameter of } \kappa_i$$

and

$$\rho_{\kappa_i} = 2 \min_{1 \leq j \leq 4} \{\text{diameter of circle inscribed in } S_j\}.$$

The regularity of the cell is measured by

$$\sigma_{\kappa_i} = \frac{h_{\kappa_i}}{\rho_{\kappa_i}}.$$

Let $\hat{\kappa}$ denote the canonical square $[-1,+1]^2$ in $(\hat{x}_1, \hat{x}_2)$ space with vertices $\hat{\mathbf{x}}_1$, $\hat{\mathbf{x}}_2$, $\hat{\mathbf{x}}_3$, and $\hat{\mathbf{x}}_4$. There is a unique invertible bilinear mapping F_{κ_i} which maps $\hat{\kappa}$ onto κ_i such that $F_{\kappa_i}(\hat{\mathbf{x}}_j) = \mathbf{x}_j$, $j = 1, \ldots, 4$, see Fig. 1. The finite element approximation spaces are defined on $\hat{\kappa}$ and in particular we shall be interested in the sets $Q_1(\hat{\kappa})$ and $Q_0(\hat{\kappa})$ of bilinear and constant functions respectively defined on $\hat{\kappa}$. The trial and test spaces for the cell vertex method are defined in terms of the sets

$$S^h = \{w \in H^1(\Omega): \ w = \hat{w} \circ F_{\kappa_i}^{-1}; \ \hat{w} \in Q_1(\hat{\kappa}), \ \kappa_i \in T^h\}$$

and

$$T^h = \{p \in L^2(\Omega): \ p = \hat{p} \circ F_{\kappa_i}^{-1}; \ \hat{p} \in Q_0(\hat{\kappa}), \ \kappa_i \in T^h\}$$

and set $S_E^h = S^h \cap H_E^1(\Omega)$. Since the flux functions for the compressible Euler and Navier-Stokes equations are nonlinear functions of $\mathbf{w}$ their boundary integration would normally have to be achieved by numerical quadrature. However, a computationally simpler approach is to use the so called product approximation [4] where the nonlinear fluxes are replaced by low order interpolating polynomials. For the cell vertex method the fluxes are also assumed to be isoparametric bilinear functions. Therefore, we introduce the interpolation operator $I^h : (H^1)^4 \longrightarrow (S^h)^4$ and we finally arrive at the discrete weak formulation which is to find $\mathbf{w}^h \in (S_E^h)^4$ satisfying

$$(\nabla .(I^h \mathbf{f}(\mathbf{w}^h), I^h \mathbf{g}(\mathbf{w}^h)), p) = 0 \qquad \forall p \in T^h. \tag{9}$$

The set of nonlinear equations for the nodal values of $\mathbf{w}^h$ are identical to those formed from the finite difference formulation (8). However, it is from this framework that we shall develop an a posteriori error estimator in section 5.

4.3 Approximation of the viscous fluxes

Since the flux functions $\mathbf{f}$ and $\mathbf{g}$ in the Navier-Stokes equations depend on the gradient of the solution vector, a prime requirement is to approximate these gradients at primary cell vertices. This has to be done carefully to match the accuracy of the inviscid flux balance.

There is considerable evidence, both theoretical and numerical, that the cell vertex approximation enjoys a *superconvergence* property of its gradient at cell centroids. That

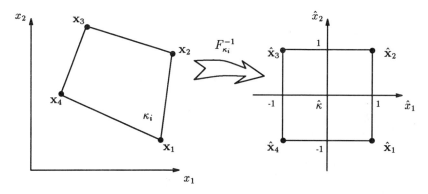

Figure 1: Transformation of a general quadrilateral to the canonical square.

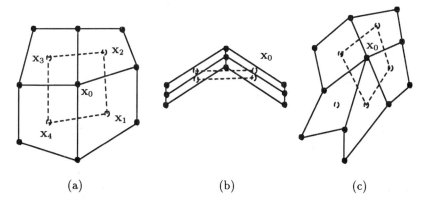

Figure 2: Recovery of gradients at $\mathbf{x}_0$ from local superconvergent points.

is the gradient of the isoparametric bilinear variation of $\mathbf{w}^h$ is a second order approximation at the cell centroid whereas it is only first order elsewhere in the cell. In particular the gradient at cell vertices is discontinuous across cell edges. The problem is then to *recover* second order accurate approximations of partial derivatives at cell vertices from the superconvergent points. This can be achieved by the notion of a recovery operator which should have the following properties: firstly the operator should be able to produce an approximation of the gradient which is of the same order of accuracy of the gradient at the superconvergent points; secondly the recovery operator should only use local information to produce a local approximation of the gradient; and finally since the first two conditions do not identify a recovery operator uniquely it is desirable that the recovery operator be optimal in the sense of minimising interpolation error.

For the cell vertex approximation the most obvious choice of recovery operator is bilinear interpolation. With reference to Fig. 2(a) the interpolation of the partial derivatives $\partial w^h/\partial x_i$ at the point $\mathbf{x}_0$ is given by the formula

$$\left.\frac{\partial w^h}{\partial x_i}\right|_{\mathbf{x_0}} = \sum_{j=1}^{4} \frac{\partial w^h(x_1^j, x_2^j)}{\partial x_i} \phi_j(\hat{x}_1^0, \hat{x}_2^0), \tag{10}$$

where $\phi_j(\hat{x}_1^0, \hat{x}_2^0)$, $j = 1, \ldots, 4$ is the set of interpolatory basis functions corresponding to the four cell centroids. The coordinates $(\hat{x}_1^0, \hat{x}_2^0)$ are computed by solving the system of two non-linear equations

$$x_i^0 = \sum_{j=1}^{4} x_i^j \phi_j(\hat{x}_1, \hat{x}_2), \qquad i = 1, 2 \tag{11}$$

by Newton's method. This system of equations can be solved in a very small number of iterations and the values of $(\hat{x}_1^0, \hat{x}_2^0)$ can be stored before the start of the main iteration.

The above procedure would be expected to be close to the optimal recovery operator if the central vertex $\mathbf{x}_0$ lay within the quadrilateral formed by joining the four cell centroids of the cells meeting at $\mathbf{x}_0$. For most structured grids this would be the case but there is one major exception which occurs for grids used to solve the Navier–Stokes equations for very high Reynolds number flows. For these flow conditions it is not unusual to use grids with aspect ratios of the order of 10^3. In regions where the mesh also has a reasonable level of curvature the central node of four cells may not lie within the quadrilateral formed from the four cell centroids see Fig. 2(b). In fact the central vertex may be some way from the four cell centroids. The straightforward use of (10) leads to an *extrapolation* of gradients, precisely in regions where a large variation of gradient would be expected. This can lead to very inaccurate results and can even lead to divergence of the main nonlinear iterative solver. A solution to this problem lies in the requirement of the recovery operator to be local. Therefore, in the troublesome situation described above the answer is to interpolate from the nearest four cell centroid values to the vertex in question. This procedure already includes the standard case described by (10) and can be used in other non-standard cases. For instance, this procedure can be used at grid singularities found in multiblock meshes, where more than four blocks may meet at one vertex, see Fig. 2(c). In this case the gradient at the singular vertex is interpolated from the closest four cell centroids.

5 An a posteriori error indicator

We begin the investigation for a posteriori error estimators by considering the model problem

$$\begin{aligned} \nabla.(\mathbf{a}w) &= f & x \in \Omega \\ w &= g & x \in \partial_-\Omega. \end{aligned} \tag{12}$$

Here $\Omega = (0,1)^2$ with inflow boundary $\partial_-\Omega$ and outflow boundary $\partial_+\Omega$. The velocity field $\mathbf{a} = (a_1, a_2)$ is a two component vector function with continuously differentiable entries defined on $\bar{\Omega}$ and for simplicity we shall assume that a_1 and a_2 are strictly positive. Furthermore, we shall assume that

$$\min_{\mathbf{x} \in \Omega} \nabla.a = c_\Omega > 0. \tag{13}$$

The weak form of (13) is to find $w \in H_E^1$ such that

$$(\nabla.(\mathbf{a}w), p) = (f, p) \qquad \forall p \in L^2(\Omega), \tag{14}$$

where $(\cdot, \cdot)$ denotes the L^2 inner-product on Ω. The cell vertex approximation of (14) is to find w^h in S_E^h, such that

$$(\nabla . I^h(aw^h), p) = (f, p) \qquad \forall p \in T^h, \tag{15}$$

where the trial and test spaces are defined analogously as before.

We now define the *residual* r of the approximation w^h as

$$r = \nabla . (aw^h) - f, \tag{16}$$

and the error in the cell vertex solution as $e = w - w^h$. We are now in a position to derive the local error estimate.

Theorem 5.1 *Suppose that condition 13 holds for the problem (13). Then the cell vertex solution satisfies the following local estimate:*

$$\|e\|_{0,2,\kappa_i}^2 + \frac{1}{c_{\kappa_i}} \int_{\partial_+\kappa_i} (\mathbf{a}.\mathbf{n}) e^2 ds \leq -\frac{1}{c_{\kappa_i}} \int_{\partial_-\kappa_i} (\mathbf{a}.\mathbf{n}) e^2 ds + \frac{1}{c_{\kappa_i}^2} \|r\|_{0,2,\kappa_i}^2, \tag{17}$$

where $0 < c_{\kappa_i} := \min_{x \in \kappa_i} \nabla . a$.

Proof From the definition of the residual (16) and the weak form (14) we have

$$(\nabla . (ae), p)_{\kappa_i} = -(r, p)_{\kappa_i} \qquad \forall p \in L^2(\kappa_i). \tag{18}$$

By substituting $p = e$ we have

$$(\frac{1}{2}(\nabla . a)e, e)_{\kappa_i} + \frac{1}{2}\int_{\partial_+\kappa_i}(\mathbf{a}.\mathbf{n})e^2 ds = -\frac{1}{2}\int_{\partial_-\kappa_i}(\mathbf{a}.\mathbf{n})e^2 ds - (r, e)_{\kappa_i}. \tag{19}$$

Using the Cauchy-Schwarz inequality we finally have

$$\|e\|_{0,2,\kappa_i}^2 + \frac{1}{c_{\kappa_i}}\int_{\partial_+\kappa_i}(\mathbf{a}.\mathbf{n})e^2 ds \leq -\frac{1}{c_{\kappa_i}}\int_{\partial_-\kappa_i}(\mathbf{a}.\mathbf{n})e^2 ds + \frac{1}{c_{\kappa_i}^2}\|r\|_{0,2,\kappa_i}^2. \tag{20}$$

□

Remark 1 This theorem tells us that we can bound the L^2 error within a cell and a weighted integral of the error on the outflow boundary by the L^2 norm of the residual within the cell and a weighted integral of the error on the inflow boundary. Since the error on $\partial_-\Omega$ is zero and we have assumed that a_1 and a_2 are both positive this allows us to estimate the error in all the cells by sweeping through the cells from bottom left to top right.

Remark 2 As it stands the theorem tells us very little about the efficiency of the error estimate. From the convergence theory of Suli [13] we know that the global L^2 error converges at $O(h^2)$ and that the weighted boundary integral of the error converge at $O(h^{3/2})$. However, it is not clear at what rate, if any, the L^2 norm of residual converges. The definition of the residual for this problem is

$$r = (\nabla . a)w^h + \mathbf{a}.\nabla w^h - f,$$

and hence we can see that the pacing term in terms of convergence rate is the accuracy of the gradient. As mentioned in section 4.3, the gradient of w^h is second order accurate only at cell centroids. Therefore if a quadrature rule is used to approximate $\|r\|_{0,2,\kappa_i}$ then it would produce a first order quantity unless the quadrature rule is the one point product Gaussian rule. This is obviously extremely beneficial when we turn to larger problems where the approximation of the norm of the residual may have been expected to involve many function evaluations.

Remark 3 The practical use of this theorem requires the use of the error at the inflow boundary to bound the error in the cell and at the outflow boundary. This is an unattractive prospect for the extension of this theorem to the compressible Euler equations since a characteristic decomposition would be necessary to determine the appropriate wave directions. The obviously next question is whether it is possible to use just use the residual as a local error estimator. Numerical evidence will be given in section 7 which gives credence to this proposition.

5.1 Relationship of the residual and truncation errors

It is worth noting here the duality of the finite element residual r and the local truncation error of a finite difference method. A finite difference approximation of (13) is simply

$$L^h w^h = f^h, \tag{21}$$

where L^h is an $n \times n$ matrix representing the difference operator and w^h and f^h are n-component real vectors. The local truncation error is defined as

$$\tau = L^h(R^h w(x)) - f^h, \tag{22}$$

where R^h is a restriction operator such that $(R^h w(x))_i = w(x_i)$. Notice the similarity between the (22) and (16). Stated simply, the finite element residual represents the inability of the approximate solution from the trial space to satisfy the differential equation, whereas the truncation error of finite difference methods represents the inability of the restriction of the exact solution to satisfy the difference equations. Both the residual and truncation errors are important in establishing error bounds for their respective methods but for adaptive calculations one requires these terms to easily calculated from the present approximate solution. This is easily done with the finite element residual whereas the truncation error requires the unknown exact solution. For finite difference methods, an approximation of the truncation error is usually obtained by either estimating high order derivatives of the exact solution or by comparing solutions on different grids, as in the Richardson-type estimates of Berger and Oliger [1] or the multilevel methods of Brandt [2].

5.2 Calculation of the residual for conservation laws

We now consider the calculation of the residual of the cell vertex finite volume approximation of the system of non-linear conservation laws

$$\mathbf{f}_{x_1}(\mathbf{w}) + \mathbf{g}_{x_2}(\mathbf{w}) = 0. \tag{23}$$

The residual is simply

$$\begin{aligned}\mathbf{r}(x_1, x_2) &= (\mathbf{f}(\mathbf{w}^h))_{x_1} + (\mathbf{g}(\mathbf{w}^h))_{x_2}, \\ &= A(\mathbf{w}^h)\mathbf{w}^h_{x_1} + B(\mathbf{w}^h)\mathbf{w}^h_{x_2}, \end{aligned} \tag{24}$$

where

$$A = \frac{\partial \mathbf{f}}{\partial \mathbf{w}} \quad \text{and} \quad B = \frac{\partial \mathbf{g}}{\partial \mathbf{w}}$$

are the flux Jacobian matrices. If we decide to measure $\mathbf{r}$ in the L^2 norm we need to calculate

$$\|\mathbf{r}\|_{0,2,\Omega} = \left\{ \int_\Omega (\mathbf{r}(x_1,x_2))^2 d\mathbf{x} \right\}^{1/2} = \left\{ \sum_i \int_{\kappa_i} (\mathbf{r}(x_1,x_2))^2 d\mathbf{x} \right\}^{1/2} = \left\{ \sum_i \|\mathbf{r}\|_{0,2,\kappa_i}^2 \right\}^{1/2}. \tag{25}$$

The approximate calculation of (25) is achieved using one-point Gaussian quadrature. The only problem is to calculate $\mathbf{w}_{x_1}^h$ and $\mathbf{w}_{x_2}^h$ at the cell centroid. Remembering that $\mathbf{w}^h$ is an isoparametric bilinear element, its derivatives have to calculated in the transformed space as functions of $\hat{x}_1$ and $\hat{x}_2$. This gives

$$\mathbf{w}_{x_1}^h = \frac{\mathbf{w}_{\hat{x}_1}^h x_{2_{\hat{x}_2}} - \mathbf{w}_{\hat{x}_2}^h x_{2_{\hat{x}_1}}}{J} \quad \text{and} \quad \mathbf{w}_{x_2}^h = \frac{\mathbf{w}_{\hat{x}_2}^h x_{1_{\hat{x}_1}} - \mathbf{w}_{\hat{x}_1}^h x_{1_{\hat{x}_2}}}{J}, \tag{26}$$

where J is the Jacobian of the bilinear mapping. It can easily be shown that the cell centroid of the cell in physical space corresponds to the origin in canonical space and hence the one-point quadrature may be achieved very simply. Numerical experiments using the cell vertex residual will be given in section 7 but before we go to the numerical results we briefly examine a very common error indication technique.

6 Interpolation errors

Most adaptive methods for compressible fluid dynamic problems at present use first or second differences of physical flow quantities like the density, Mach number or the pressure as a measure of local flow activity. The use of differences of the calculated solution as an adaptivity criteria is usually argued on intuitive physical principles. However, there is a theoretical sense in which this procedure may be partially justified by considering interpolation errors.

Let $k \geq 0$, $m \geq 0$ be integers and $p \geq 1$, $q \geq 1$ be reals such that

$$W^{k+1,p}(\hat{\kappa}) \hookrightarrow W^{m,q}(\hat{\kappa}).$$

Then we introduce the interpolation operator $\hat{\pi} \in \mathcal{L}(W^{k+1,p}(\hat{\kappa}); W^{m,q}(\hat{\kappa}))$ such that

$$\hat{\pi} t = t \quad \forall t \in Q_k(\hat{\kappa}). \tag{27}$$

We are now in the position to state the interpolation theorem.

Theorem 6.1 *Let κ_i be a convex quadrilateral and let k, m, p, q and $\hat{\pi}$ be defined as above. If we define the operator $\pi \in \mathcal{L}(W^{k+1,p}(\kappa_i); W^{m,q}(\kappa_i))$ by $\hat{\pi v} = \hat{\pi}\hat{v}$ then there exists a positive constant C, independent of the geometry of κ, such that*

$$|v - \pi v|_{0,q,\kappa_i} \leq C \sigma_{\kappa_i}^{2/p} h_{\kappa_i}^{k+1+2/q-2/p} |v|_{k+1,p,\kappa_i} \quad \forall v \in W^{k+1,p}(\kappa_i), \tag{28}$$

and

$$|v - \pi v|_{m,q,\kappa_i} \leq C \sigma_{\kappa_i}^{4m-2+2/p} h_{\kappa_i}^{k+1-m+2/q-2/p} |v|_{k+1,p,\kappa_i} \quad m \geq 1 \quad \forall v \in W^{k+1,p}(\kappa_i). \tag{29}$$

Proof See Cartan [3] □

Remark 1 If for instance we are interested in measuring errors in the maximum norm and we assume that $|v|_{2,\infty,\kappa_i}$ exists then we have

$$|v - \pi v|_{0,\infty,\kappa_i} \leq Ch_{\kappa_i}^2 |v|_{2,\infty,\kappa_i}. \tag{30}$$

Now

$$|v|_{2,\infty,\kappa_i} := \max_{|\alpha|=2} \left\{ \max_{\mathbf{x} \in \kappa_i} |D^\alpha v| \right\}$$

$$= \max_{\mathbf{x} \in \kappa_i} \left\{ \left| \frac{\partial^2 v}{\partial x^2} \right|, \left| \frac{\partial^2 v}{\partial x \partial y} \right|, \left| \frac{\partial^2 v}{\partial y^2} \right| \right\}, \tag{31}$$

which may be approximated by divided second differences of the approximate solution at grid vertices, if a finite difference type of method were being used. It is clear then that by combining the $h_{\kappa_i}^2$ scaling with the divided differences approximating (31), that one can be led in a rather heuristic way, to using undivided second differences as a local error indicator.

Remark 2 For some problems the solution may not lie in $W^{2,\infty}(\kappa_i)$ but may lie in $W^{1,\infty}(\kappa_i)$. In this case Theorem 6.1 yields the bound

$$|v - \pi v|_{0,\infty,\kappa_i} \leq Ch_{\kappa_i} |v|_{1,\infty,\kappa_i}. \tag{32}$$

As before

$$|v|_{1,\infty,\kappa_i} = \max_{\mathbf{x} \in \kappa_i} \left\{ \left| \frac{\partial v}{\partial x_1} \right|, \left| \frac{\partial v}{\partial x_2} \right| \right\},$$

which when combined with the h_{κ_i} scaling and use of divided differences to approximate the derivative results in the use of an undivided first difference as a local error indicator.

7 Numerical experiments

7.1 Two-dimensional linear advection

The first test case considered is linear advection in two dimensions. This is a sensible starting point to investigate the reliability of the error estimator due to the availability of an exact solution and is a viable testing ground for Theorem 5.1. The model problem considered is

$$\begin{aligned} \nabla \cdot (\mathbf{a} w) &= f & \mathbf{x} \in \Omega \\ w &= g & \mathbf{x} \in \partial_- \Omega, \end{aligned} \tag{33}$$

where $\Omega = (0,1)^2$. The convective velocity field is given by

$$\mathbf{a} = \left(\frac{(1+x)^{1+\beta}}{2\beta}, \frac{(1+y)^{1+\beta}}{2\beta} \right)^T,$$

which for $\beta > 0$ has both components being strictly positive. For simplicity the problem will be approximated on a uniform grid and in each cell $\kappa_{jk} = [x_{j-1}, x_j] \times [y_{k-1}, y_k]$ we have

$$\nabla \cdot \mathbf{a} \geq \frac{(1+\beta)((1+x_{j-1})^\beta + (1+y_{k-1})^\beta)}{2\beta} = c_{\kappa_{jk}} > 0.$$

With a forcing function given by

$$f = \frac{(\beta+1)((1+x)^\beta + (1+y)^\beta) \exp^{-(1+x)^{-\beta} - (1+y)^{-\beta}}}{(2\beta - 1)},$$

the solution of (33) with boundary conditions

$$w(x,y)|_{\partial_-\Omega} = \begin{cases} \exp^{-(1+y)^{-\beta} - 1}, & x = 0 \\ \exp^{-(1+x)^{-\beta} - 1}, & y = 0 \end{cases}$$

is

$$w(x,y) = \exp^{-(1+x)^{-\beta} - (1+y)^{-\beta}}.$$

Fig. 3(a) shows a contour plot of the exact solution when $\beta = 5$. The computed cell vertex solution looks identical to this at plotting accuracy on grid sizes of 32×32 and finer. Fig. 3(b) shows the error of the cell vertex solution at the nodes of a 32×32 cell grid. The areas of highest nodal error, corresponding to the lighter areas, are in regions were the convective velocity is closest to being parallel to a set of grid lines, which is consistent with the error analysis of the cell vertex method for this problem. Figs. 3(c) and (d) are of $\|e\|_{0,2,\kappa_{jk}}^2$ and

$$\|\hat{e}\|_{0,2,\kappa_{jk}}^2 := \sum_{k=1}^{N} \sum_{j=1}^{N} \frac{\|r\|_{0,2,\kappa_{jk}}^2}{c_{\kappa_{jk}}^2},$$

which corresponds to ignoring the effect of the boundary integrals of the error as discussed in remark 3 in section 5. Both of the norms have been calculated using a one point quadrature formula . From these figures it is clear that there is a very good correlation between the error estimator and the actual error. The agreement is quantified in table 1 which indicates not only the second order convergence of the nodal errors, labelled here as $\|e\|_{0,\infty,\Omega^h}$, and the approximate L^2 norm of the error but more importantly that approximate L^2 norm of the error estimator is also second order convergent. The final column of table 1 shows the asymptotic convergence of the efficiency index θ_h of the error estimator which is defined as $\theta_h = \|\hat{e}\|_{0,2,\Omega} / \|e\|_{0,2,\Omega}$. What is very encouraging is that we can observe an $O(h)$ convergence rate to the ideal value of unity.

7.2 Transonic Euler flow over a NACA0012 aerofoil

We now consider the performance of the cell residual as an error indicator for inviscid transonic flow over a NACA0012 aerofoil with flow conditions $M_\infty = 0.85$ at an angle of attack of $\alpha = 1°$. The grid used for this calculation has 128×32 grid cells and was chosen deliberately to be coarse at the leading edge of the aerofoil which can be seen in Fig. 4(a). This type of grid is too coarse at the leading edge and would normally be rejected, however the luxury of choosing a suitable initial grid is not always possible especially in three dimensions. The computed Mach contours are shown in Fig. 4(b)

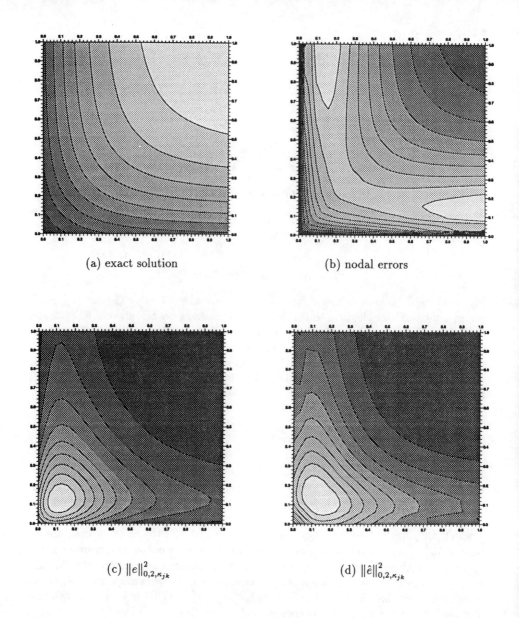

Figure 3: (a) contour plot of the exact solution of the 2d linear advection test problem, (b) contour plot of the nodal errors, (c) shows the contours of $\|e\|^2_{0,2,\kappa_{jk}}$ and finally (d) shows the contours of $\|\hat{e}\|^2_{0,2,\kappa_{jk}}$.

Table 1: Performance of residual error estimator for 2D linear advection.

N	$\|e\|_{0,\infty,\Omega^h}$	$\|e\|_{0,2,\Omega}$	$\|\hat{e}\|_{0,2,\Omega}$	θ
4	3.0151E-02	1.4180E-02	2.2188E-02	1.5647
8	6.8083E-03	3.3300E-03	4.1726E-03	1.2530
16	1.3051E-03	8.2551E-04	9.2016E-04	1.1147
32	2.9218E-04	2.0609E-04	2.1676E-04	1.0518
64	7.2376E-05	5.1506E-05	5.2638E-05	1.0220
128	1.8076E-05	1.2875E-05	1.2972E-05	1.0075

which clearly indicate the insufficient grid resolution at the leading edge and trailing edges of the aerofoil which can be deduced from the erroneous entropy generation on the upper and lower surface of the aerofoil upstream of both shocks. Fig. 5 shows contour plots of the four components of $\|\mathbf{r}\|_{0,2,\kappa_i}^2$ for this case, where the highest error is indicated by the darker shading. From these plots we can see that all components of the error indicator are large at the leading edge of the aerofoil as expected. However, what is surprising is the absence of any error indication at both shock locations. It is highly likely that the shocks are being indicated but at a far lower level of error, and will come into view after the quality of the grid at the leading edge region is improved. To test this hypothesis Fig. 6 shows the contour plots of the four components of the residual for a calculation using a grid with improved resolution at the leading edge of the aerofoil. The error at the leading edge has been reduced and the error indicator is now drawing attention to the error caused by the inadequate resolution of the strong shock on the upper surface.

For comparison Fig. 7 shows contours of of first and second *undivided* differences of pressure and Mach number for the initial grid. These plots clearly indicate the solution features which one would expect to pose problems; the leading and trailing edges and shocks. The differences in Mach number also seems to pick up the slip line in the wake region. However, one should expect this sort of flow sensor to be quite indiscriminate in indicating problem areas and could quite easily overpredict the extent of solution error. Nevertheless, the expectation of errors at shocks and their subsequent identification with this method ensures this is still the most popular type of adaptivity criteria.

8 Error estimates and adaptive mesh generation

The development of a posteriori error estimates is just one half of an adaptive algorithm. The other half is the design of an effective mesh selection strategy. How this strategy is formulated depends on the problem that the adaptive algorithm is proposed to solve. Two typical examples of such problems are the following: first of all, given a fixed number of grid cells, calculate the grid on which the norm of the error estimator is minimal. The second problem is to find the grid with the minimal number of grid cells such that the norm of the error estimator is less than a given tolerance. If solutions to these two problems are to be found then the error estimators must have sufficient structure to characterise the target meshes and also to suggest an iterative strategy which converges to the target mesh.

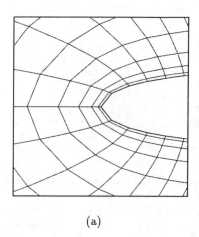 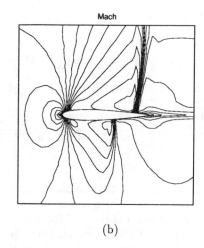

(a) (b)

Figure 4: (a) close up of leading edge region; (b) shows the computed Euler solution over a NACA0012 with $M_\infty = 0.85$ and $\alpha = 1°$.

The majority of local error indicators are based upon truncation error estimates, residuals, defects, solutions of local problems or just crude estimation of derivatives of the solution. The local error however is made up a nonlinear contribution of these error estimators where the local error estimator may or may not be a significant factor. For elliptic problems a large local residual should indicate large local error. For hyperbolic problems the local error is a non-local function of the residual. Therefore 'upstream' residuals are to be believed and corrected first which should have the effect of purifying the 'downstream' values. Elliptic problems which are convection dominated behave much the same as hyperbolic problems in the interior of the domain, however there is the added complication of boundary layers. Merely equidistributing the residual normal to the boundary fails to resolve the boundary layer in the error.

Adaptive mesh movement algorithms come under the category of methods which are used to solve the first problem outlined above. For problems and numerical methods which produce error estimates of a particular form, the target mesh can be characterised by having the error equidistributed throughout the domain. Mesh movement methods use a weight function related to an error indicator or some other adaptivity criteria, as the driving force to move the points within the mesh. The iterative method is made more robust by under-relaxing the effect of the weight function before use. Firstly, it is smoothed to avoid producing local oscillations in the mesh size; secondly it is rescaled such that the ratio of the largest to smallest weight values lies between specified limits. This has the desirable effect of retaining points in the regions of high accuracy. The final adapted mesh should be independent of these two under-relaxation procedures. When a sophisticated error estimator is not available, mesh movement routines optimise the smoothness of the grid, the distribution of the weight function and local orthogonality. The implicit belief here is that the accuracy of most numerical methods is adversely effected by non-orthogonal and highly stretched grids. However, analysis of the cell vertex method has shown no direct link between errors and non-orthogonality or non-smoothness of the mesh. The main requirement of a computational grid when using

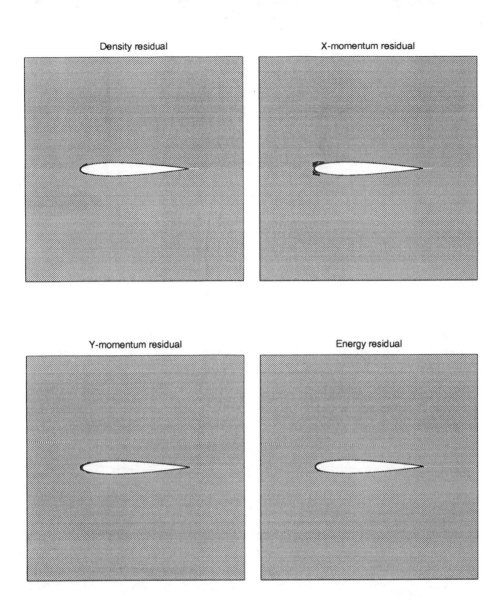

Figure 5: Contour plots of $\|\mathbf{r}\|^2_{0,2,\kappa_i}$ for flow over a NACA0012 aerofoil with $M_\infty = 0.85$ and $\alpha = 1°$.

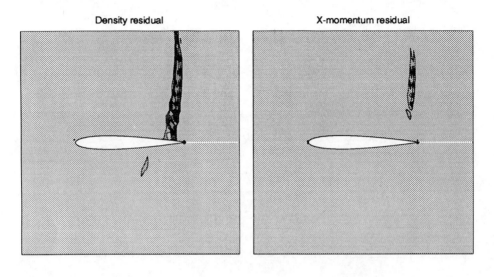
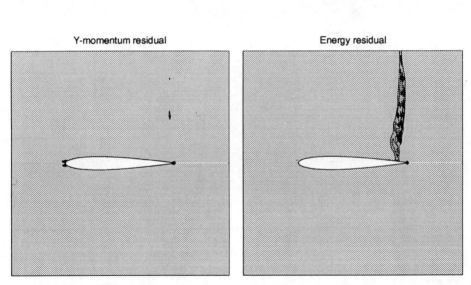

Figure 6: Contour plots of $\|\mathbf{r}\|^2_{0,2,\kappa_i}$ for flow over a NACA0012 aerofoil with $M_\infty = 0.85$ and $\alpha = 1°$ on an improved grid.

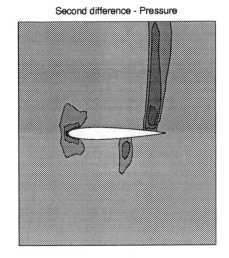

Figure 7: Plots of first and second undivided differences of pressure and Mach number for flow over a NACA0012 aerofoil with $M_\infty = 0.85$ and $\alpha = 1°$.

the cell vertex method is that the cells are close to being parallelograms [11] [13]. A new algorithm has been developed which uses bisection within a multigrid framework to quickly redistribute the mesh nodes and which also utilises the particular qualities of the cell vertex method [7].

A mesh enrichment algorithm is a method used to attempt to solve our second adaptive problem. In a typical mesh enrichment algorithm the cells are bisected if the error indicator exceeds some predescribed value. Additional flexibility can be gained if the cells are only bisected in one direction to reduce errors which are highly anisotropic e.g. in boundary layers or at shocks. Having the facility to derefine cells as well as refine cells increases the possibility of obtaining a suitable grid using the minimal number of grid cells. Derefinement can also help in the case for hyperbolic problems where initial poor refinements may be corrected by later adaptions. A mesh enrichment algorithm has been developed which works well within a multigrid framework [8]. Care must be taken when designing the data structures for such methods or the data structure management system can make large refinement prohibitive. A description of a data structure which has been designed such that the linkage does not depend linearly upon the number of unknowns can be found in Mayfield [9].

9 Conclusions

The finite element residual is the finite element equivalent of the local truncation error of finite difference methods. For hyperbolic problems the local truncation error may be an ineffective error estimator and some form of directional integration of the truncation errors is needed. Most of the interest in hyperbolic problems lies in the possibility of non-smooth solutions. It can be shown that for such problems the residual measured in the L^2 norm diverges whereas the approximate solution converges in this norm. The divergence of the residual should however be localised to the area of non-smoothness and the residual would then be used as a local error indicator. However, this is far from an ideal situation.

The problem with non-smooth solutions could be better dealt with by measuring the residual in the more appropriate dual norm. However, the calculation of the dual norm is not straightforward and can be approximated by the solution of local projection equations. The cost of such a procedure for large multidimensional problems is at this stage unclear.

Other commonly used error estimation techniques include bounds on local interpolation error and the Richardson extrapolation method.

Richardson extrapolation techniques were originally used for elliptic problems discretised on rectangular domains on uniform meshes. Their use also assumes that the underlying approximation scheme has a convergence rate of 2^p for some integer p. The use of these techniques in aerodynamic calculations can be questioned on many counts. First of all almost all meshes for fluid dynamic calculations are non-uniform and a careful choice of coarse and fine grids is required. Secondly as mentioned before a major interest is in problems which have non-smooth solutions. Most approximation schemes will not be convergent at the same rate as for smooth problems for which the theory is applicable. Finally, most solution domains for aerodynamic problems are not rectangular and this leads to difficulties around leading and trailing edges of aerofoils.

The attractiveness of local interpolation error bounds is the relative cheapness of the error calculation and the application to non-smooth flows. Also, local interpolation bounds are strongly dependent on the local geometry of the computational grid. However, there is still a major problem of relating the local interpolation error to the global error and it is at this point that the locality and cheapness of the interpolation estimate can be expected to be lost. On the theoretical side it is unclear for hyperbolic problems whether estimating derivatives of the exact solution via the approximate solution is reliable and efficient. However, the use of undivided differences will still continue to be the most popular error sensor due to its simplicity and its "believability" for large aerodynamic calculations.

For realistic external aerodynamic calculations we are still some way from developing reliable and efficient error indicators. However, many error indicators are available and the cell vertex residual may be added to the list. This paper has demonstrated the validity of such an approach.

References

[1] M.J. Berger and J. Oliger. Adaptive mesh refinement for hyperbolic partial differential equations. *Journal of Computational Physics*, 53:484–512, 1984.

[2] A. Brandt. Multi-level adaptive solutions to boundary value problems. *Mathematics of Computation*, 31:333–390, 1977.

[3] H. Cartan. *Calcul Différentiel*. Herrmann, 1967.

[4] I. Christie, D. F. Griffiths, A. R. Mitchell, and J. M. Sanz-Serna. Product approximation for non-linear problems in the finite element method. *IMA. J. Numer. Anal.*, 1:253–266, 1981.

[5] P.I. Crumpton, J.A. Mackenzie, and K.W. Morton. Cell vertex algorithms for the compressible Navier-Stokes equations. Technical Report NA91/12, Oxford University Computing Laboratory, 11 Keble Road, Oxford, OX1 3QD, 1991. Submitted for publication.

[6] R.G. Jones. Error estimators for adaptive methods. Technical Report NA90/17, Oxford University Computing Laboratory, 11 Keble Road, Oxford, OX1 3QD, 1990.

[7] A.J. Mayfield. Discussion and presentation of a new method for mesh movement. Technical Report NA92/11, Oxford University Computing Laboratory, 11 Keble Road, Oxford, OX1 3QD, 1992.

[8] A.J. Mayfield. Design and implementation of an adaptive multigrid algorithm. D.Phil thesis, Oxford University, 1992.

[9] A.J. Mayfield. A new efficient data structure for mesh enrichment algorithms. Technical Report NA92/12, Oxford University Computing Laboratory, 11 Keble Road, Oxford, OX1 3QD, 1992.

[10] T.A. Manteuffel and A.B. White Jr. The numerical solution of second-order boundary value problems on nonuniform meshes. *Mathematics of Computation*, 47(176):511–535, 1986.

[11] K.W. Morton and M.F. Paisley. A finite volume scheme with shock fitting for the steady Euler equations. *Journal of Computational Physics*, 80:168–203, 1989.

[12] H. Schlichting. *Boundary–Layer Theory*. McGraw–Hill, 1955.

[13] E. Süli. Finite volume methods on distorted meshes: stability, accuracy, adaptivity. Technical Report NA89/6, Oxford University Computing Laboratory, 11 Keble Road, Oxford, OX1 3QD, 1989.

MULTIGRID METHODS FOR THE ACCELERATION AND THE ADAPTATION OF THE TRANSONIC FLOW PROBLEMS

A.E.Kanarachos, N.G.Pantelelis and I.P.Vournas
National Technical University of Athens,
P.O.Box 64078, 15710, ATHENS, GREECE

ABSTRACT

In this paper the two of the three parts of a study towards the Block Adaptive Multigrid (BAM) method are presented. The BAM method is specially designed for implicit schemes with the aim to handle finite volume discretizations with the optimum accuracy at the minimum CPU-time and storage, points very critical when industrial problems are to solved. The first part is a thorough study of the multigrid schemes and their implementation into the single grid code, an implicit upwind code using a characteristic flux extrapolation scheme is exhibited. This results into a very efficient code capable to handle complex transonic viscous flow problems even by small workstations. As the correct implementation of the multigrid theory is essential whereas the adaptive method will rely on, whereas the grid independent convergence rate is the best proof for any inconsistencies.

With respect to the error analysis the prediction of the truncation error of the solution is considered. The calculation of the truncation error, although based theoretically on the multigrid procedure for the present approach, it can be used also without multigrid. Examples of the effectiveness of the prediction of the solution error are presented for common test cases, where a correlation between the predicted errors and the actual errors of the solution is established. Additionally, in order to prove the validity of the error prediction, crude adaptions with grid refinement procedures are used for regions with high truncation errors whereas the accuracy of the solution is improved considerably. The efficiency achieved by the combination of multigrid and the adaptive refinement procedure is extraordinary, but further studies on the adaptive strategies and the refinement grid solution are required. Moreover, the extension of the above methods for turbulent three dimensional problems is straight forward and very promising.

1. INTRODUCTION

The multigrid methods have been initiated from the early seventies and nowadays have been established as a standard acceleration technique for all the kind of flow problems not only due to the acceleration that is attained but mainly due to the linear increase of the required work with respect to the number of volumes (O(N) convergence rate). On the other hand, higher grid resolution is required only in certain regions of the domain, thus it is important that the discretization and solution process do not demand a considerable increase in the overall computation due to these local phenomena. Unfortunately, the objective of efficiently adapting the solution to the local features is often in conflict with the solution procedures moreover when data structures for irregular grids, demanding big overheads, may negate the advantages of the adaptive process. Thus, the development of an adaption technique should take full advantage of the multigrid methods thus, the ultimate target should be to construct a single method combining multigrid and adaptive strategies both tight together in such a way that the grid independence convergence rate is still valid in such a way that in order to attain a certain accuracy the totally minimal CPU time is required, taking also under consideration the computer software and hardware.

The adaptive strategies can be grouped into global and local strategies either of them with advantages and drawbacks. After determining the erroneous regions or both strategies the regions the global strategy keeps the number of grid points constant rearranging the grid density when the local strategy keeps the initial grid distribution the same but enrichment takes place in order to increase the grid resolution. Both methods have limitations approaching the optimal solution thus it is important to realize that the solution errors can not be driven to zero by either of the strategies. For the global methods this is due to the inability of the addition of grid points which normally are required and for the local methods grid alignment with the physical features is impossible. Referring to the local adaptive strategies that have been studied up till now there can be classified into three categories. The first group uses the truncation error estimation as a prediction of the solution error without implementing multigrid acceleration, although this is left as an option. Such schemes have been implemented successfully in [1,3,4] mainly for compressible inviscid flow problems with the additional common feature that the refinement grids are taken as rectangular blocks. The second group has implemented multigrid solution to solve the refinement grids choosing for adaption criterion several solution gradients. Although this is a more conventional approach, it is considered to be the most applicable and covers a wide range of problems [2,7,10]. It should be underlined that at this approach a more unstructured refinement is constructed in order to adapt exactly to the flow gradients. The most radical approach is a combination of the multigrid acceleration and the truncation error prediction, which has been proposed from the very early development of the multigrid theory, but its implementation has been kept mainly for the solution of elliptic equations [1,5,6,13,14,20,21].

The objectives of the study were to establish the multigrid efficiency and to explore the multigrid mechanisms as a first step for the adaption process (chapter 3). Afterwards, the development of a realistic and reliable error sensor is essential in order to define the regions that the error solution needs to be reduced (chapter 4). These regions should be also the regions that propagate the errors that dominate the computational domain. The processing of the above information should be included in order that the refinement regions are compatible with the multigrid process. The final and most critical step is the correct incorporation of the refinement levels within the multigrid cycle, which still is under investigation [17] and only limited it was used at the present study in order to have a global understanding of the capabilities of the method. Common test cases were chosen in order to evaluate the proposed method among the other adaptive methods considered in the project. The results which are presented in chapter 5 show the capabilities of the multigrid acceleration together with the validity of the error sensor.

2. GOVERNING EQUATIONS

The basic equations are the time-dependent Reynolds-Averaged Navier

Stokes equations in two dimensions. Conservation laws are used with body-fitted coordinates ξ, η, and cartesian velocity components u,v :

$$U_t + E_\xi + F_\eta = 0 \qquad (2.1)$$

and the steady state solution is given for $U_t = 0$. The fluxes normal to $\xi = $ const. and $\eta = $ const. faces are:

$$E = J \cdot (\bar{E}\,\xi_x + \bar{F}\,\xi_y) \quad \text{and} \quad F = J \cdot (\bar{E}\,\eta_x + \bar{F}\eta_y) \qquad (2.2)$$

the solution vector is:

$$U = J \cdot \bar{U} = J \cdot [\,\varrho \quad \varrho u \quad \varrho v \quad e\,]^T \qquad (2.3)$$

The flux-vectors E and F are taken as the summation of the inviscid and the viscous terms:

$$\bar{E} = \bar{E}_{inv} - \bar{E}_{vis} \qquad \bar{F} = \bar{F}_{inv} - \bar{F}_{vis}$$

The inviscid and viscous cartesian fluxes are given by:

$$\bar{E}_{inv} = \begin{bmatrix} \varrho u \\ \varrho u^2 + p \\ \varrho u v \\ (e+p)u \end{bmatrix} \qquad \bar{F}_{inv} = \begin{bmatrix} \varrho v \\ \varrho u v \\ \varrho v^2 + p \\ (e+p)v \end{bmatrix}, \qquad (2.4)$$

$$\bar{E}_{vis} = \begin{bmatrix} 0 \\ \sigma_{xx} \\ \sigma_{xy} \\ \sigma_{xx} u + \sigma_{xy} v + q_x \end{bmatrix} \qquad \bar{F}_{vis} = \begin{bmatrix} 0 \\ \sigma_{xy} \\ \sigma_{yy} \\ \sigma_{xy} u + \sigma_{yy} v + q_y \end{bmatrix} \qquad (2.5)$$

where $\sigma_{xx} = \frac{2}{3}\mu(2u_x - v_y)$, $\sigma_{xy} = \mu(u_y + v_x)$ and $\sigma_{yy} = \frac{2}{3}\mu(2v_y - u_x)$ and $q_x = -k\,T_x$, $q_y = -k\,T_y$. The e is the total energy: $e = \frac{p}{(\gamma-1)} + \frac{1}{2}\varrho(u^2 + v^2)$, p and ϱ are the pressure and the density respectively. J, the Jacobian of the inverse mapping, is the area of each cell for the two-dimensional case. For the calculation of the viscosity coefficient for the laminar flow is computed via the Sutherland law.

In order to solve eq.(2.1) a finite-volume scheme with an implicit backward Euler solver for the evolution in time is used, which allows high CFL numbers. The governing method is described in detail by Eberle [8] and its principle, modified for the present method, is summarized below. As only the steady-state solution is required, a scheme first order accurate in time is used and the discretized implicit form of eq.(2.1) is given as:

$$\frac{U^{n+1} - U^n}{\Delta t} + E^{n+1}_\xi + F^{n+1}_\eta = 0 \qquad (2.6)$$

By linearizing the fluxes around the time-level n,

$$E^{n+1} = E^n + \frac{\partial E}{\partial U}^n \Delta U^{n+1} \qquad F^{n+1} = F^n + \frac{\partial F}{\partial U}^n \Delta U^{n+1} \qquad (2.7)$$

a delta-formulation for eq(2.5) with $\Delta U = U^{n+1} - U^n$ can easily be found:

$$\frac{\Delta U}{\Delta t} + (A^n \Delta U)_\xi + (B^n \Delta U)_\eta = -(E^n_\xi + F^n_\eta) = -\text{Res}(U^n) \qquad (2.8)$$

where A and B are the Jacobians of the fluxes E and F, respectively, including inviscid and viscous terms:

$$A = \frac{\partial E}{\partial U} = \frac{\partial E_{inv}}{\partial U} - \frac{\partial E_{vis}}{\partial U} = A_{inv} - A_{vis}$$

$$B = \frac{\partial F}{\partial U} = \frac{\partial F_{inv}}{\partial U} - \frac{\partial F_{vis}}{\partial U} = B_{inv} - B_{vis}.$$

When inviscid flows are calculated, the viscous terms are discarded:
$$E_{vis} = F_{vis} = A_{vis} = B_{vis} = 0.$$

Upwind differencing of the flux vectors is the natural way to reach a diagonal dominant system and also to introduce numerical dissipation, which both are crucial for the efficiency and robustness of the solver. For the flux calculations at the edges of the cells a linear one-dimensional Riemann solver (Godunov approach) is employed at the finite-volume faces which guarantees the homogeneous property of the Euler fluxes [8]. The mean values of the conservative variables at both sides of the cell faces are used as flow variables at the volume faces for the Riemann solver while for the viscous fluxes central differencing at the cell faces is performed. For the laminar flow the viscosity coefficient is computed via the Sutherland law.

Depending on the sign of the eigenvalues λ^i the conservative variables are extrapolated up to third order in computational space to each face of the volumes (MUSCL-type interpolation). Sensing functions are used to guarantee monotonic behavior of the flow field except at shocks, where the scheme reduces to first order accuracy in order to avoid oscillations. With the divergence of the characteristically extrapolated fluxes on the right-hand-side (RHS), equation (2.8) has to be solved approximately at each time step. At the i,j cell the variation in time can be calculated by:

$$A \Delta U^{n+1}_{i-1,j} + B \Delta U^{n+1}_{i,j-1} + C \Delta U^{n+1}_{i,j} + D \Delta U^{n+1}_{i,j+1} + E \Delta U^{n+1}_{i+1,j} = \omega \cdot RHS \quad (2.9)$$

where the RHS and $\Delta U_{i,j}$ are 4·1 matrices given from eq.(2.8) and **A,B,C,D,E** are 4·4 matrices which emerge from a Godunov-like first-order flux-splitting scheme and form a block pentadiagonal system if eq.(2.9) is applied to the entire computational domain. The ω is an underrelaxation factor which compensates the different spatial order of accuracy between the RHS and the LHS (the LHS is always first order accurate) and for the present implementation was from 0.45 to 0.60. Because of the pseudo-time approach we can accelerate the procedure by advancing in time with different time step Δt for each cell, keeping the CFL number constant varying from 100 to 200 (local time-stepping). In order to avoid the manipulation of the LHS terms at the boundaries, simple conditions are prescribed on U and ΔU in phantom cell rows all around the boundaries. For the outer boundary conditions no special care should be taken on the free stream condition as the code extracts automatically such information.

3. IMPLEMENTATION OF THE MULTIGRID METHODS FOR THE ACCELERATION OF THE SOLUTION

3.1 COMMON FEATURES OF THE MULTIGRID SCHEMES

In order to develop a multigrid scheme it is essential to define several features, common for linear and nonlinear problems. Thus, is necessary to define a sequence of grids in such a way that a coarse volume is constructed by four finer volumes, deleting the internal edges. This cellwise coarsening is essential for the proper representation of the fine grid solution, through the finite volume theory, to the coarser levels and vice versa. Supposing that the fine grid equations (2.8) have the general formulation (by 1 is denoted the finest grid-level and with 2,3,...,N the coarser levels are denoted):

$$L^1 \cdot \Delta U_1 = -\text{Res}^1(U_1) \tag{3.1}$$

in contrast to:

$$L^1 \cdot \Delta U_1 + \text{Res}^1(U_1) = 0 \tag{3.2}$$

which was used in [11], the multigrid concept is implemented correctly. Taking under consideration that the LHS operator is calculated for each grid-level and the RHS operator is taken from the previous level by conservative interpolation, the difference between (3.1) and (3.2) is that in eq.(3.2) recalculation of Res(U) at lower levels introduces non-vanishing fluxes which mislead the solution. Contrarily, when only changes of variables are transferred and moreover fluxes and residuals are conserved, automatically the fine grid solution cannot be misled or corrupted by the multigrid process.

To obtain efficient smoothing a selection should be made from an abundance of available relaxation schemes. Although the existing Collective Gauss-Seidel relaxation in lexicographic order behaves satisfactorily the Symmetric Collective Gauss-Seidel relaxation is adopted. Thus, effectiveness and robustness are gained without implementing a computationally expensive and complex scheme like a line or a block relaxation scheme.

As the coarsening is performed cellwise the coarse grid points are not a subset of the fine grid points, hence, special care should be taken at the intergrid transfers of the physical variables. Instead, as the external faces of a coarse volume are common with those of the fine one, the restriction of the fluxes and residuals is straight. Because of the fact that two different kind of variables are restricted, thus two restriction operators have to be defined in such a way that four fine volumes results equivalently in a single coarse volume. For the physical variables the simplest interpolation scheme, denoted by I^n_{n-1}, is the weighted volume average over the corresponding four fine volumes. Supposing that $J^{i,k}$ is the area of the (i,k) volume of the n-th grid level, we get for the restriction of the U variables:

$$U^{i,k}_{n+1} = \frac{(JU)^{i,k}_n + (JU)^{i+m,k}_n + (JU)^{i,k+m}_n + (JU)^{i+m,k+m}_n}{(J^{i,k} + J^{i+m,k} + J^{i,k+m} + J^{i+m,k+m})}$$

$$= I^{n+1}_n U^{i,k}_n. \tag{3.3}$$

The integer m gives the increment so that the neighboring points of the (i,k) volume at the n-th grid level are found immediately. Thus, using the same subroutines for all grid levels can be used with the additional input of m which for all directions equals $2^{(n-1)}$ (i.e. m=1, 2 and 4 for the levels n=1,2,3, respectively).

To obtain the restriction operator for the generalized fluxes, Res and τ, it is necessary to consider that these are the summation of integrals over the faces of the volumes, hence a simple summation, denoted by Σ^{n+1}_n, over the four volumes cancels the integrals of the inner common faces, leaving only the outer fluxes, so conservation is preserved. Thus, for the restriction of the fluxes Res(U) it holds:

$$\text{Res}_{n+1}^{i,k} = \text{Res}_{n}^{i,k} + \text{Res}_{n}^{i+m,k} + \text{Res}_{n}^{i,k+m} + \text{Res}_{n}^{i+m,k+m} = \Sigma_{n}^{n+1}\text{Res}_{n+1}^{i,k}. \tag{3.4}$$

With respect to the prolongation operator a relatively accurate interpolation should be used as no relaxation sweeps are performed to smooth out the prolongation errors. Thus, instead of using constant (zero order) prolongation, cellwise bilinear interpolation is preferred. To avoid complexity, this prolongation is independent of the cell volumes, demanding four coarse volumes to prolong to one fine volume.

$$\Delta U_{n}^{i,k} = \frac{1}{16}\left(9 \cdot \Delta U_{n+1}^{i,k} + 3 \cdot \Delta U_{n+1}^{i-m,k} + 3 \cdot \Delta U_{n+1}^{i,k-m} + \Delta U_{n+1}^{i-m,k-m}\right)$$
$$= I_{n+1}^{n} \Delta U_{n+1}^{i,k}. \tag{3.5}$$

The prolongation operator, because it relates numerous volumes inevitably requires calculations over boundary volumes, which is done simply by taking advantage of the phantom volumes at the boundaries, already introduced for the fine grid boundary conditions. These volumes are also used to set the coarse grid boundary conditions.

With respect to the multigrid strategy two basic ideas are implemented in order to achieve the maximum acceleration [15]. Firstly, when no relaxation sweeps are performed on the way from the coarse to fine levels apart from the prolongation interpolation, the multigrid cycle will be more effective and faster. Without relaxation sweeps on the way up it is likely that a few extra relaxation sweeps are required at the end of the multigrid cycle on the finest grid to smooth out the interpolation errors, taking under consideration that the order of the prolongation interpolation is low. In practice one post relaxation sweep is enough and only for the viscous flows which are more sensitive to the interpolation errors. This extra sweeps is performed after the recalculation of the fluxes, considering the more recent values of the U variables. The second idea concerns the number of relaxation sweeps at each level in such a way that going from fine to coarse levels the number of sweeps should be increasing, thus the relaxation sweeps are taken equal to the grid level (i.e. one for the finest level, two for the next coarser e.t.c.). With reference to the grid generation there are some restrictions to the choice of the grid size. Thus if no limits to the coarsest grid of the multigrid sequence are desired then the number of volumes of each direction of the finest grid should be a power of two for both directions, on the other hand if the coarsest grid level is set the finest grid level should only be a multiply of two of the size of the coarsest grid, hence no strict limitations are posed at the grid generation process even when a multiblock structure is utilized.

3.2 FULL APPROXIMATION AND CORRECTION SCHEMES

The Full Approximation Scheme (FAS) was developed by Brandt [1] with the aim to cope with nonlinear problems. Although the FAS was developed as an extension of the Correction Scheme (CS) which was built for linear problems, at this study there are analyzed from the reverse point of view. As the linear case is a special case of the nonlinear problems it is likely to describe the nonlinear multigrid scheme and proceed through the linear assumptions to the linear problem.

Formulating FAS method with the "alternate point of view" (analyzed in [5,14]) the finer grids are considered mainly as levels to increase the spatial accuracy of the solution where the coarser grids are responsible for the solution process where the relaxation sweeps are much cheaper. The problem which is solved at the coarser levels will have the same formulation with that of the finest grid but additionally on the RHS of eq.(3.1) the fine-to-coarse defect correction τ will be added. Thus, eq.(3.1) can be written in a general formulation for the n-th grid-level solution:

$$L^n \cdot \Delta U_n = -Res^n(U_n) + \tau^n \qquad \text{taking} \quad \tau^1 = 0 \,. \qquad (3.6)$$

The use of the multigrid defect correction formulation underlines the different solution strategy that should be followed and offers a convenient environment for advanced techniques such as adaptive techniques to be included. Following the proposed strategy only few relaxations should be spent on the fine grids, in order to compute the correct defect correction) while most of the relaxation sweeps should be performed on the coarser levels. The formulation of the fine-to-coarse defect correction at the n-th grid level, (n-1 is the finer level) can be given by:

$$\tau^n_{n-1} = L^n \cdot (I^n_{n-1} \Delta U_{n-1}) - I^n_{n-1}(L^{n-1} \cdot \Delta U_{n-1}) \qquad \text{with} \quad \tau^1_0 = 0 \qquad (3.7)$$

denoted as relative local truncation error between n and n-1 levels, or by:

$$\tau^n = \Sigma^n_{n-1} \tau^{n-1} + \tau^n_{n-1} \qquad \text{with} \quad \tau^1 = 0 \qquad (3.8)$$

denoted as total local truncation error between levels 1 and n. Although eq. (3.8) should be adopted eq.(3.7) is also studied so the proposed formulations for the multigrid cycle are:

$$L^n \cdot \Delta U_n = -Res^n(U_n) + \tau^n_{n-1} \quad \text{or} \quad L^n \cdot \Delta U_n = -Res^n(U_n) + \tau^n. \qquad (3.9)$$

The two formulations yield only different convergence rates (the steady state solution is not affected) and only when more than two grid levels are used. Analyzing these formulations, equation (3.8) results in a more accurate and correct representation of the finest solution to the coarser ones so it is used for the viscous flows and equation (3.7) introduces additional numerical dissipation which is useful for the inviscid flows. With respect to the LHS coarse grid operator instead of the Galerkin approximation which could be used, the direct approach is adopted so the LHS operator (L in eq.(3.1)) is the same for all grid levels yielding a more natural approximation for the coarse grid equations. Thus, the single grid subroutines for the calculation of the LHS (eq.(2.8)) are used but with the extra variable of the increment m (see below eq.(3.3), above). Finally, referring to the reverse direction from the coarser to finer grids only the prolongation of the ΔU variables is needed as no relaxation sweeps are performed at this direction. Thus, for the prolongation step it holds:

$$\Delta U_{n-1} = \Delta U_{n-1} + I^{n-1}_n (\Delta U_n - I^n_{n-1}\Delta U_{n-1}). \qquad (3.10)$$

In order to determine the best scheme to apply for the multigrid acceleration it is necessary to choose between CS and FAS, a choice which initially seems simple as it emerges from the kind of the solver that it is used. Certainly when such a solver does not exist then the question is much more complicated as there are two possibilities, to linearize the solver and apply the CS scheme or to solve the original nonlinear problem with the aid of the FAS scheme [5,12]. In order to explore the links between these schemes it is necessary to quote the Correction Scheme (CS).

Using for a starting point the FAS formulation (equations (3.6) to (3.8)) and supposing that the differential operator (L) is linear the cancellation of two operators occurs because it holds:

$$L^n \cdot \Delta U_n - L^n \cdot (I_{n-1}^n \Delta U_{n-1}) = L^n \cdot (\Delta U_n - I_{n-1}^n \Delta U_{n-1}) = L^n \cdot \Delta V_n \quad (3.11)$$

$$\Delta V_n = \Delta U_n - I_{n-1}^n \Delta U_{n-1} . \quad (3.12)$$

Thus instead of using the ΔU variable as coarse grid variable a new one is introduced in order to avoid unnecessary operations at the coarse grid levels, thus in accordance to the FAS scheme the τ formulation takes the form:

$$\tau_{n-1}^n = -I_{n-1}^n (L^{n-1} \cdot \Delta U_{n-1}) \quad \text{taking} \quad \tau_0^1 = 0 \quad (3.13)$$

where the restriction and the prolongation operators are defined by equations (3.3) to (3.5) using the new coarse grid variable ΔV. The only difference with reference to the FAS scheme is that by using a different multigrid variable (ΔV), a different prolongation operator should be also used. So instead of the formulation of equation (3.10), taking under consideration equation (3.12), the prolongation step is defined as:

$$\Delta V_{n-1} = \Delta V_{n-1} + I_n^{n-1} \Delta V_n \quad (3.14)$$

and for the finest level a new approximation to the ΔU variable is found:

$$\Delta U_1 = \Delta U_1 + I_2^1 \Delta V_2 . \quad (3.15)$$

Displaying both schemes a clear disadvantage of the CS is that a new system of variables should be introduced only for the coarse grid levels which results in extra code complexity and deviation from the physical variables of equation (2.1) although there is a gain in the number of operations with respect to the FAS. In contrast to the CS, the FAS is much more general making possible the implementation of the multigrid either for linear or for nonlinear problems with great simplicity [11]. With reference to the additional operations this can easily be accomplished simultaneously within the relaxation step as the L operator is computed at each grid level. Additionally, new features can easily be entered the solution method like adaptation strategies, composite grid solutions and defect correction techniques. Hence, although CS theoretically is a little faster than FAS for the linear problems, FAS can handle all kind of problems and normally should be considered as the best multigrid scheme.

4. IMPLEMENTATION OF THE MULTIGRID METHODS FOR THE ADAPTATION OF THE SOLUTION

4.1 APPROXIMATION OF THE SOLUTION ERROR

There are essentially two approaches in order to determine the problematic regions of a computational domain. The first one is the physically based information of the problem (solution gradients and similar features), while the second one is based on the determination of the local discretization error which is numerically motivated and as a result more reliable. The latter case takes into account features of the discretization and the solution procedures like systems of equations, higher order discretizations and nonlinear problems while the former does not. On the other hand, the former scheme may be preferable and simpler to be used but this may lead to implicating the solution process with grids which are refined due to purely local errors without having actual indications of the global errors. The truncation error estimator, in contrast, is expressed in terms of the global effects of the local truncation error, i.e. the error in the equation which is exactly the type of errors that should be confronted by the solution process itself, especially when grid refinement techniques are applied. Thus, essentially it is not attempted to compute the total error of the solution but the part of the total error of the solution which has to do with the grid size which is exactly the error that can be handled by a refinement procedure compatible with the multigrid methods. The grid can be refined until the solution error estimate is below some desired level. The solution is computed on the intermediately coarse grid and from these two solutions, the solution error can be estimated by Richardson extrapolation. Of course such an error prediction can be applied approximately also when different adaptive techniques, like moving point techniques, are used.

A critical point concerns the choice of the grid level solutions to compare with i.e. to compare the solution of the present finest grid with a coarser or with a finer one, whereas the most efficient procedure (Full or Nested Multigrid Scheme) starts from a coarse grid and proceed to the finer ones, via refinement procedures. Although it is more reliable to compare directly the current solution with the solution of the immediate finer grid the cost of this comparison is four times the integration time of the currently finest grid not mentioning the required expensive prolongation interpolations. The alternative approach, considering that the solution is smooth enough, computes the corresponding truncation error of the current solution and this of the next coarser grid and using the Richardson Extrapolation to approximate the truncation error of the required grid level. This approach is less expensive (one forth of the integration time of the currently finest grid), less sensitive to relaxation errors and as a result more attractive, although a hybrid scheme which would use the former and more expensive scheme for the coarse grids and the latter one for the fine grids may be in some cases more appropriate. At the present study only the second scheme will be considered which gives fairly good results even for very coarse grids. According to the truncation error definition, it is determined by:

$$Q(h) \, u = \tau \, h^p \tag{4.1}$$

where Q is the differential operator, u is the physical solution, h is the mesh size, p is the order of accuracy of the method and τ is a factor which together with p are typically h- independent. Comparing solutions of two different grids predictions can be made in the way a change of the mesh size will affect the accuracy of the solution. The following holds [3,4,13,20]:

$$\tau = \frac{Q(2h) - Q(h)}{h^p \, (2^p - 1)} = \frac{Q(h) - Q(h/2)}{h^p \, (1 - 2^{-p})} \, . \tag{4.2}$$

Thus the prediction of the truncation error of the grid level $h/2$ is:

$$Q(h) - Q(h/2) = \frac{(1 - 2^{-p})}{(2^p - 1)} \, [\, Q(2h) - Q(h) \,] \, . \tag{4.3}$$

Although the FAS scheme through the τ formulation (eq.(3.7)) has an inherent information of the solution error where it holds for the n-th grid level (n-1 and n+1 are the finer and the coarser levels, resp.):

$$\tau^n_{n-1} = \frac{(1 - 2^{-p})}{(2^p - 1)} \, \tau^{n+1}_n \, . \tag{4.4}$$

Due to the Newton linearization that it is employed, the LHS has a first order delta formulation while the accuracy of the solution is determined by the computation of the fluxes at the RHS of eq.(2.9). Additionally, the usage of the local truncation error of eq.(3.7) is connected only with the relaxation errors occurred at each grid level dealing mainly with the convergence of the linearization step but also having converged to the steady state the ΔU vector should be almost zero.

Fortunately, there is a very interesting alternative operator for the truncation error prediction which shows none of the previous disadvantages but, in contrast requires some additional computations. This sensor although it is based on the multigrid process and on the discretization error motivation does not rely strictly on the multigrid cycle but only uses its restriction interpolation. This means that some additional computations have to be performed independently of the multigrid cycle but on the other hand this feature enables the proposed sensor to be used even if multigrid techniques are not included in the code, thus the proposed sensor is applicable to all the kind of problems. The basic idea is to use the same truncation error estimator as in eq.(3.7) but using the Res(U) operator of eq.(3.1) instead of the Q differential operator, in the following way:

$$T_n^{n+1} = \Sigma_n^{n+1} \text{Res}_n(U_n) - \text{Res}_{n+1}(I_n^{n+1} U_n) \qquad T_{n-1}^n = \frac{(1 - 2^{-p})}{(2^p - 1)} T_n^{n+1}. \qquad (4.5)$$

Moreover when an approximation of the truncation error is normally adequate only T is required. Otherwise, complexities in order to specify p might occur as the order p is not always known exactly *a priori*, especially for higher order methods where a reduction of the accuracy occurs near discontinuities to avoid oscillations. This approach has several advantages as the truncation error is of the order of accuracy of the steady state solution while in order to take advantage of the multigrid acceleration adequate approximations are available from the initial time steps, thus no wasted computational work is performed in order to define the problematic regions. This point is very critical as employing the full (nested) multigrid scheme very fast solutions in a few work units can be achieved.

4.2 SOLUTION OF THE COMPOSITE GRID STRUCTURE

The major drawback that the local adaptive methods face with respect to the global ones is the solution of grids with different mesh size and "hanging" nodes. For the solution of such grids additional features have to be introduced in order to optimize the data structure. A state of the art approach is to use blocks (or else rectangles) of refined grids in such a way that the data management is minimized as it is used in [1,3,4]. The data management although it is a crucial decision, it has been studied very thoroughly and as it is almost independent of the flow solver, its choice is only a strategic matter. These blocks could be identified with the blocks of the multiblock mesh generation or not, taking under consideration that the blocks belonging to the adaptive process should have a size big enough to be compatible with the multigrid process. Hence, if the multiblock grid generation is performed efficiently with the usage of the truncation error sensor simply the finest grid level of each block is imposed in order to attain the maximum efficiency of the problem.

On the other hand the most individual characteristic of a composite grid implementation is the manipulation of the "hanging" nodes in order to maintain the fine grid accuracy. Mistreating of the integration process at the internal boundaries of two grids with different meshsize can be degrade considerably the accuracy of the adaptive scheme. Thus exact simulation of the uniform grid logic has to adopted at these artificial interfaces in order to depress the propagation of errors not consistent with the solution method. The problem gets more difficult when shocks and other discontinuities are encountered, thus conservation and a properly chosen interpolation method are a necessity. Additionally, in order to facilitate the handling of the different grid levels and to maintain the fast convergence the multigrid method has to be incorporated with the adaptive procedure. In the present study the composite grids are used to test the efficiency of the prediction of the truncation errors thus, only crude adaptions are performed. Additionally, the convergence of the block adaptive multigrid method needs to be tested extensively the dependence of the efficiency of the method on numerous factors is expected and will be presented in a future study.

5. RESULTS

In order to validate the multigrid acceleration and the correct prediction of the solution error for real life problems, four test cases (those which were come up from the EUROMESH project) are investigated. From these test cases which are commonly accepted, there are two Euler cases: case 1 (denoted by E1) is a NACA-0012 airfoil with Mach number 0.85 and angle of attack 1.0 degree while the second case (denoted by E2) is the RAE-2822 airfoil with 0.73 Mach number and 2.79 degrees angle of attack. For the laminar cases there are two similar cases, both NACA 0012 airfoil at 10 degrees angle of attack, but the first (denoted by L1) is transonic (Mach number is 0.8 and Reynolds number is 73) and the other one (denoted by L2) is supersonic (Mach number is 2.0 and Reynolds number 108). For all the test cases good agreement with other published results were achieved, although for the laminar cases difficulties occurred in order to converge the problem when very fine grids were used near the airfoil. In order to validate the error prediction of the solution extra grids with "ad hoc" false regions were also tested.

At first, the efficiency and the robustness of the presented FAS and CS schemes are compared together with the results from the single grid code which uses a collective point Gauss-Seidel relaxation method. For all the test cases, the solvers gave exactly the same steady-state solutions so, separate results for each solver are not given. In order to compare the efficiency of the solvers, a Work Unit (W.U.) is defined by the CPU-time (of an 1-MFlop serial machine) that is needed for the global single grid calculation to perform a relaxation sweep (not symmetric). As it was expected only small differences occurred at the performance of the two schemes (CS and FAS) which their sources should be attributed to minor differences on the strategies that were followed by the implementation of each scheme. For the Euler cases FAS implementation seems to be a little faster than CS although grid independent convergence rates are established, together with acceleration from 5 to 12 times for moderate size grids. The same holds for the laminar cases where for the first case (L1) the lift and drag coefficients convergence rates of the schemes are shown in fig.1. A superiority of the multigrid schemes over the single grid solution with respect to the required CPU-time is apparent as even higher acceleration than that of the Euler cases were achieved, ranging from 10 to 34 times. Extensive results for the present test cases and additional results for turbulent problems for each of the multigrid schemes can be found in [15,18,19].

Concerning the truncation error sensor proposed, it is necessary to clearly demonstrate its capabilities for real problems where a substantial need for adaptive methods exists. In such complex cases neither analytical nor prescribed solutions exist, hence the introduction of an error estimation of the exact solution is essential in order to evaluate the proposed error prediction sensor. A common solution is to accept the finest grid solution as the exact one in such a way that a comparison between this solution and that of the coarser levels give an estimation of the actual error of the solution [15]. As a system of four variables exists, the choice of the pressure as the comparative variable is best suited because it combines all the unknown variables of a volume under a physical law and it is the required variable from the analysis of the flow. Thus, the actual error indicator is the difference of the pressure computed at the finest level via the finest grid solution with the pressure at the same level computed via the variables of the next coarser grid after prolongating them to the finest grid by injection. This estimator is computed after individual solution and total convergence of the couple of comparative grid levels. Further results on the evaluation of the truncation error prediction are presented in [23].

In order to study the behaviour of the sensor for different grids, a very fine algebraic grid (256x56) and its nested coarser grids are tested thoroughly with an elliptic generated grid (128x32) (provided by the BAe) having on purpose reduced number of points around the airfoil, especially at the leading edge. For the elliptic grid very good accordance is achieved as the truncation error prediction is very similar to the pressure error contours both indicating the leading edge area as an error propagating region as can be seen from fig. [3 a,b]. Naturally the upper and lower shocks are also producing errors but at the same levels as those of the leading edge which unavoidable destroys the solution giving lift and drag coefficients of 0.25 and 0.09 respectively. When an even coarser grid (64x16, nested in the fine grid) is tested the error at the leading edge dominates completely the flow field and underlines the need for refinement at this region. In contrast, the algebraic grid where the leading edge resolution is much better, the regions of the shocks dominate the error distribution of the algebraic grid solution whereas at the leading edge only a slight error indication exists (fig.[3 c,d]). It is worth mentioning that the truncation error levels are also lower at the algebraic grid than the elliptic (for the same number of volumes the maximum values of the truncation error are 0.012 and 0.019, respectively).

In order to have a better idea of the concept of the adaption it is essential to employ the error truncation sensor into a composite grid structure. This structure is based on the finest grid and all the nested grids, derived by deleting appropriate grid lines from the finest grid. The very coarse grids are spread globally, in contrast to the finer ones which their limits are defined by the truncation error distribution. Thus, taking under consideration equation (4.5) with p equals two it is expected that introducing one refinement level the truncation error will be diminished approximately by one forth, hence based on these predictions *a priori* refinements are made, two crude adaptive grids and one more strict are tested with the intention to explore the accuracy and the behavior of the error prediction. The crude adaption combines only two grid levels (128x28 globally and 256x56 locally) refining only a wide region around the airfoil and gave almost exact results comparing to the corresponding fine grid results. The more strict adaption combines three grid levels (64x14 globally with 128x28 and 256x56 locally) and the finest level spread around the shock regions and the leading edge. This approach is more efficient as appears on fig.(2). The superiority of the adaption schemes is apparent with respect to the global grids, in addition to the parallel convergence rates of the drag coefficients and the truncation error values with respect to the number of the required volumes. On the other hand the multigrid efficiency and grid independent convergence are apparent (fig.[3e]) for either global or adaptive grids. The quality of the adapted solution is the same achieved by the finest global grid and the Mach number contours are shown together with the composite grid structure in fig.[3.f].

For the other Euler case (E2) two algebraic grids with different grid distribution are tested, one with 256x56 volumes and one with 192x28 (provided by RAE), where the latter is better adapted to the flow features at the leading and trailing edges, thus more accurate results are achieved. Due do better resolution of the leading edge the RAE grid presents globally smaller truncation error and pressure error distributions with respect to the former grid (fig.[4 a,c]) which in contrast presents an increase of the truncation error and the pressure error at the leading edge (fig [4 b,d]). With respect to the multigrid implementation is observed that the convergence rate of the solution is almost independent from the grid size not only for the global grids but also for the adaptive grids (fig.(4e)) where again the efficiency of the three-level adaptive solution is also demonstrated. Thus, in conclusion it was established a robust and reliable estimation of the solution error which still works for nonsmooth fields and the adaptive grids that are based on this are far more efficient from the global multigrid solution, not to mention the single grid code where a total acceleration of twenty is easy to achieve.

For the laminar test cases it was observed that the flow features were very insensitive with reference to the grid generation, especially the supersonic case (L2). For the subsonic case (L1) the truncation error contours and the Mach contours are shown in figures 5.a and 5.b, respectively. With respect to the changes of the grid resolution taking under consideration the truncation error distribution, a reduction of the maximum of the truncation error of order of two it was achieved, although the Cl and Cd coefficients were not affected considerable. For the supersonic case L2, which was proved almost insensitive to the grid changes, the Mach contours are given in fig.6.b, where the bow shock dominates the flow and was predicted by the truncation error sensor (fig.6.a). It must be underlined that at this case the level of the truncation error prediction is considerably high in comparison with the other test cases and this should be referred to the strongly supersonic flow. Further results should be conducted for the turbulent Navier Stokes equations in order to evaluate the influence of the viscous fluxes on the proposed error sensor.

CONCLUSIONS

At the present study the multigrid methods together with the truncation error prediction were developed with very promising results. The incorporation of the considerable multigrid acceleration together with a fast and reliable error indicator, sets the basis for a very efficient solution method for inviscid and viscous compressible flows (BAM method). It is expected that the developed methods can be very easily, with the corresponding changes for the additional third direction, extended to three dimensional problems were the need of acceleration and adaption solutions are essential. As it is shown the Correction Scheme has no real existence as it exhibits much less advantages than the FAS scheme which when it is applied with the correct concept can be faster than CS scheme even for linearized problems. Moreover, composite grid techniques can be developed easily in accordance to the FAS theory, as the handling of different grid levels can be done very efficiently.

The implementation of the truncation error of the solution as an established method to predict the actual error of the solution is demonstrated. The truncation error prediction enables the construction of an adaptive scheme which has the potential, first to move from one grid to another finer when the error indication is high till this is reduced to the desired level or to coarsen the grid when so fine resolution is not required. Additionally the required work can be connected directly to the desired accuracy thus a fully artificial intelligent adaptive scheme can be considered. The introduction of the composite grid refinement in block structuring is very well suited either to the multigrid methods or to the truncation error prediction and also for parallel computer architectures in contrast to other refinement techniques which are strongly serial. The local grid refinement process takes full advantage of the multigrid construction as it is fully compatible with this logic. Additionally, without implementing complicated interpolations, the introduction of new refinement regions does not cause restart of the iterative process but a Full Multigrid scheme can be proceeded.

As future aspects are considered, first the implementation of a very efficient integrator that can handle different grids that have the blockwise logic but are strictly adapted to the error prediction without a reduction on the multigrid efficiency. The great disadvantage of the local refinement methods is that grid alignment with a flow feature like a shock can not be achieved in order to reduce the false diffusion of the solution. Thus there is a requirement of a global adaptive scheme [23] that should work concurrently with the presented concept in order to reduce the solution error totally to zero in just a few work units.

ACKNOWLEDGEMENTS

This work was supported by CEC BRITE/EURAM Area 5 Aeronautics Project Aero 0018 Euromesh. The authors would like also to thank MBB GmbH for providing the NsFlex code.

REFERENCES

[1] D.Bai and A.Brandt, Local mesh refinement multilevel techniques, SIAM J. Sci. Stat. Comput.,8 (1987) pp. 109
[2] F.Bassi, F.Grasso and M.Savini, A local multigrid strategy for viscous transonic flows around airfoils, Notes in Num.Fluid Mechanics, Vol.20 (Vieweg Verlag, Braunschweig,1987) pp.17u
[3] M.J.Berger and A.Jameson, Automatic Adaptive Grid Refinement for the Euler equations, AIAA J., 23 (1985) pp. 561
[4] M.J.Berger and P.Collela, Local adaptive mesh refinement for shock hydrodynamics, J.Comp.Phys., 82 (1989), pp. 64-84.
[5] A.Brandt, Multigrid Techniques: 1984 Guide with Applications to Fluid Dynamics, GMD -Studien No 85, (GMD, Sankt Augustin, 1984).
[6] Q.Chang and G.Wang, Multigrid and adaptive for solving the nonlinear Schrodinger equation, J.Comp.Physics, 88, (1990) pp.362-380
[7] J.F.Dannenhoffer, A comparison of adaptive-grid redistribution and embedding for steady transonic flows, Int.J.Num.Meth.Engng, 32 (1991) pp. 653
[8] A.Eberle, Characteristic flux averaging approach to the solution of the Euler equations, VKI Lecture series 1987
[9] R.E.Ewing,R.D.Lazarov and P.S.Vassilevski, Local Refinement techniques

for elliptic problems on cell-centered grids I.Error analysis, Math.Comp., **56** (1991) pp.437

[10] L.Fuchs, An adaptive multigrid scheme for simulation of flows, Multigrid methods II, Proceedings, Cologne 1985, Lecture notes in Mathematics 1228 ed. W.Hackbusch and U.Trottenberg (Springer Verlag, Berlin,1985).

[11] D.Hanel, M.Meinke and W.Schroder, Application of the multigrid method in solutions of the compressible Navier-Stokes equations, in Proceedings of the fourth Copper Mountain conference on multigrid methods, eds. J.Mandel et al (SIAM, Philadelphia,1989) pp.234-254.

[12] D.C. Jespersen, Recent Developments in Multigrid Methods for the Steady Euler Equations, Lecture Series in Computational Fluid Dynamics, von Karman Institute, Rhode St. Genese, Belgium (1984).

[13] W.Joppich, A multigrid algorithm with time-dependent, locally refined grids for solving the nonlinear diffusion equation on a nonrectangular geometry, in: Proceedings of the 3rd European Conference on Multigrid methods, Int. Ser. of Num. Math, Vol.98 (Birkhauser Verlag, Basel,1991) pp. 241

[14] M.Heroux, S.McCormick, S.McKay, J.W.Thomas, Applications of the fast Adaptive Composite grid Method, in Multigrid methods Theory, Applications and Supercomputing, ed. S.F.McCormick (M.Dekker,1988,N.York).

[15] A.Kanarachos and N.Pantelelis, An alternative multigrid method for the Euler and Navier Stokes equations, Submitted in the J.Comp. Phys.

[16] A.Kanarachos, N.Pantelelis and C.Provatidis, Recent developments in the Multiblock-Multigrid method for the compressible Euler equations, Submitted in the Comp. Meths Appl.Mech.Engng.

[17] A.Kanarachos and N.Pantelelis, Block Full multigrid adaptive scheme for the compressible Euler equations, presented at the 13th Intl Conf.Num. Meth.Fluid Dynamics, Rome, 1992.

[18] A.Kanarachos and I.Vournas, Multigrid techniques applied to the compressible Euler equations, Acta Mechanica, to appear.

[19] A.Kanarachos and I.Vournas, Multigrid solution for the compressible Euler equations by an implicit characteristic flux averanging, Proceedings of the 1st European Computational Fluid Dynamics Conference, ed.C.Hircsh, (Elsevier, North Holland, 1992).

[20] S.F.McCormick, Multilevel Adaptive Methods for Partial Differential Equations (SIAM, Philadelphia,1989)

[21] M.C.Thompson and J.H.Ferziger, An adaptive multigrid technique for the incompressible Navier-Stokes Equations, J.Comp.Phys, **82** (1989) pp. 94-121.

[22] D.Lee and Y.M.Tsuei, A formula for estimation of truncation errors of convection terms in a curvilinear coordinate system, J.Comp.Phys, **98**, (1992) pp. 90-100.

[23] N.G.Pantelelis and I.P.Vournas, Final report, EUROMESH project, (NTUA, Athens, 1992).

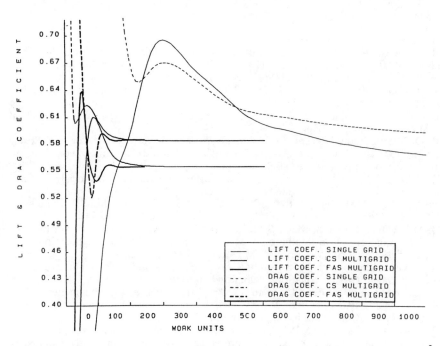

FIG 1 (above) CASE T1: Comparison of the lift and drag convergence of the single grid and CS, FAS multigrid schemes.
FIG 2 (below) CASE E1: Convergence of the truncation error and the Drag coefficient with respect to the number of volumes.

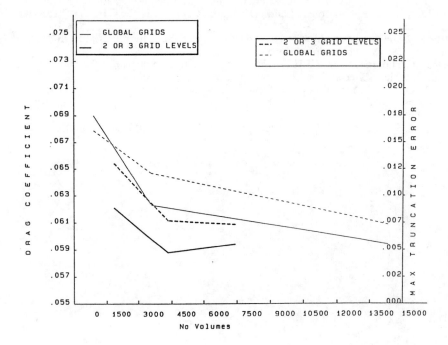

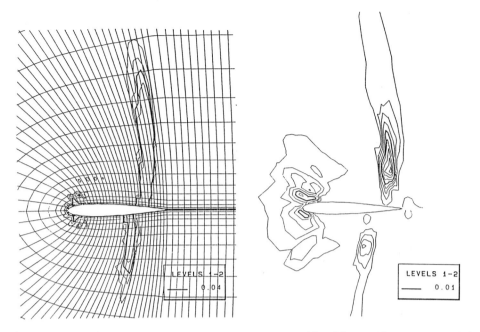

FIG 3 a & b (above) CASE E1: Elliptic grid with actual pressure error contours (LEFT) and the predicted truncation error contours (RIGHT).
FIG 3 c & d (below) CASE E1: Algebraic grid with actual pressure error contours (LEFT) and the predicted truncation error contours (RIGHT).

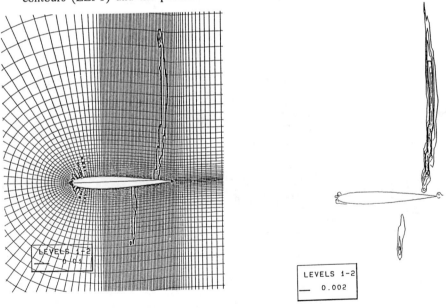

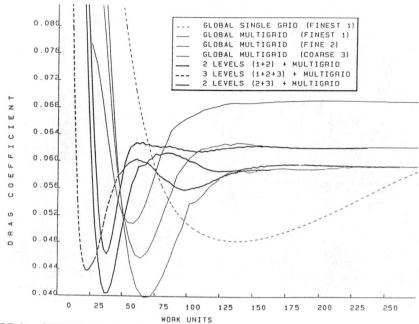

FIG 3 e (UPPER) CASE E1: Comparison of the convergence of the Drag coefficient for the single grid, and multigrid for the three global grids and the three adaptive grids.

FIG 3 g (LOWER) CASE E1: Adaptive solution. Adaptive grid with three grid levels with MACH contours.

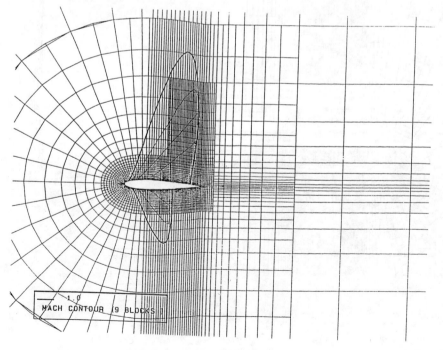

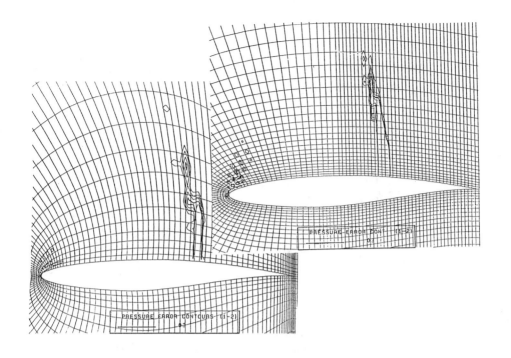

FIG 4 a & b (LEFT) CASE E2: Elliptic grid with actual pressure error contours (above) and the predicted truncation error contours (below).
FIG 4 c & d (RIGHT) CASE E2: Algebraic grid with actual pressure error contours (above) and the predicted truncation error contours (below).

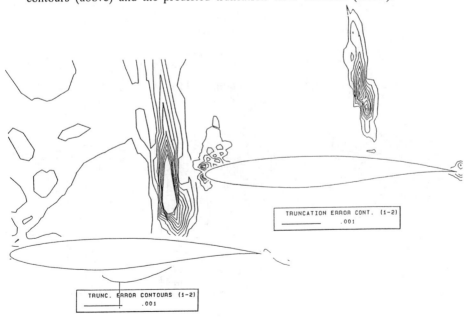

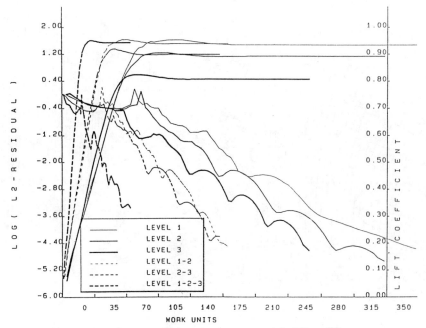

FIG 4 e (UPPER) CASE E2: Comparison of the convergence of the Lift coefficient and the logarithm of the euclidean norm of the residual using multigrid for the three global grids and the three adaptive grids.

FIG 4 g (LOWER) CASE E2: Adaptive solution. Adaptive grid companing three grid levels with MACH contours.

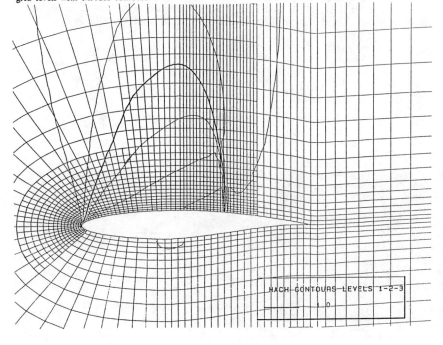

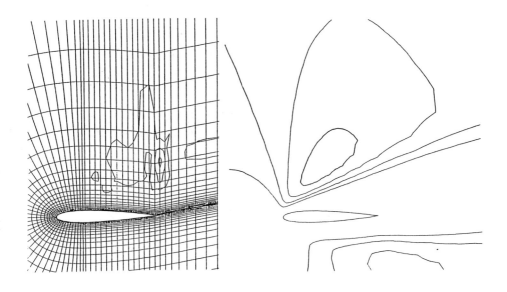

FIG 5 a,b (above) CASE L1: Algebraic grid with the predicted truncation error contours (left) and the Mach contours of the solution (right).
FIG 6 a,b (below) CASE L2: Algebraic grid with the predicted truncation error contours (left) and the mach contours of the solution (right).

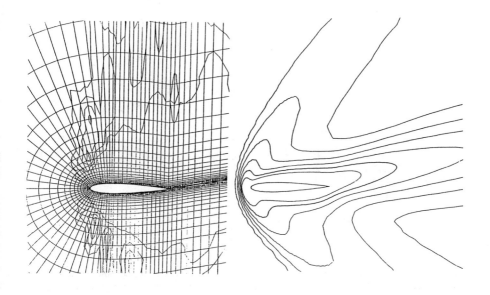

Addresses of the Editors of the Series "Notes on Numerical Fluid Mechanics"

Prof. Dr. Ernst Heinrich Hirschel (General Editor)
Herzog-Heinrich-Weg 6
D-85604 Zorneding
Federal Republic of Germany

Prof. Dr. Kozo Fujii
High-Speed Aerodynamics Div.
The ISAS
Yoshinodai 3-1-1, Sagamihara
Kanagawa 229
Japan

Prof. Dr. Bram van Leer
Department of Aerospace Engineering
The University of Michigan
Ann Arbor, MI 48109-2140
USA

Prof. Dr. Keith William Morton
Oxford University Computing Laboratory
Numerical Analysis Group
8-11 Keble Road
Oxford OX1 3QD
Great Britain

Prof. Dr. Maurizio Pandolfi
Dipartimento di Ingegneria Aeronautica e Spaziale
Politecnico di Torino
Corso Duca Degli Abruzzi, 24
I-10129 Torino
Italy

Prof. Dr. Arthur Rizzi
FFA Stockholm
Box 11021
S-16111 Bromma II
Sweden

Dr. Bernard Roux
Institut de Mécanique des Fluides
Laboratoire Associé au C.R.N.S. LA 03
1, Rue Honnorat
F-13003 Marseille
France

Brief Instruction for Authors

Manuscripts should have well over 100 pages. As they will be reproduced photomechanically they should be typed with utmost care on special stationary which will be supplied on request.
In print, the size will be reduced linearly to approximately 75 per cent. Figures and diagrams should be lettered accordingly so as to produce letters not smaller than 2 mm in print. The same is valid for handwritten formulae. Manuscripts (in English) or proposals should be sent to the general editor, Prof. Dr. E. H. Hirschel, Herzog-Heinrich-Weg 6, D-85604 Zorneding.